Practicing Circular Economy

Practicing Circular Economy

Authored by

Prasad Modak

CRC Press is an imprint of the
Taylor & Francis Group, an **informa** business

First edition published 2021
by CRC Press
6000 Broken Sound Parkway NW, Suite 300, Boca Raton, FL 33487-2742

and by CRC Press
2 Park Square, Milton Park, Abingdon, Oxon, OX14 4RN

Library of Congress Cataloging-in-Publication Data

Names: Modak, Prasad, author.
Title: Practicing circular economy / Prasad Modak.
Description: First edition. | Boca Raton : CRC Press, 2021. | Includes bibliographical references and index.
Identifiers: LCCN 2021000712 (print) | LCCN 2021000713 (ebook) | ISBN 9780367619534 (hardback) | ISBN 9781003107248 (ebook)
Subjects: LCSH: Sustainable development. | Recycling (Waste, etc.)
Classification: LCC HC79.E5 M6233 2021 (print) | LCC HC79.E5 (ebook) | DDC 338.9/27–dc23
LC record available at https://lccn.loc.gov/2021000712
LC ebook record available at https://lccn.loc.gov/2021000713

ISBN: 978-0-367-61953-4 (hbk)
ISBN: 978-0-367-61957-2 (pbk)
ISBN: 978-1-003-10724-8 (ebk)

Typeset in Palatino
by KnowledgeWorks Global Ltd.

Contents

List of Figures

List of Tables

List of Boxes

List of Activities

Preface

Circular Economy (CE) is considered as one the most important strategies to address the Sustainable Development Goals. Many countries have come up with national road maps on CE and have enforced legislation. More recently, investors have started showing their preference for circular projects.

The canvas of CE is, therefore, rapidly evolving. Universities have started offering courses on CE at the postgraduate level. Several continuing education programs have emerged and have been gaining traction. Policymakers, investors, and entrepreneurs have shown interest in knowing more about CE. Unfortunately, very few books are available as a resource to understand the evolution, policy, and, more importantly, the practice experience in a circular economy.

This book is expected to meet this gap. It is targeted to the needs of mid and senior-level managers, policymakers, consultants, investors, entrepreneurs, researchers, and academic students.

"Practicing Circular Economy" provides an overview of CE, covering its evolution describing the key concepts, programs, policies, and regulations. It illustrates several business opportunities over a hundred hand-picked case studies covering numerous sectors, various scales of operations, and geographies. Another unique feature of the book is the activities listed in each chapter to invoke thoughts, frame assignments, and generate discussions. The book can thus be used as a textbook for running a full course on CE at postgraduate level. In addition, it will serve as a supplement to the continuing education programs on CE. The book will also help policymakers, investors, and entrepreneurs as a valuable reference and for an immersion to this exciting subject.

The book is organized into 11 chapters.

Chapter 1 introduces the global challenges faced with regard to the rise in population and overconsumption of resources. Concepts such as virtual water, biocapacity, ecological footprint, and decoupling have been introduced to gain a deeper understanding of the impact of human activities on the environment. The chapter ends with introducing the term "Circular Economy" and its relevance.

Chapter 2 acts as a foundation to key concepts and programs that led to the emergence of circular economy. Those who are "late entrants" to the subject of CE will find this chapter very useful. Chapter 3 describes key methodologies, tools, and knowledge bases that help in putting circular economy in practice.

Chapter 4 presents the 12Rs of a circular economy. Each of the Rs are described though case studies, handpicked from different geographies, scales of operation and representing diverse sectors. It also describes the importance of collaboration between different stakeholders to promote, deliver and sustain circular solutions. At the end of the chapter, case studies are described that emphasize mapping of Rs across the life cycle and involving key stakeholders at every stage.

Chapter 5 discusses the importance of making products last longer in an economy through strategies such as reuse, repair, and refurbishing. Case studies that portray how society perceives second-hand goods and how policy makers and companies can bring in behavioral change are included.

Chapter 6 focuses on strategies to close the loop of an economy. This chapter discusses recycling infrastructure, related policies, and economic instruments taking examples from various sectors and waste streams. It also highlights the informal sector's role to promote a circular economy, especially in developing economies.

Chapter 7 presents an overview of opportunities in circular economy in three sectors – viz. textile, steel, and agriculture and food. For each sector, opportunities are mapped across the lifecycle with case studies and international initiatives.

Chapter 8 introduces readers to 8 business models operating in a circular economy mapping with the life cycle. Each business model is described with case studies. The chapter ends with hybrid business models. Chapter 9 discusses the importance of innovation in a circular economy and its efforts to promote innovations and entrepreneurship. Funds active in the space of circular economy are also described. This chapter is especially relevant for entrepreneurs in circular economy as well as to circular businesses who are seeking finance.

Chapter 10 provides a global overview of policies, roadmaps, and legislations to promote circular economy. It further describes various economic, market-based, and information-driven instruments. The need for policy impact assessments supported by tools such as Life Cycle Sustainability Assessments and System Dynamic Modelling is emphasized. This chapter is extremely relevant to the policymakers.

Chapter 11 builds on the previous chapters and discusses the steps ahead, emphasizing the importance of regional circular economy. It introduces a new framework CIRCULAR, that can help to prepare action plans on circular economy. Important areas such as sustainable packaging, circular supply chains, sustainable public procurement, and behavioral change are described. The chapter ends with impact of the COVID-19 pandemic on the UN Sustainable Development Goals and how circular economy could provide an opportunity to restructure and revitalize the economy.

I hope that "Practicing Circular Economy" will serve as a useful resource for every circular economy practitioner and especially to those who aspire to make a career in circular economy.

Acknowledgement

This book has been a culmination of my four decades of global experience in policy and practice in circular economy. I would like to thank all my friends and colleagues who gave me opportunities to work on some exciting projects across the globe and generously shared their resources and experiences.

I must thank Aneesa Patel, who supported me in writing the book, providing key research inputs, proofing all the drafts, and ensuring publisher's requirements. Her constant support and hard work played an important role in completing this demanding project.

I would also like to record my appreciation to Chandra Mouli and Dr. Anand Palkar for a thorough technical review of the drafts. Credit also goes to Sakshi Gore and Bhushan Bhaud for editing the visuals inserted in this book and Kashyapi Dave and Neeraj Borgaonkar for building an initial repository of case studies.

Finally, all book writing projects take time, where working hours extend, and home becomes another office. No words would be adequate to express my gratitude to my wife, Kiran, who supported and tolerated me and patiently so, during the COVID-19 times.

Author Biography

Dr. Prasad Modak was a Professor at Centre for Environmental Science & Engineering at IIT Bombay and later as Professor (Adjunct) at the Center for Technology Alternatives in Rural Areas. He is currently the Executive President of Environmental Management Centre LLP and Director of Ekonnect Knowledge Foundation.

Dr Modak has worked with almost all key UN, multi-lateral and bi-lateral developmental institutions in the world. Apart from Government of India and various State Governments, Dr Modak's advice is sought by Governments of Bangladesh, Egypt, Indonesia, Mauritius, Thailand, and Vietnam.

Dr Modak is currently member of Indian Resources Panel at Ministry of Environment, Forests and Climate Change and Member of Task Force on Sustainable Public Procurement at the Ministry of Finance of Government of India.

Dr Modak serves as a member of the National Committee on Circular Economy of the Federation of Indian Chamber of Commerce and Industry (FICCI). He is also on the three-member advisory committee of the Circular Economy Club that is a worldwide initiative with more than 8000 members.

Dr Modak prepared Global Status Report on Cleaner Production for UNEP DTIE. He has also co-authored the first Global Waste Management Outlook and served as Chief Editor of Asia Waste Management Outlook for the United Nations Environment.

List of Abbreviations

AUD	Australian Dollars
CAD	Canadian Dollars
CAGR	Compound Annual Growth Rate
CE	Circular Economy
CO_2	Carbon Dioxide
COVID-19	Coronavirus disease 2019
D4D	Design for Disassembly
D4R	Design for Repairability
D4S	Design for Sustainability
DMC	Domestic Material Consumption
DRS	Deposit and Refund Schemes
EIP	Eco-Industrial Park
e-LCA	Environmental Life Cycle Assessment
EMS	Environmental Management System
EPA	Environmental Protection Agency
EPD	Environmental Product Declarations
EPR	Extended Producer Responsibility
EU	European Union
EUR	Euro
GBP	Great Britain Pounds
GDP	Gross Domestic Product
GFN	Global Footprint Network
GHG	Greenhouse Gas
GP	Green Productivity
ILO	International Labor Organization
INR	Indian Rupees
ISO	International Organization for Standardization
LCA	Life Cycle Assessment
LCSA	Life Cycle Sustainability Assessment
LCT	Life Cycle Thinking
MFA	Material Flow Analysis
MMT	Million Metric Tons
MRF	Material Recovery Facilities
MSW	Municipal Solid Waste
NCPC	National Cleaner Production Centres
OECD	Organisation for Economic Co-operation and Development
o-LCA	Organisational Life Cycle Assessment
PaaS	Product as a Service

PET	Poly-Ethylene Terephthalate
PPP	Public-Private Partnerships
PRO	Producer Responsibility Organisations
RECP	Resource Efficient Cleaner Production
SCP	Sustainable Consumption and Production
SDG	Sustainable Development Goals
s-LCA	Social Life Cycle Assessment
SME	Small and Medium sized Enterprises
SRM	Secondary Raw Material
TERI	The Energy Research Institute
UK	United Kingdom
UN	United Nations
UN IRP	United Nations International Resource Panel
UNCED	United Nations Conference on Environment and Development
UNCSD	United Nations Conference on Sustainable Development
UNCTAD	United Nations Conference on Trade and Development
UNDESA	United Nations Department of Economic and Social Affairs
UNEP	United Nations Environment Program
UN-HABITAT	United Nations Human Settlement Programme
UNIDO	United Nations Industrial Development Organization
USA	United States of America
USD	United States Dollar
WBCSD	World Business Council for Sustainable Development
WEF	World Economic Forum
WRAP	Waste & Resources Action Program
WWF	World-Wide Fund

1

Challenges We Face Today

1.1 Rising Population and Urbanization

Resource consumption and waste generation have become critical concerns questioning the sustainability of the planet. Population growth has been one of the major drivers. It is estimated that the world population would more than double by 2100 compared to 1990[1]. Do we have sufficient resources to meet this challenge? Figure 1.1 and Box 1.1 highlight some important statistics about global population growth.

Population growth has significantly contributed to urbanization. The urban population of the world has grown rapidly from 751 million in 1950 to 4.2 billion in 2018. Studies conducted by the United Nations show that 68% of the world population is projected to live in urban areas by 2050.[1] At present, 22 of the world's 33 megacities, each having 10 million or more people, are in Asia and Africa. It is predicted that by 2030, there will be 10 more megacities and 9 of them will be on these two continents[2].

While cities act as economic hubs, they are plagued by issues such as overcrowding, traffic congestion, waste generation, quality of energy supply, air pollution, and income inequality. With rapid rise in urbanization, there will be a strain on the existing housing and transportation infrastructure, and essential services such as health and education. It is estimated that cities consume more than two-thirds of the world's energy and account for more than 70% of global CO_2 emissions. With the majority of the world's urban areas situated on coastlines, cities are at high risk from some of the devastating impacts of climate change such as rising sea levels and powerful coastal storms[3].

When cities expand, there is a threat to agricultural land, ecosystem services, and biodiversity, thereby affecting food production and climate resilience. This could also result in spatial inequalities and put a strain on economies. It is estimated that urban land area could increase by 80% globally between 2018 and 2030, assuming a Compound Annual Growth Rate (CAGR)[4]. This is alarming.

Apart from rise in population and urban area expansion, the "in-migration" of people from rural to urban areas causes rapid urbanization. In several Asia

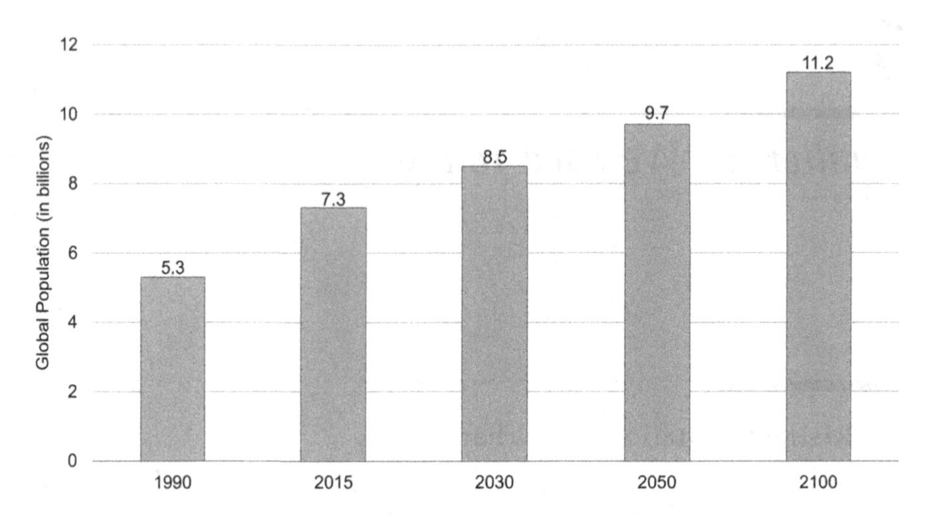

FIGURE 1.1
Global population until 2100[1].

BOX 1.1 GLOBAL POPULATION TRENDS[1]

- The world population is projected to reach 8.5 billion in 2030 and increase further to 9.7 billion in 2050 and 11.2 billion by 2100.
- More than half of the global population growth between 2020 and 2050 is expected to occur in Africa.
- Around 2027, India is projected to overtake China as the world's most populous country, while China's population is projected to decrease by 31.4 million between 2019 and 2050.

Pacific countries, in-migration from rural areas to cities has occurred due to natural resource depletion, calamities, civil conflicts, poverty, and other pressures on rural areas[5]. Economies in rural and urban contexts need to be supportive or collaborative to reduce the in-migration to cities.

1.2 Rising Consumption and Impact of Trade Flows

With the rise in the consumers' affordibility, the demand for many goods and services is increasing rapidly. Rising consumer demand and infrastructure investments in prospect will fuel the world economy's growth. However, in

ACTIVITY 1.1 IDENTIFICATION OF REGIONS WITH HIGH URBANIZATION

The International Institute for Environment and Development has created an interactive data portal covering all cities with 500,000-plus inhabitants. It shows the urban population's growth and the rise in cities' formation between the years 1800 and 2030.

Visit: https://www.iied.org/cities-interactive-data-visual to access the portal.

Instructions to use the portal[6]:

1. To access the data, click on the dateline button at the top and slide the scale to the right and advance through the years. This will enable you to see how and where cities of 500,000+ inhabitants have grown since 1800, plus predictions for future growth up to 2030. A summary of the wider global changes taking place can be found at the top.
2. By hovering the cursor over individual cities, precise data regarding those cities could be obtained. Alternatively, click on the "grid" view toward the visual's top right to see a chart of the individual cities and where they rank in terms of population.
3. How the populations have changed over time, and the predictions until 2030, can be seen by sliding the scale to the right.

Use the portal described above to identify the main regions that show rapid urbanization. Identify the decade in which there was a significant increase in urbanization. Also, identify the regions where a decrease in urbanization can be noticed. Can you think of the socioeconomic reasons leading to such a decline?

this process, there will be a mounting strain on global natural and capital resources. The overall rise in the per capita income is leading to increased consumption and generation of wastes and residues. The middle class in the urban population is contributing significantly to resource consumption and waste generation.

Notwithstanding the recent economic disruptions due to the COVID-19 pandemic, reports have forecasted growth in every sector. According to a European Commission report, income growth in emerging economies has led to an increase in the consumption of higher-value products such as meats and dairy products[7]. The global sales of smartphones in 2020 were predicted to increase 3% year on year[8], while the global white goods market (home and kitchen appliances) is projected to record a substantial CAGR of 7.9% in the forecast period of 2019–2026[9]. This increase in demand has contributed in the global logistics sector's growth, which is predicted to grow at 3.48% CAGR to reach USD 12.26 trillion by 2022[10].

Global trade plays an essential role in the movement of materials that consist of resources (e.g., coal), products (e.g., automobiles), and wastes and residues. Over the years, the material flows across the globe are increasing and getting skewed. Advancements in communications, logistics and supply chain technologies, improved geopolitical relations, and more trade openness have been responsible factors.

For instance, agricultural produce is traded from regions where it is available in abundance or surpassing the national production needs. Sometimes the resources are obtainable at a lower price through inter-state trade, even when transportation costs are added. In some cases, trade is inevitable as the materials are not available locally. Often the products made in some developed countries travel back to the very countries that provide the raw resources but do not have local manufacturing facilities. Wastes are also transported and sold as secondary materials. Before the imposition of the "green fence" in China, nearly 45% of the world's wastes were traded to China[11].

There has been a shift in the products that are getting traded. The international trade of "exotic" products under the fresh vegetables, fruits, and flowers categories has significantly increased during the past decade, in addition to the boom in the exports of by-products like biofuel made from crops. Such activities have led to a series of ecological issues in dry and semiarid regions. In such regions, high-water-consuming crops (such as Jatropha) compete for water and land with local communities, deplete and degrade local ecosystems, worsen local and national food sovereignty, and alter existing production modes and income distribution. Consider the following instances[12]:

1. The export of fruits and vegetables from Peru's dry desert coast increased tenfold from 2001 to 2015.
2. During the same period, Ecuador tripled its flower export from the fragile drought-prone Andean hill slopes North of Quito and is now the third flower-exporting country globally.
3. In the desert of North Coast of Peru, multinational and national companies have acquired land and water rights to start large-scale sugarcane production for biofuels, showing a sharp increase in export since 2008.

Trade flows have resulted in drivers of consumption across geopolitical boundaries, hence raising issues related to resource sufficiency, livelihoods, and sustainability.

When we think of an increase in material flows and resource insufficiency due to a rise in trade, we should pay attention to the concept of "virtual

BOX 1.2 GLOBAL RESOURCE CONSUMPTION

- Resource extraction increased 12-fold between 1900 and 2015[15].
- In the past 40 years alone, the global use of materials has almost tripled, from 26.7 billion tons in 1970 to 84.4 billion tons in 2015. It is expected to double again to between 170 and 184 billion tons by 2050[9].
- A quantitative analysis of global resource requirements carried out by the United Nations International Resource Panel (UN IRP)[16] estimates that the world's cities' material consumption will grow from 40 billion tons under business-as-usual circumstances in 2010 to about 90 billion tons by 2050.
- According to UN IRP, during the 20th century, the annual extraction of ores and minerals grew by a factor of 27, construction materials by a factor of 34, fossil fuels by a factor of 12, and biomass by a factor of 3.6. In total, material extraction increased by a factor of about 8[10].

water". The idea of virtual water refers to the volume of water used in producing food and fiber and non-food commodities, including energy production. For example, the production of a ton of wheat requires about 1300 m^3 of water and the production of a ton of beef requires 16,000 m^3 of water. Production of a cotton T-shirt weighing about 250 g requires about 2.7 m^3 of water[13]. Today, prominent discussions on virtual water address global water resource scarcity vis-à-vis commodity production and consumption[14].

The combined effects of urbanization, industrialization, and globalization of trade have thus deeply influenced global resource consumption. See Box 1.2.

Due to the rising intensity of resource consumption, biocapacities of the regions are exceeded. Biocapacity is defined as the ecosystems' capacity to produce useful biological materials and absorb waste materials generated by humans, using current management schemes and extraction technologies[17].

The stress on biocapacity due to resource consumption is described as the "Ecological Footprint". Ecological footprint is the impact of human activities measured in terms of the area of biologically productive land and water required to produce goods consumed and to assimilate the wastes generated[18]. More simply, it is the amount of the environmental resource necessary to produce the goods and services required to support a particular lifestyle or consumption pattern.

Both, ecological footprint and biocapacity are expressed in global hectares. Each city, state, or nation's ecological footprint can be compared to its biocapacity. If a population's ecological footprint exceeds the region's biocapacity,

ACTIVITY 1.2 IDENTIFYING REGIONAL BIOCAPACITY AND ECOLOGICAL FOOTPRINT

The Global Footprint Network (GFN) released a report called "The Ecological Wealth of Nations". This report explains the concepts as well as documents country-wise ecological footprint and biocapacity until the year of 2006[19].

Building on its work, GFN has developed an interactive visualization, documenting the current global ecological footprint and biocapacity. Visit: http://data.footprintnetwork.org/#/? to access the visualization.

How to use the visualization:

1. Choose from the tabs below the map to view different ecological footprint indicators per person and total biocapacity.
2. Make use of the labels to identify the difference in the representation of each color on the map.
3. Click on a country map to view an in-depth analysis of how the country's ecological footprint and biocapacity have changed over the years.

The GFN report is periodically updated. Can you conduct research on the latest information on the biocapacity of your country and the Ecological Footprint? Discuss why you find a change, and what could be done to address the ecological deficit if so observed.

then that region is in an ecological deficit. This means that its demand for goods and services, e.g., fruits and vegetables, meat, fish, wood, cotton for clothing, and carbon dioxide absorption – exceeds what the region's ecosystems can renew. A region in ecological deficit then meets demand by importing, i.e., trade and liquidating its own ecological assets (such as overfishing)[11].

1.3 Decoupling

In the past several decades, a coupling has been observed between economic growth and waste generation and Greenhouse Gases (GHG) emissions. The more we progress economically, greater are the risks to the ecosystems being damaged, loss of biodiversity, and global warming. Figure 1.2 shows the coupling between waste generation per capita and Gross Domestic Product (GDP) per capita.

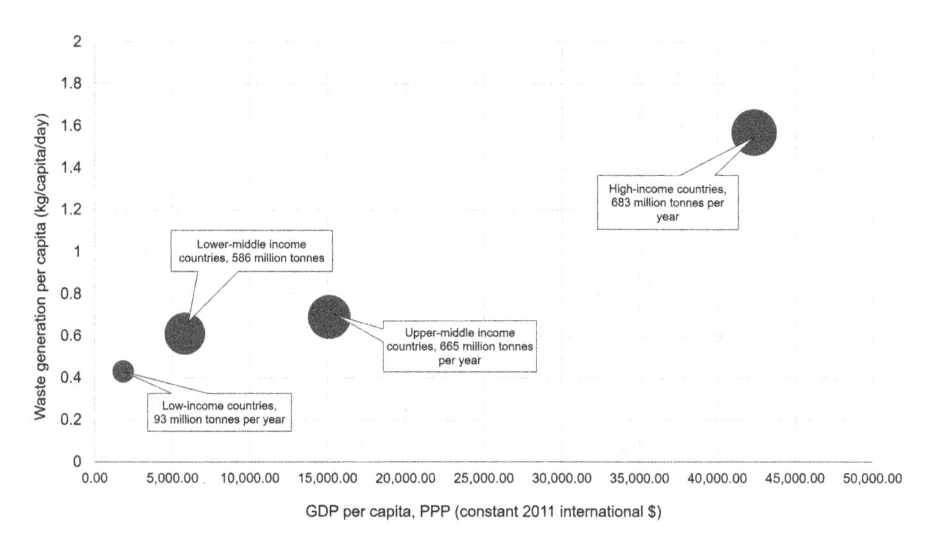

FIGURE 1.2
Waste generation per capita vs GDP per capita, by income group[20].

Decoupling occurs when the growth rate of environmental pressure (for example, GHG emissions) is less than that of its economic driving force (for example, GDP per capita) over a given period[21].

Box 1.3 highlights the findings of the Organisation for Economic Co-operation and Development (OECD) Report on waste generation and GDP decoupling.

BOX 1.3 OECD REPORT ON WASTE GENERATION AND GDP DECOUPLING[22]

- Most countries continue to generate increasing amounts of waste; only a few have managed to decouple total waste generation (i.e., all sources of waste) from population and economic growth (such as France, Hungary, Japan, Slovak Republic, and Spain).
- The developments for municipal solid waste (MSW) are more positive as their growth rate appears to have peaked after the year 2000. A person living in the OECD area generates, on average 525 kg of municipal waste per year; this is 20 kg more than in 1990, but 35 kg less than in 2000.
- Waste is increasingly being recycled but landfilling remains the primary disposal method in many OECD countries.

A study was undertaken by International Global Environmental Strategies (IGES) in 2019 to assess regional sustainable development by decoupling waste generation from economic growth in Asian countries[23]. The research included the following:

- The occurrence of decoupling of MSW generation from economic growth.
- It examined six representative countries for the period 1970–2015 as an example to determine the relationship between MSW generation and related driving factors.
- The logarithmic mean Divisia index technique was used to disassemble MSW generation into four factors: population, economic activity (GDP divided by population), material intensity (Domestic Material Consumption divided by GDP), and waste rate (MSW generation divided by Domestic Material Consumption).

The results show that the decoupling situation and its extent varied among countries:

- In Vietnam and Cambodia, decoupling of MSW generation from economic growth has not occurred.
- China has shifted from coupling (during 1981–1990) to relative decoupling since 1991.
- Japan has shifted from relative decoupling (during 1971–2000) to absolute decoupling since 2001.
- Singapore shifted from absolute decoupling (during 2001–2005) to relative decoupling since 2006.
- Thailand has shown relative decoupling during 2011–2015.

As reported by IGES, Japan is at the forefront of such decoupling by demonstrating that waste can be reduced even with economic development. Figure 1.3 shows the reduction in average waste emission per capita per day in Japan over the years.

Box 1.4 shows the case of decoupling in the high-income city of Kitakyushu, Japan, and South Korea.

Apart from per capita waste generation, GHG emissions are considered good indicators of decoupling. Countries with policy frameworks supportive of renewable energy and climate change mitigation tend to show greater decoupling between GHG emissions and GDP. The relationship between GHG emissions and GDP trends in such countries has also become much weaker in the last two decades than the preceding decades[27]. Figure 1.4 illustrates relationship between CO_2 emissions per capita and GDP per capita. Box 1.5 explains the case of decoupling of GHG emissions in the United States.

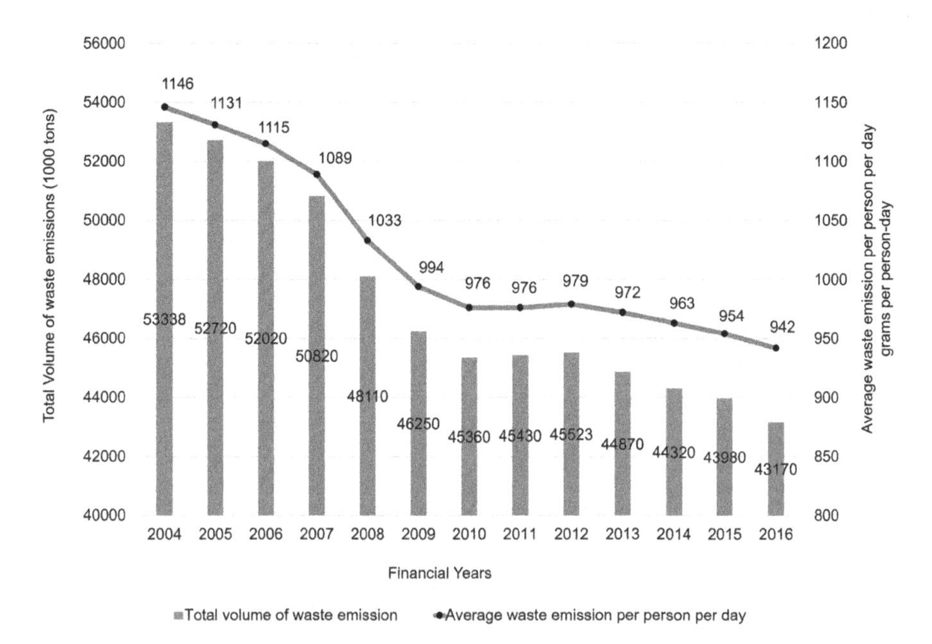

FIGURE 1.3
Average waste emission per capita per day in Japan[24].

BOX 1.4 DECOUPLING CASE IN THE CITY OF KITAKYUSHU, JAPAN[25], AND SOUTH KOREA[26]

Kitakyushu is an industrial city in Japan that was suffering from significant waste pollution. To manage the waste, it pursued an environmentally sound approach rather than a disposal-focused one. Their efficient waste management system's main drivers cover all steps from the start to endpoints: sorting the waste at source, composting widely at the household level, recycling, and engaging citizens.

These measures were complemented with financial incentives/disincentives. The waste handling user fees were based on volume rather than on flat fees per household. Over time, Kitakyushu even built up an "Eco-Town" to increase environmental awareness and recover materials from many types of waste, including cars and appliances.

In South Korea, solid waste generation rose rapidly due to rising incomes and changes in consumption. In 1995, the government introduced a system where people had to pay for bags to dispose off garbage. The price varied based on the municipality and reflected the local cost of disposal. Despite the charges levied not covering the full cost, waste generation fell from 1.3 kg per person per day in 1994 to 0.95 kg in 2014. The recycling rate rose from 15.4 to 59% over the same period.

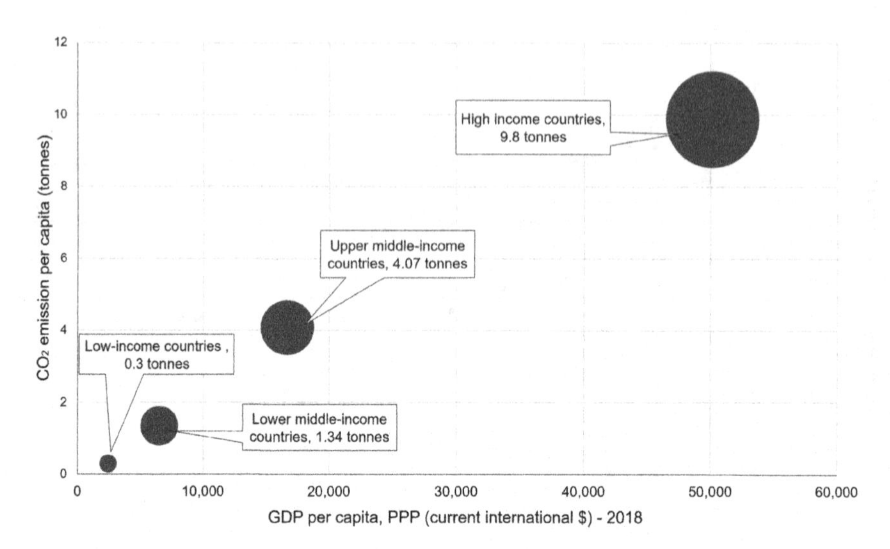

FIGURE 1.4
Relationship between CO_2 emissions per capita and GDP per capita[28].

BOX 1.5 DECOUPLING OF GHG EMISSIONS IN CERTAIN STATES OF UNITED STATES[29]

Between 2005 and 2017, 41 states of United States and the District of Columbia reduced their energy-related GHG emissions while increasing real GDP, debunking the myth that economic growth can only occur by compromising the security of planet's ecosystem. These states included both large and small states located in every part of the country such as Maine and New York in the Northeast, Alabama and Georgia in the South, Indiana and Ohio in the Midwest, and Alaska and Nevada in the West. Of these 41 states, Maryland reduced its GHG emissions the most (38%), followed by New Hampshire (37%), the District of Columbia (33%), Maine (33%), Alaska (29%), and Georgia (28%). Given reversals of various climate policies at the federal level, this finding offers hope that the United States can make significant progress toward tackling climate change at the state level regardless of federal policy.

1.4 Emergence of Circular Economy

Decoupling is not easy to achieve. It will require well-designed and concerted policy packages with well-chosen and coordinated sustainability actions, particularly resource efficiency, sustainable consumption, product life extension, and integrated waste management that help recycle and recover used resources.

To support these interventions, there is a need to come up with business models and leapfrogging technologies. These technologies should allow economic output to be achieved with fewer resource inputs, waste reduction, and cost savings to further expand the economy or improve resource security. The business models should involve multiple stakeholders, e.g., both from formal and informal sectors and operate based on partnerships to achieve a wider social and economic impact. However, their uptake and upscaling will require policies that remove barriers and promote a transition towards greater resource productivity and a regenerative system to close the loop of material flows.

Economies often do not naturally adjust to these changes and can suffer from blocks to transition where current consumption and production patterns remain "locked-in". The legacy of past policy decisions and technological, behavioral, organizational, and institutional biases can present significant barriers. Although much of the policy "know-how" exists in the form of legislation, incentive systems, administrative measures, and institutional reform, additional policy options need to be "unlocked" for a yet more strategic and long-term avenue toward ecologically sound and smart sustainable growth.

One such strategy that has emerged in the last two decades is the circular economy. The circular economy shows a great promise to address the challenges described, especially related to decoupling. To put the circular economy in practice, you need to conceive and manage economies at a regional level in a systems perspective. The context and scope of the set of instruments that drive a circular economy will depend on the region's context. In this process, multiple benefits can be achieved by coordinating with other national plans, e.g., for climate change mitigation and adaptation, as well as with national plans for the protection, conservation, and sustainable use of natural resources.

The circular economy strategy is like a conductor of an orchestra who guides the musicians to produce sweet music, getting the best of their expertise. It is the conductor who decides the role of every musician engaged and does coordination.

There are more than a hundred definitions of the term circular economy[30]. Most of the definitions emphasize reduce, reuse, and recycle activities, and often do not highlight that circular economy requires a systemic shift in the economy and behavioral change. Box 1.6 depicts the most used circular economy definitions. The definition by China is more operational while the European Union (EU) definition emphasizes how life cycle thinking spells out the outcomes or benefits. Decoupling is not explicitly stated.

The principal drivers and strategies stated in the definitions of the circular economy are not new. In the last five decades, several initiatives have been taken across the world on sustainable and responsible management of

BOX 1.6 DEFINING CIRCULAR ECONOMY

- **Definition by EU**[31] – In a circular economy, the value of products and materials is maintained for as long as possible. Waste and resource use are minimised, and when a product reaches the end of its life, it is used again to create further value. This can bring major economic benefits, contributing to innovation, growth, and job creation.
- **Definition by Ellen MacArthur**[32] – A circular economy is based on the principles of designing out waste and pollution, keeping products and materials in use, and regenerating natural systems.
- **Definition as per Circular Economy Law in China**[33] – The circular economy herein is the general term for the activities of decrement, recycling and resource recovery in production, circulation, and consumption.
    - Decrement herein means the reduction of the resource consumption and waste generation in production, circulation, and consumption.
    - Recycling herein means the direct use of wastes as products, or the use of wastes as products after the repair, renovation or reproduction of them, or the use of wastes, wholly or partly, as parts of other products.
    - Resource recovery herein means the direct use of wastes as raw material, or waste regeneration.

wastes and resources. These initiatives have led to positive outcomes at the facility level, across supply chains, over product life cycles and in some cases influencing the regional economy and the policy frameworks. The evolution of a circular economy is perhaps better understood if we familiarize with some of the key concepts. Chapter 2 explains these concepts in a narrative and unfolds the canvas of circular economy.

1.5 Key Takeaways

- Population growth, urbanization, and global trade have been the major factors toward the rising consumption of resources.
- Ecological footprints of many cities and regions have exceeded their biocapacities.

- Economic development and the generation of wastes and GHG emissions are coupled. That gives a challenge and a threat to the limited resources we have.
- Decoupling between economic development and wastes and emissions is, therefore, necessary in the interest of planet's sustainability.
- There is evidence of decoupling, however, more systematic efforts are needed to reduce consumption of virgin material resources, improve resource efficiency, extend product life, and practice reuse, recovery, and recycling of used resources.
- Practicing a circular economy could be such a systematic strategy which would address the aforementioned challenges.

ADDITIONAL READING

1. Rising population and urbanization
   a. **Urban world: Cities and the rise of consuming class:** This report from 2012 focuses on the economic power of cities, the expansion of the consuming class, and the implications of the increase in consumption on the infrastructure and natural resources.

Source: McKinsey Global Institute. 2012. *Urban World: Cities and the Rise of Consuming Class.* https://www.mckinsey.com/~/media/McKinsey/Featured%20Insights/Urbanization/Urban%20world%20Cities%20and%20the%20rise%20of%20the%20consuming%20class/MGI_Urban_world_Rise_of_the_consuming_class_Full_report.ashx [Accessed 2 October 2020].

   b. **IPAT tool to measure resource consumption:** As a result of the rise in population and the subsequent resource consumption, scientists Ehrlich, John Holdren, and Barry Commoner in the early 1970s devised a formula to measure our rising impact. This formula is called IPAT, in which (I)mpact equals (P)opulation multiplied by (A)ffluence multiplied by (T)echnology. The IPAT formula can help us realize that our cumulative impact on the planet is not just in population numbers but also in the increasing amount of natural resources each person uses.

Source: Dimick, D., 2014. *As world's population booms, will its resources be enough for us?* https://www.nationalgeographic.com/news/2014/9/140920-population-11billion-demographics-anthropocene/ [Accessed 2 October 2020].

2. Rising consumption and impact of trade flows
    a. **Globalization in transition: The future of trade and value chains:** This article focuses on the future of global trade. It emphasizes the shift in the trade of goods to the trade of services. It sheds light on how technology is playing a role in enabling the change in trade patterns.

Source: McKinsey Global Institute. 2019. *Globalization in transition: the future of trade and value chains.* https://www.mckinsey.com/featured-insights/innovation-and-growth/globalization-in-transition-the-future-of-trade-and-value-chains [Accessed 2 October 2020].

    b. **Evolution of the global virtual water trade network:** This paper focuses on the virtual water trade network associated with international food trade. The evolution of this network from 1986 to 2007 is analyzed and linked to trade policies, socioeconomic circumstances, and agricultural efficiency. It is found that the number of trade connections and the volume of water associated with global food trade has more than doubled in 22 years.

Source: Dalin, C., Konar, M., Hanasaki, N., Rinaldo, A., and Rodriguez-Iturbe, I., 2012. Evolution of the global virtual water trade network. *Proceedings of the National Academy of Sciences,* 109(16), pp. 5989–5994.

3. Decoupling
    a. **Research on eco-towns in Japan: Implications and lessons for developing countries and cities:** This report focuses on identifying the key lessons learned in the setting up of the eco-towns. Eco-towns have enabled several developmental objectives such as stimulating the local economy, securing employment, disposing of waste in an environmentally sound manner and protecting air and water resources. A number of lessons have been learned in setting up such eco-towns in Japan, not only within these eco-towns but also in the cities where they are located.

Source: UNEP. 2005. *Research on eco-towns in japan: implications and lessons for developing countries and cities.* https://www.unenvironment.org/resources/report/research-eco-towns-japan-implications-and-lessons-developing-countries-and-cities [Accessed 2 October 2020]

b. ***What a Waste* report (2018):** According to the World Bank's *What a Waste 2.0* report, the world generates 2.01 billion tons of MSW annually, with at least 33% of that not managed in an environmentally safe manner. The report projects that rapid urbanization, population growth, and economic development will push global waste to increase by 70% over the next 30 years – to a staggering 3.40 billion tons of waste generated annually.

Source: World Bank. 2018. *What A Waste: an updated look into the future of solid waste management.* https://www.worldbank.org/en/news/immersive-story/2018/09/20/what-a-waste-an-updated-look-into-the-future-of-solid-waste-management [Accessed 2 October 2020].

c. **Global Resource Outlook report (2019)** – The report attempts to understand the impacts of our growing resource use and develop coherent scenario projections for resource efficiency and sustainable production and consumption that decouple economic growth from environmental degradation.

Source: UN International Resource Panel. 2019. *Global Resources Outlook.* https://www.resourcepanel.org/reports/global-resources-outlook#:~:text=IRP%20Global%20Resources%20Outlook%20 2019,-Download%20the%20Full&text=Since%20the%201970s%2C%20 global%20population,has%20brought%20across%20the%20 globe [Accessed 2 October 2020].

d. **Global Waste Management Outlook** – The document establishes the rationale and the tools for taking a holistic approach towards waste management and recognizing waste and resource management as a significant contributor to sustainable development and climate change mitigation. The Outlook is primarily focused on the "governance" issues which need to be addressed to establish a sustainable solution – including the regulatory and other policy instruments, the partnerships, and the financing models.

Source: UNEP – UN Environment Programme. 2015. *Global Waste Management Outlook.* https://www.unenvironment.org/resources/report/global-waste-management-outlook [Accessed 2 October 2020].

e. **The Long-Run Decoupling of Emissions and Output: Evidence from the Largest Emitters** – This paper uses simple trend decomposition to provide evidence of decoupling between greenhouse gas emissions and output in richer nations, particularly in European countries, but not yet in emerging markets. The evidence also shows that the relationship between trend emissions and trend GDP has also become much weaker in the last two decades than in preceding decades.

Source: Cohen, G., Jalles, J., Loungani, P. and Marto, R., 2018. The long-run decoupling of emissions and output: Evidence from the largest emitters. *Energy Policy*, 118, pp. 58–68.

4. Emergence of circular economy

a. **Introduction to Circular Economy by Ellen MacArthur Foundation:** Ellen MacArthur Foundation is one of the leading institutions dealing with the research and spread of awareness about the circular economy. This presentation is a brief and straightforward overview of what circular economy means.

Source: Ellen Macarthur Foundation. 2019. *Introduction to the circular economy.* https://www.ellenmacarthurfoundation.org/assets/downloads/Introduction-to-the-circular-economy-presentation-slides.pdf [Accessed 2 October 2020].

b. **From linear to circular — Accelerating a proven concept:** This web resource by the World Economic Forum discusses the need to shift from a linear to circular economy model as well as the advantages and concepts of the circular economy.

Source: World Economic Forum. n.d. *From linear to circular—accelerating a proven concept.* https://reports.weforum.org/toward-the-circular-economy-accelerating-the-scale-up-across-global-supply-chains/1-the-benefits-of-a-circular-economy/ [Accessed 2 October 2020].

Notes

1. United Nations Department of Economic and Social Affairs. 2020. *68% of the World Population Projected to Live in Urban Areas by 2050, Says UN | UN DESA | United Nations Department Of Economic And Social Affairs.* https://www.un.org/development/desa/en/news/population/2018-revision-of-world-urbanization-prospects.html [Accessed 1 October 2020].
2. World Economic Forum. 2019. *10 Cities are Predicted to Gain Megacity Status by 2030.* https://www.weforum.org/agenda/2019/02/10-cities-are-predicted-to-gain-megacity-status-by-2030/ [Accessed 1 October 2020].
3. C40. n.d. *A Global Opportunity for Cities to Lead.* https://www.c40.org/why_cities [Accessed 2 October 2020].
4. World Resources Institute. 2019. *Too Many Cities Are Growing Out Rather Than Up. 3 Reasons That's A Problem.* https://www.wri.org/blog/2019/01/too-many-cities-are-growing-out-rather-3-reasons-s-problem [Accessed 1 October 2020].
5. Anh, D., (2003). Internal migration policies in the ESCAP region. *Asia-Pacific Population Journal,* 18/3. https://doi.org/10.18356/f5ba72c7-en
6. International Institute for Environment and Development. n.d. *Cities: An Interactive Data Visual.* <https://www.iied.org/cities-interactive-data-visual> [Accessed 19 October 2020].
7. European Commission. 2019. *Global Food Consumption Growth and Changes In Consumer Preferences.* https://ec.europa.eu/info/news/global-food-consumption-growing-faster-population-growth-past-two-decades-2019-sep-10_en [Accessed 2 October 2020].
8. ETtech.com. 2020. *Global Sales of Smartphones to Increase 3% This Year: Gartner.* https://tech.economictimes.indiatimes.com/news/mobile/global-sales-of-smartphones-to-increase-3-this-year-gartner/73698988#:~:text=Worldwide%20sales%20of%20smartphones%20to,smartphone%20market%20experienced%20a%20decline. [Accessed 2 October 2020].
9. Globe Newswire News Room. 2020. *White Goods Market Growth Analysis, Business Demand with Growing CAGR of 7.9% By 2026.* https://www.globenewswire.com/news-release/2020/06/11/2047036/0/en/White-Goods-Market-Growth-Analysis-Business-Demand-with-growing-CAGR-of-7-9-by-2026-Leading-Players-like-Whirlpool-Corporation-Johnson-Controls-IFB-Industries-Samsung.html [Accessed 2 October 2020].
10. PRN News Wire. 2020. *Logistics Market Size Is Projected to Reach USD 12,256 Billion By 2022 - Valuates Reports.* https://www.prnewswire.com/news-releases/logistics-market-size-is-projected-to-reach-usd-12-256-billion-by-2022—valuates-reports-301065076.html [Accessed 2 October 2020].
11. National Geographic. 2018. *China's Ban On Trash Imports Shifts Waste Crisis To Southeast Asia.* https://www.nationalgeographic.com/environment/2018/11/china-ban-plastic-trash-imports-shifts-waste-crisis-southeast-asia-malaysia/ [Accessed 1 October 2020].
12. Vos, J. and Boelens, R., 2016. The Politics and Consequences of Virtual Water Export. *Eating, Drinking: Surviving,* pp. 31–41.
13. Eni.com. n.d. *The Concept Of Virtual Water.* <https://www.eni.com/en-IT/low-carbon/concept-of-virtual-water.html> [Accessed 18 October 2020].

14. Antonelli, M. and Sartori, M., 2015. Unfolding the potential of the virtual water concept. What is still under debate. *Environmental Science & Policy*, 50, pp. 240–251.
15. World Economic Forum. 2018. *Circular Economy In Cities: Evolving The Model For A Sustainable Urban Future*. http://www3.weforum.org/docs/White_paper_Circular_Economy_in_Cities_report_2018.pdf [Accessed 1 October 2020].
16. UNEP. 2011. Decoupling natural resource use and environmental impacts from economic growth, A Report of the Working Group on Decoupling to the International Resource Panel. Fischer-Kowalski, M., Swilling, M., von Weizsäcker, E.U., Ren, Y., Moriguchi, Y., Crane, W., Krausmann, F., Eisenmenger, N., Giljum, S., Hennicke, P., Romero Lankao, P., Siriban Manalang, A., Sewerin, S. (format not to be changed – citation rules mentioned in the report)
17. World Wide Fund. n.d. *Biocapacity*. https://wwf.panda.org/knowledge_hub/all_publications/living_planet_report_timeline/lpr_2012/demands_on_our_planet/biocapacity/ [Accessed 1 October 2020].
18. World Wide Fund. n.d. *Ecological Footprint*. https://wwf.panda.org/knowledge_hub/teacher_resources/webfieldtrips/ecological_balance/eco_footprint/ [Accessed 1 October 2020].
19. Footprint Network. n.d. *The Ecological Wealth of Nations*. <https://www.footprintnetwork.org/content/images/uploads/Ecological_Wealth_of_Nations.pdf> [Accessed 4 December 2020].
20. Kaza, Silpa; Yao, Lisa C.; Bhada-Tata, Perinaz; Van Woerden, Frank. 2018. What a Waste 2.0: A Global Snapshot of Solid Waste Management to 2050. Urban Development;. Washington, DC: World Bank. © World Bank. https://openknowledge.worldbank.org/handle/10986/30317 License: CC BY 3.0 IGO
21. Office for National Statistics. 2019. The Decoupling Of Economic Growth From Carbon Emissions: UK Evidence - Office For National Statistics. https://www.ons.gov.uk/economy/nationalaccounts/uksectoraccounts/compendium/economicreview/october2019/thedecouplingofeconomicgrowthfromcarbonemissionsukevidence> [Accessed 18 October 2020].
22. OECD. 2020. *Circular Economy - Waste And Materials*. https://www.oecd.org/environment/environment-at-a-glance/Circular-Economy-Waste-Materials-Archive-June-2020.pdf [Accessed 1 October 2020].
23. Liu, C. and Fei, J., 2018. *Decoupling Waste Generation From Economic Growth In Asian Countries*. IGES. https://www.iges.or.jp/en/pub/decoupling-waste-generation-economic-growth/en [Accessed 1 October 2020].
24. Data Source: Japan Industrial Waste Information Center. 2018. *Waste Management In Japan - Rules And Figures*. <https://www.jwnet.or.jp/assets/pdf/en/20190322133536.pdf> [Accessed 31 October 2020].
25. Wahba, S., Kaza, S. and Ionkova, K., 2019. *A New Phenomenon – Realizing Economic Growth While Cutting Waste. HOW?* https://blogs.worldbank.org/sustainablecities/new-phenomenon-realizing-economic-growth-while-cutting-waste-how [Accessed 1 October 2020].
26. World Economic Forum. 2020. *Which Countries Produce The Most Waste?* <https://www.weforum.org/agenda/2020/02/waste-global/#:~:text=Denmark%20tops%20the%20list%20of,produced%2C%20with%20a%20significant%20reduction> [Accessed 23 October 2020].

27. UNEsP. 2014. Decoupling 2: Technologies, opportunities and policy options. A Report of the Working Group on Decoupling to the International Resource Panel. von Weizsäcker, E.U., de Larderel, J., Hargroves, K., Hudson, C., Smith, M., Rodrigues, M. (Format not to be changed. Citation rules mentioned in report)
28. Data sources: GDP per capita – World Bank. n.d. *GDP Per Capita, PPP (Current International $).* <https://data.worldbank.org/indicator/NY.GDP.PCAP.PP.CD?end=2018&start=2018> [Accessed 31 October 2020].
    CO_2 emissions - Our World in Data. n.d. *Where In The World Do People Emit The Most CO_2?.* <https://ourworldindata.org/per-capita-co2> [Accessed 31 October 2020].
29. World Resources Institute. 2020. *Ranking 41 US States Decoupling Emissions and GDP Growth* <https://www.wri.org/blog/2020/07/decoupling-emissions-gdp-us> [Accessed 23 October 2020].
30. Kirchherr, J., Reike, D. and Hekkert, M., 2017. Conceptualizing the circular economy: An analysis of 114 definitions. *Resources, Conservation and Recycling,* 127, pp. 221–232.
31. Internal Market, Industry, Entrepreneurship and SMEs – European Commission. n.d. *Circular Economy – Internal Market, Industry, Entrepreneurship and SMEs – European Commission.* https://ec.europa.eu/growth/industry/sustainability/circular-economy_en#:~:text=Circular%20economy-,Circular%20economy,again%20to%20create%20further%20value. [Accessed 1 October 2020].
32. Ellen Macarthur Foundation. n.d. *What Is The Circular Economy?* https://www.ellenmacarthurfoundation.org/circular-economy/what-is-the-circular-economy [Accessed 1 October 2020].
33. FDI Gov China. n.d. *Circular Economy Promotion Law of The People's Republic Of China.* http://www.fdi.gov.cn/1800000121_39_597_0_7.html#:~:text=Article%202%20The%20Circular%20Economy,in%20production%2C%20circulation%20and%20consumption. [Accessed 1 October 2020].

2

Key Concepts, Strategies, and Programs

This chapter describes some of the key concepts, strategies, and programs that have helped to put a circular economy into practice. These instruments can be categorized as:

- Facility/corporate focused – that are implemented at the facility or corporate level.
- Regional/across life cycle/across supply chains – that focus on product lifecycles, or across supply chains or are implemented on a regional scale, and not limited to a facility.
- Policy/strategy – high-level strategies and programs that influence regional, national, and international economies.

However, many of these instruments have commonalities, overlaps, and interlinkages and should not be considered in silo.

2.1 Key Concepts and Strategies

Figure 2.1 shows the evolution of some of the key concepts over the last five decades. Many of these concepts have played a significant role in the process of decoupling between economic growth and waste generation. These concepts have also further matured, expanded, and have been later introduced in environment management practice in the form of programs. However, we limit only to some concepts and strategies that have had a significant influence on global and regional levels for building the path toward a circular economy.

2.1.1 Waste Minimization

Waste minimization was one of the early concepts that initially drew attention at the facility and corporate levels. Later the concept was expanded to the neighborhood, regional, and national scales.

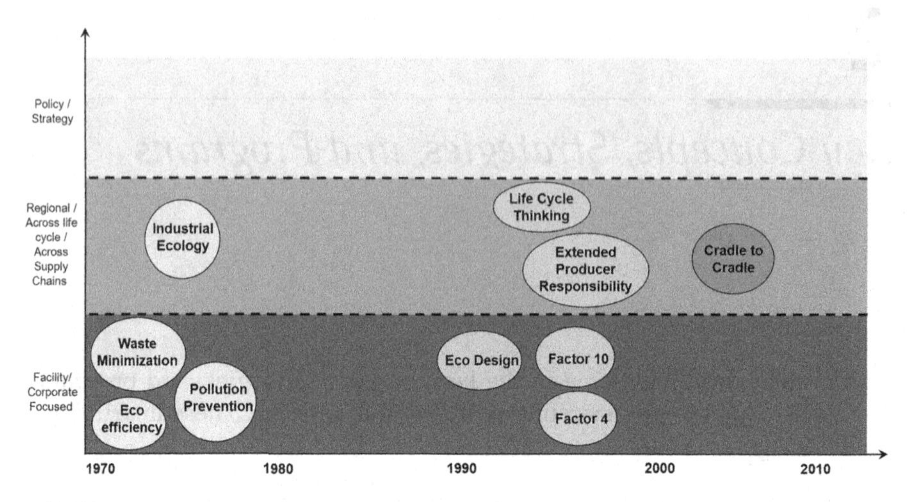

FIGURE 2.1
Evolution of key concepts and strategies.

2.1.1.1 Waste Minimization at a Facility/Corporate Level

In the late 1980s, the concept of waste minimization emerged in the United States in response to the rising costs of waste treatment and disposal. Regulatory standards on wastes and emissions were getting stricter, requiring more investments on the "end of pipe" treatment.

In 1984, the Hazardous and Solid Wastes Amendments to the Resource Conservation and Recovery Act led to unprecedented increases in waste management costs. In response, the industry started exploring and prioritizing waste minimization. Practicing waste minimization helped reduce the waste-related liabilities under the Comprehensive Environmental Response, Compensation, and Liabilities Act (CERCLA, or superfund). In this context, waste minimization became an attractive proposition to the industry. Box 2.1 defines waste minimization that had then a focus on hazardous waste.

BOX 2.1 DEFINITION OF WASTE MINIMIZATION[1]

The reduction, to the extent feasible, of hazardous waste that is generated or subsequently treated, stored, or disposed of. It includes any source reduction or recycling activity undertaken by a generator that results in either:

- The reduction of total volume or quantity of hazardous waste.
- The reduction of toxicity of hazardous waste, or both, so long as such reduction is consistent with the goal to minimize present and future threats to human health and environment.

2.1.1.2 Waste Minimization at the Neighborhood Level and in Clusters

Soon, the concept of waste minimization was used to build waste minimization at a program, regional and national levels not limiting to a facility or hazardous waste. Box 2.2 illustrates two interesting waste minimization programs that operated at the national level.

BOX 2.2 PROGRAMS TO PROMOTE WASTE MINIMIZATION AT THE NATIONAL LEVEL

WASTE MINIMIZATION CLUBS IN THE UNITED KINGDOM[2]

The concept of waste minimization club (WMC) was first developed in the Netherlands in the early 1990s to encourage industries to reduce pollution. WMCs in the United Kingdom (UK) consisted of a group of industries within the same sector or in a given geographical area or neighborhood. In addition to providing varying levels of training and consultancy support to participating industries, WMCs provided members an opportunity to share knowledge and experience through club meetings and events.

Based on data from the firm Envirowise, it has been estimated that 150 WMCs operated across the UK from 1992 to 2004, working with 5222 industries. This figure has to be treated with caution as it only includes those clubs within Envirowise's database. In 2001, Envirowise concluded that 138 clubs and projects had generated savings of GBP 45 million between 2005 and 2008. In addition, 70 resource efficiency clubs were set up by Envirowise covering an additional 1330 industries, leading to an additional GBP 25 million of savings over three years.

WASTE MINIMIZATION CIRCLES IN INDIA[3]

National Productivity Council (NPC), India, saw the need for a new mechanism that could overcome various barriers faced by Small and Medium Enterprises (SMEs) in waste minimization efforts and developed the Waste Minimization Circle concept in cooperation with the Ministry of Environment and Forests (MoEF).

WMC is a group of SMEs that manufacture similar products or employ similar processes and meet regularly to learn about waste minimization options to increase profitability and reduce waste generation.

The program started in 1994 and underwent three phases. Phase II that lasted from July 1997 to 2005, WMC members made investments of USD 5.56 million in waste minimization efforts and reaped annual savings of USD 3.7 million, bringing total savings to USD 9.26 million by the end of 2005. The environmental benefits included reductions in water, energy, and material consumption of an average 10–30%; reduction in solid waste generation of 5–20%; and yield improvement of 2–5%. As of 2013, 158 WMCs had been established in 17 Indian states, representing more than 40 industrial sectors.

Visit the link below to view different case studies on industries that have successfully implemented the WMC initiative:

https://www.indiawaterportal.org/articles/case-studies-waste-minimisation-circle-initiative-ministry-environment-and-forests

ACTIVITY 2.1 FUNDING FOR WASTE MINIMIZATION CLUBS

Many waste minimization initiatives similar to WMCs did not include financing/investments for their implementation. Conduct research on WMC programs that involved a fund or financing institution. Could this arrangement have made a difference? What could have been the challenges?

Could you think of a revolving fund that is built on the profits or savings made by the members of the WMC? How could such a fund be structured on a co-operative basis?

2.1.1.3 Waste Minimization at City/Regional Level – Concept of Zero Waste

Possibly the first mention of the term "zero waste" came from Daniel Knapp's concept of Total Recycling[4]. Box 2.3 defines the term "zero waste".

In the 1980s, Daniel Knapp and his wife founded a company in Berkeley, California, called "Urban Ore" for salvaging waste operation and marketing Waste Minimization. Urban Ore could divert a variety of wastes from landfill and reuse it within the community. Urban Ore had a contract with the City of Berkeley for exclusive salvage rights of reusable materials from the city's transfer station tip floor. Urban Ore salvagers spotted and reported hazardous wastes to city staff to avoid illegal disposal. Its success is still evident today.

Like Urban Ore, there have been multiple zero-waste initiatives that promote awareness and offer solutions toward reducing and recycling waste through "zero-waste" shops. See Box 2.4.

Upon Urban Ore's success in the community at Berkeley, Daniel Knapp proceeded to spread his "Total Recycling" concept globally. In 1995, Knapp travelled to Australia for the first series of talks with governments, businesses, and citizens in major cities on how to maximize materials recovery and minimize through reuse, recycle, and composting[4]. During that time, he

BOX 2.3 DEFINITION OF ZERO WASTE[37]

The Planning Group of the Zero Waste International Alliance adopted the first peer-reviewed internationally accepted definition of Zero Waste on November 29, 2004. The Zero Waste International Alliance defines Zero Waste as "the conservation of all resources by means of responsible production, consumption, reuse, and recovery of products, packaging, and materials without burning and with no discharges to land, water, or air that threaten the environment or human health".

discovered a resource recovery company in Australia called Revolve: that led the activism of "No Waste by 2010". Revolve was supported by the government. Daniel acted as a consultant and helped the founder of Revolve expand its operations to build a no-waste transfer facility[6].

While Daniel was touring Australia, that same year[7] in the USA, Lynn Landes set up the Zero Waste USA[8], a website to change waste habits on an individual level, and Bill Sheehan founded the GrassRoots Recycling Network[9].

In 2003, Vaughan Levitzke established Zero Waste South Australia[10]. The Office of Zero Waste SA was proclaimed a statutory authority by the South Australian Government, and in February 2004, legislation to create Zero Waste SA was passed[11]. South Australia's waste management achievements have been recognized in the United Nations Human Settlement Programme (UN-HABITAT) publication "Solid Waste in the World's Cities that assessed waste management and recycling systems of more than 20 cities worldwide"[12]. Today the Zero Waste SA is transformed into Green Industries SA[13] that has taken leadership in circular economy in Australia and internationally[14]. The National Action Plan 2019, in Australia aims to influence the businesses and households to realize the full value of recyclable materials and work toward more sustainable resource use.

BOX 2.4 ZERO-WASTE SHOPS[5]

Zero-waste shops operate in many countries such as Germany (Original Unverpackt), Austria (Lunzers), Spain (Granel), Italy (Effecorta), and the UK (The Zero Waste Shop). In the United States, one of the first zero-waste shops called "In.gridients" was opened in Austin, Texas, in 2012. However, it permanently closed in April 2018 as a result of low sales. Despite the financial risks, similar businesses have appeared across the United States. Zero Market opened in Denver in 2017, and the Refill Shoppe in Ventura, California, opened in 2010.

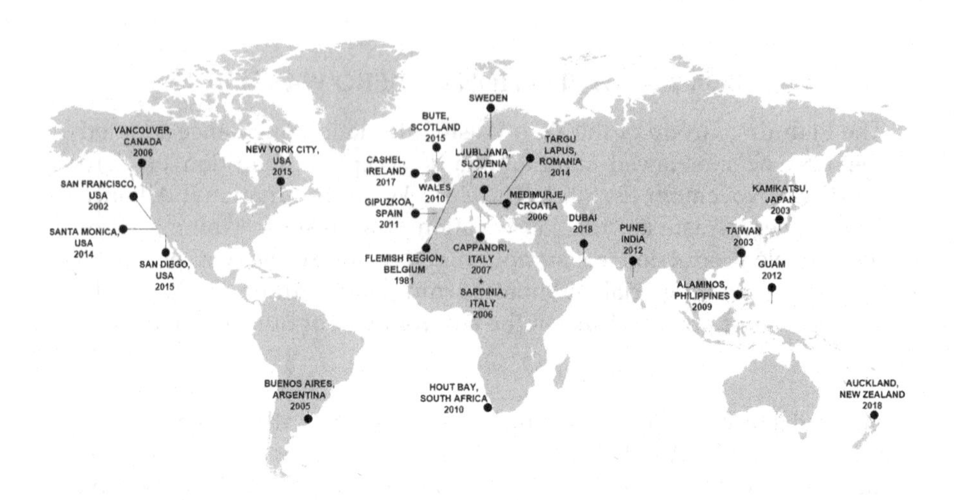

FIGURE 2.2
Twenty-five cities that have committed to going zero-waste[15].

Various countries today have prepared plans and set up policies to shift to a zero-waste economy. For instance, Scotland and Singapore aim to convert into a zero-waste society by following a circular economy. These case studies have been discussed in Chapter 10. These countries have set ambitious targets for reducing waste and increasing recycling through various policy reforms for encouraging reuse and recycling. Multiple cities have committed to a zero-waste economy. Figure 2.2 shows twenty-five cities across the world that have pledged for going zero waste.

2.1.2 Pollution Prevention

Pollution prevention is a term used interchangeably with waste minimization. However, there is a difference between the two terms.

Pollution prevention reduces or eliminates waste at the source by modifying production processes, promoting the use of non-toxic or less toxic substances, implementing conservation techniques, and reusing materials rather than putting them into the waste stream[18]. United States Environmental Protection Agency (US EPA) coined this definition in 1992 through a memorandum under the Pollution Prevention Act (See Box 2.5).

Pollution prevention was introduced in ISO 14001 Environmental Management System (EMS). The definition of pollution prevention in the standard was, however, too broad. It defined "prevention of pollution" as the "use of processes, practices, materials, or products that avoid, reduce or control pollution, which may include recycling, treatment, process changes, control mechanisms, efficient use of resources and materials substitution". The ISO 14001 standard thus does not distinguish between pollution prevention and pollution control.

BOX 2.5 DEFINITION OF POLLUTION PREVENTION[16]

The Pollution Prevention Act defines "source reduction" to mean any practice which:

- Reduces the amount of any hazardous substance, pollutant, or contaminant entering any waste stream or otherwise released into the environment (including fugitive emissions); prior to recycling, treatment, or disposal and
- Reduces the hazards to public health and the environment associated with the release of such substances, pollutants, or contaminants.

The term pollution prevention includes equipment or technology modifications; process or procedure modifications; reformulation or redesign of products; substitution of raw materials; and improvements in housekeeping, maintenance, training, or inventory control.

Under the Pollution Prevention Act, however, recycling, energy recovery, treatment, and disposal are not included within the definition of pollution prevention.

2.1.3 Green Chemistry

Use of chemicals in manufacturing of products has intensified over last five decades. These chemicals can have significant implications to the environment across the life cycle. The aim of green chemistry is to reduce chemical related impacts on human health and ecosystems through avoidance of hazardous chemicals by substitution of safer alternatives and the same time improve process and product performance. Application of green chemistry also helps in the technological and commercial viability of recycling and recovery of used products. Paul T. Anastas and John C. Warner laid out 12 principles of green chemistry. (See Box 2.6)

2.1.4 Industrial Ecology

Robert Frosch and Nicholas E. Gallopoulos popularized the concept of industrial ecology in a Scientific American article in 1989. Frosch and Gallopoulos' vision was, "why would not our industrial system behave like an ecosystem, where the wastes of a species may be a resource to another species?

BOX 2.6 TWELVE PRINCIPLES OF GREEN CHEMISTRY[17]

1. Prevention of waste to avoid treating or cleaning up waste after it has been created.
2. Atom economy[18] through new synthetic methods designed to maximize the incorporation of all materials used in the process into the final product.
3. Less hazardous chemical syntheses designed to use and generate substances that possess little or no toxicity to human health and the environment.
4. Design of safer chemicals able to carry out the desired function while minimizing their toxicity.
5. Avoiding wherever possible or minimizing the use of auxiliary substances (e.g., solvents, separation agents, and others), and introducing safer solvents and auxiliaries that are innocuous when they must be used.
6. Design for energy efficiency of chemical processes to minimize their environmental and economic impacts and if possible, to introduce synthetic methods to be conducted at ambient temperature and pressure.
7. Promotion of the use of renewable raw materials or feedstock instead of depleting ones whenever technically and economically practicable.
8. Reduce derivatives through minimizing or avoiding the use of blocking groups, protection/deprotection, and temporary modification of physical/chemical processes that require additional reagents and can generate waste.
9. Use catalytic reagents as selective as possible.
10. Design for degradation of chemical products at the end of their function into innocuous degradation products not persisting in the environment.
11. The development of analytical methodologies needed to allow real-time analysis for pollution prevention, in-process monitoring and control prior to the formation of hazardous substances.
12. Inherently safer chemistry for accident prevention substances and the form of a substance used in a chemical process to be chosen to minimize the potential for chemical accidents, including releases, explosions, and fires.

Why would not the outputs of an industry be the inputs of another, thus reducing the use of raw materials, pollution, and saving on waste treatment?"[19] Box 2.7 provides a definition of industrial ecology. In many ways the concept of industrial ecology is like a "biomimicry" that will be discussed later in this chapter.

Industrial ecology promoted a "systems thinking" approach. It was also recognized as a strategy for industrial symbiosis, where industries can gain a competitive advantage through the physical exchange of materials, energy, water, and by-products, thereby fostering inclusive and sustainable development[22]. While revolutionary, this approach needed a mechanism where planning of industrial clusters was carried out on such principles. A good example of putting industrial ecology in practice was the design and implementation of Eco-Industrial Park (EIP). EIPs will be discussed in Chapter 11, illustrating case studies.

BOX 2.7 DEFINITION OF INDUSTRIAL ECOLOGY

Industrial ecology conceptualizes industry as a human-made ecosystem that operates similarly to natural ecosystems. The waste or by product of one process is used as an input into another process. Industrial ecology interacts with natural ecosystems and attempts to move from a linear to cyclical or closed loop system. Like natural ecosystems, industrial ecology is in a continual state of flux[20,21].

2.1.5 Eco-efficiency

In 1990s, Schaltegger and Sturm introduced the idea of eco-efficiency as a "business link to sustainable development"[23]. However, the concept can be traced back to the 1970s. The concept was further popularized for the business sector by the World Business Council for Sustainable Development (WBCSD), during the United Nations Conference on Environment and Development (UNCED) that was held in 1992.

Eco-efficiency is described as "the delivery of competitively priced goods and services that satisfy human needs and bring quality of life, while progressively reducing ecological impacts and resource intensity throughout the life cycle to a level at least in line with the Earth's estimated carrying capacity"[24]. In short, it means creating more value with less impact across the life cycle.

Eco-efficiency is thus, a management philosophy that encourages businesses to search for environmental improvements that yield parallel economic benefits. The definition made businesses think beyond the factory gates and address impacts due to resource consumption across the life cycle. Eco-efficiency however emphasizes economics, in addition to environmental protection and improvement.

Following are the key terms used in describing eco-efficiency that are worth noting:

Ecological	Economic
• Ecological impacts • Resource intensity • Lifecycle • Carrying capacity	• Competitive pricing of goods • Satisfaction of human needs • Quality of life

Later at the WBCSD, the eco-efficiency concept expanded and evolved over time and matured to lead to several programs and activities. Table 2.1 shows the evolution.

TABLE 2.1

Evolution of Eco-efficiency at World Business Council for Sustainable Development[25]

Year	Activity
1991	The term Eco-efficiency was coined after a search to find a phrase that captured the notion of doing more with less while being environmentally sound. It was first used in the Changing Course (analysis report on how businesses can adapt to sustainable development) and is still used today.
2002	WBSCD Young Managers Team – A program established to inculcate tomorrow's high-potential senior executives with the principles of sustainability before they reach positions of power. This was the starting block for WBCSD's acclaimed Future Leaders program, formalized in 2004. It is now known as the WBSCD Leadership Program.
2006	Inclusive Business – A term coined by WBCSD to describe expanding access to goods, services, and jobs for people at the base of the economic pyramid. This term is now used around the world by leading multinational companies as well as international development organizations and at the G20 political level.
2010	Vision 2050 – The cornerstone business plan for a sustainable world. Vision 2050 was compiled by 29 global companies from 14 industries, after 18 months of dialogue with more than 200 companies and external stakeholders across 20 countries.
2012	Business Action for Sustainable Development – A coalition of business groups built ahead of the Rio+ 20 Conference. It provided a platform for exchanging policy recommendations, business solutions and partnerships to help advance sustainable development.
2015	Low Carbon Technology Partnerships Initiative (LCTPi) – A unique, action- oriented program that was presented at COP21 in Paris and recognized as the voice of business at the event. LCPTi brings together more than 170 companies and 70 partners, to accelerate the development of low-carbon technology solutions.
2016	Natural Capital Protocol – A global framework helping businesses make better decisions by understanding their impacts and dependencies on natural capital. WBSCD was a key partner in the Natural Capital Coalition, the unique global Multi – stakeholder collaboration that developed the Protocol.
2018	Factor 10 initiative – 30 leading companies with a combined revenue of USD 1.3 trillion join forces to implement the circular economy.

2.1.6 Factor 4 and 10

Eco-efficiency is concerned with resource productivity, that is, maximizing the value added per unit of resource input. The term "Factor 4" refers to a hypothetical fourfold increase in "resource productivity", brought about by simultaneously doubling wealth and halving resource consumption.

The Factor 4 concept was introduced in 1998 in a book of the same name written by L. Hunter Lovins and Amory Lovins of the Rocky Mountain Institute, and Ernst von Weizsäcker, founder of the Wuppertal Institute for Climate, Environment & Energy. Its origins date back to 1972 when a report by the Club of Rome called "Limits to Growth" issued a stark warning that economic growth was using up resources at a rate that could not be sustained for much longer[26].

In 1989, the Factor 10 concept was developed by Schmidt-Bleek[27]. In 1994, the Factor 10 Club, an international body of senior government, non-government, industry, and academic leaders was formed. Today, the Factor 10 Club remains inactive. Box 2.8 provides a definition of Factor 10.

Factor 10 is today WBCSD's Circular Economy project[29]. It brings companies together to reinvent how business finds, uses, and disposes of the materials that make up global trade.

BOX 2.8 DEFINITION OF FACTOR 10[28]

Factor 10 stated that over the next 30–50 years (one generation) a decrease in energy use and material flows by a factor of 10 and an increase in resource productivity/efficiency by a factor of 10 is required to achieve dematerialization. That is, to attain sustainability and environmental protection we need to reduce resource turnover by 90% on a global scale within the next 50 years.

2.1.7 Life Cycle Thinking

Life Cycle Thinking (LCT) is about going beyond the traditional focus on production site or facility and manufacturing processes to address the environmental, social, and economic impacts of a product (that includes packaging) over its entire life cycle. LCT's main goals are to reduce a product's resource use and emissions to the environment and improve its socio-economic performance across the value chain through its life cycle[30]. Box 2.9 provides definition of LCT and Figure 2.3 illustrates the typical life cycle of a product.

Figure 2.4 illustrates the life cycle of a mobile phone. It may be observed that the life cycle of a product maps to the fundamental stages but is unique with some variations.

LCT's precursors emerged in the late 1960s and early 1970s from concerns about limited natural resources, particularly oil. These concerns came in the

BOX 2.9 DEFINITION OF LIFE CYCLE THINKING[31]

Life Cycle Thinking (LCT) seeks to identify possible improvements to goods and services in the form of lower environmental impacts and reduced use of resources across all life cycle stages. LCT begins with raw material extraction and conversion, then manufacture and distribution, through to use and/or consumption. It ends with re-use, recycling of materials, energy recovery, and ultimate disposal.

form of global modeling studies and energy audits at facilities. These audits were referred to as Resource and Environmental Profile Analyses and Net Energy Analyses[32].

Today, LCT has become a key tool in policy and decision-making in a circular economy. It has also inspired innovation, especially in product design and influenced consumer behavior. Thinking through the life cycle, businesses recognize that each choice sets the stage for how the product will look and function and how it will impact the environment and the community as it is manufactured, used, disposed of, or re-used and recycled[33]. Box 2.10 outlines some of the EU's key policies and instruments that have been influenced by LCT.

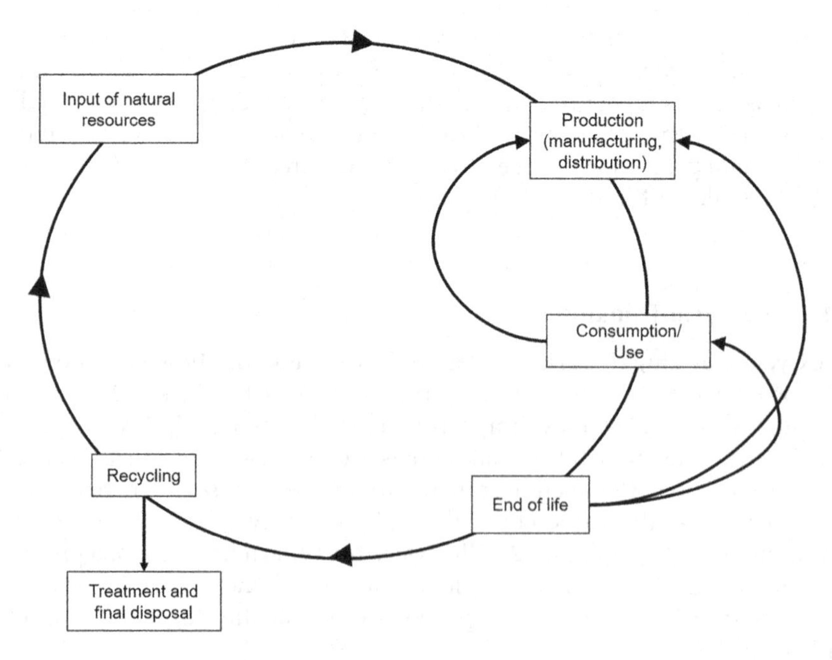

FIGURE 2.3
Typical life cycle of a product.

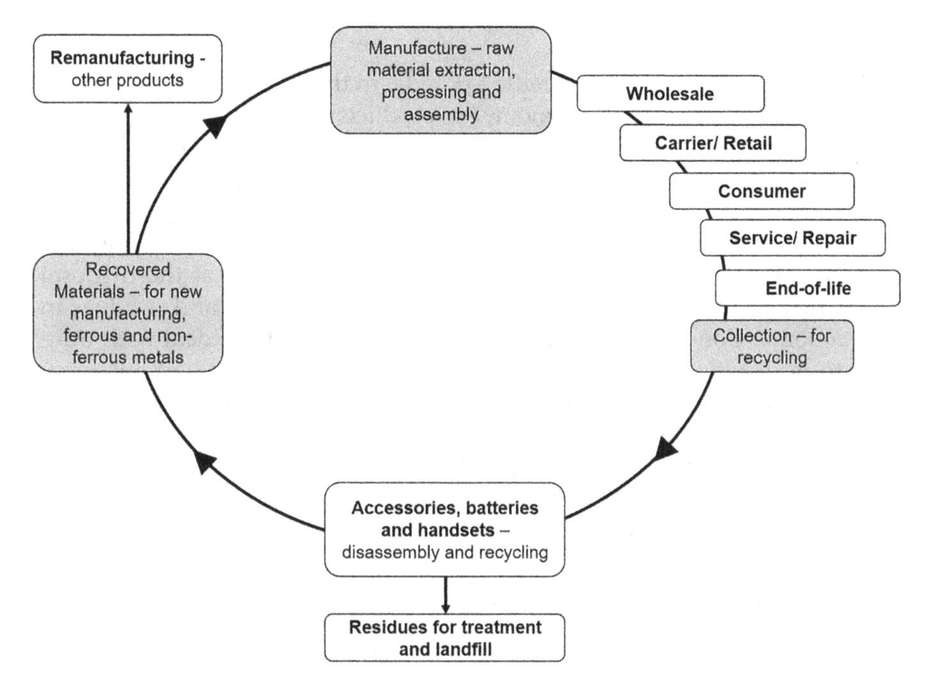

FIGURE 2.4
Life cycle of a mobile phone.

BOX 2.10 INFLUENCE OF LIFE CYCLE THINKING ON EU POLICIES[25]

In the European Union, LCT is at the heart of developing policies and instruments in areas such as:

- Integrated Product Policy, the Sustainable Consumption and Production (SCP) and Sustainable Industrial Policy Action Plan, Green Public Procurement, EU Ecolabel, EU Eco-Management and Audit Scheme, Ecodesign, Retail Forum.
- Waste – LCT in the Waste Framework Directive, used to help determine the benefits of different prevention or management options. LCT is also central to the Thematic Strategy on the prevention and recycling of waste and the Thematic Strategy on the sustainable use of natural resources.
- Eco-innovation and the EU Environmental Technologies Action Plan (ETAP).

2.1.8 Cradle to Cradle

Walter Stahel coined the term Cradle to Cradle in the 1970s. Later this concept led to the term "performance economy" as products started getting replaced by services. Together with Bill McDonough, Michael Braungart went on to develop the Cradle to Cradle™ concept and introduced a certification process in 2002[34].

In many ways, cradle to cradle concept was allied to "closing the loop" and understanding a circular economy. With the introduction of this concept, the open-ended lifecycle of a product (which ended at the disposal stage), turned into a vicious cycle of re-introducing wastes and residues in the economy. Conceptually, the idea of Cradle to Cradle is ideal with minimal to no waste generated. However, in practice, it poses multiple challenges. For instance, the chemicals introduced during the production stage could contaminate and lead to difficulties in the recycling stage and pose risks to the consumer. Additionally, the economics of recycling (high use and cost of energy and infrastructure) and the associated environmental and social impacts make this concept questionable. This challenge is called "spillovers".

2.1.9 Eco-design and Design for Sustainability

The first initiatives on eco-design started in the late 1980s in Europe and the USA. LCT provided direction to practicing eco-design. Eco-design promoted environmental aspects into the product design and development process at all life cycle stages while meeting functional product requirements and consumer needs[35].

At the beginning of the 1990s, the first range of eight eco-design demonstration projects in different industrial sectors (like furniture, automotive, and packaging) took place in the Netherlands. Based upon the experiences, a first serious attempt was made to develop an eco-design methodology and tools. In 1994, the "PROMISE" eco-design manual was published.

Later, a Design for Sustainability (DfS and later changed to D4S) program was introduced in collaboration with United Nation Environment Program (UNEP) and other Dutch partners. The concept of eco-design thus expanded to addressing sustainability, especially addressing the social aspects. Today, D4S efforts are linked to wider concepts such as product-service mixes, systems innovation, and other life cycle-based efforts. Box 2.11 provides a definition of D4S.

2.1.10 Biomimicry

The term Biomimicry was popularized by Janine Benyus in her book, *"Biomimicry: Innovation Inspired by Nature"*, published in 1997. Biomimicry is a discipline that studies nature and then imitates nature to design and

BOX 2.11 DEFINITION OF DESIGN FOR SUSTAINABILITY[36]

D4S was a natural outcome of UNEP's work on Cleaner Production (CP), eco-efficient industrial systems, and life cycle management. A broad definition of D4S would be that industries take environmental and social concerns as a key element in their long-term product innovation strategy. This implies that companies incorporate environmental and social factors into product development throughout the life cycle of the product, throughout the supply chain, and with respect to their socio-economic surroundings (from the local community for a small company, to the global market for a transnational corporations).

address problems related to sustainability. Biomimicry relies on three key principles[37]:

- Nature as a model: Study nature's models and emulate these forms, processes, systems, and strategies to solve human problems.
- Nature as a measure: Use an ecological standard to judge the sustainability of our innovations.
- Nature as a mentor: View and value nature not based on what we can extract from the natural world, but what we can learn from it.

In 2006, the Biomimicry Institute was founded by Janine Benyus and Bryony Schwan to naturalize biomimicry in the culture by promoting the transfer of ideas, designs, and strategies from biology to sustainable human systems design[38].

2.1.11 Extended Producer Responsibility

Thomas Lindquist first introduced the concept of Extended Producer Responsibility (EPR) in Sweden in 1990. He suggested that the life of products and packaging post-consumer use should be the responsibility of product manufacturers and distributors. Companies being held responsible for their products would drive them to simplify and streamline product designs to be more recyclable and/or reusable, as well as less costly to manage. This would in turn reduce waste and increase resource efficiency. EPR could also reduce costs of waste management of local administration.

Recognition and participation of the informal sector and application of tools for waste aggregation are necessary for the effective implementation of the EPR. It is also important that the consumers take a responsibility for the segregation of post-consumer products. Box 2.12 provides definition of EPR.

BOX 2.12 DEFINITION OF EXTENDED PRODUCER RESPONSIBILITY[39]

Extended Producer Responsibility (EPR) is a policy approach under which producers are given a significant responsibility – financial and/or physical – for the treatment or disposal of post-consumer products. Assigning such responsibility could in principle provide incentives to prevent wastes at the source, promote product design for the environment, and support the achievement of public recycling and materials management goals.

Today, the concept of EPR has influenced policies on a circular economy, globally. In 1991, Germany introduced the first example of EPR in Europe with a requirement that manufacturers assume responsibility for recycling or disposing of packaging material they sold. In response, the German industry set up a "dual system" for waste collection, picking up household packaging alongside municipal waste collections[40]. Business and policy perspective of EPR will be discussed in Chapters 8 and 10.

EPR led to the concepts of Design for Disassembly (D4D) and Design for Repairability (D4R) in the eco-design. D4D is a technique to design the product, such that products can be disassembled with flexible configurations. D4D helps in cost-effective separation and recovery of reusable components and materials at the end of life of the product and allows application of processes such as take back, product reuse, remanufacture, and recycling. The disassembly process has thus become an important strategy to reduce the environmental impact and increase the value of end-of-life products[41]. D4R addresses safe, efficient, and easier product maintenance and enhanced serviceability. D4R plays an important role in extending product life.

2.2 Key Programs

Many of the concepts introduced in Section 2.1 of this chapter led to the operation of programs at the national, regional, and international scale. These programs influenced the consumption and production patterns, helped in decoupling or doing more with less. Many of these programs were operated in partnerships and influenced policies toward circularity, changed corporate behavior as well as investment flows. This section highlights a few such programs that helped and continue to set the stage toward a circular economy (see Figure 2.5).

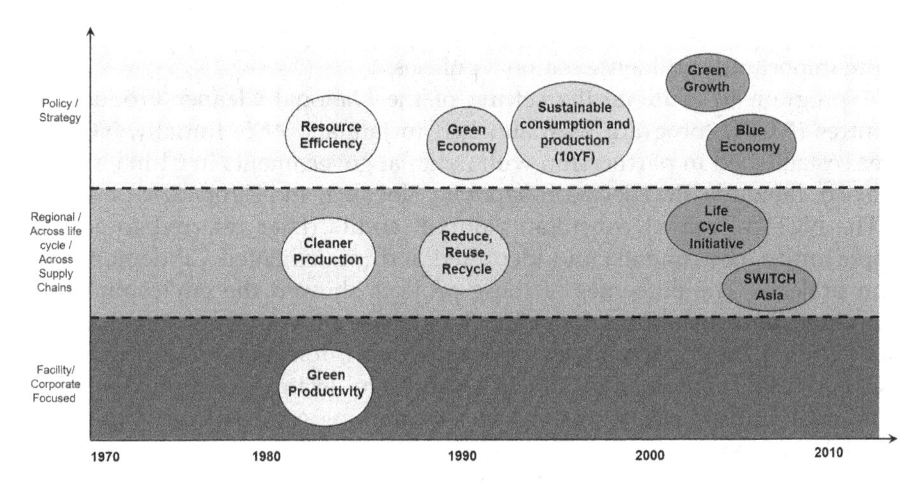

FIGURE 2.5
Some of the influencing programs that set the stage for a circular economy.

2.2.1 Cleaner Production Program

The term Cleaner Production (CP) was defined by the UNEP in 1990. Box 2.13 provides this definition.

UNEP and United Nations Industrial Development Organization (UNIDO) jointly ran the CP program. Three principal activities undertaken in this program included[43]:

- advancing the frontiers of understanding the concept of CP
- marketing the message of the "win-win", environmental and economic potentials of CP and
- enhancing the local national and regional capacities to ensure effective responses to the growing demand for CP services.

Under this program, bi-annual high-level meetings were held at different world regions for policy dialogue and sharing of practice experiences. These

BOX 2.13 DEFINITION OF CLEANER PRODUCTION[42]

"The continuous application of an integrated environmental strategy to processes, products, and services to increase efficiency and reduce risks to humans and the environment". This definition has been used as the working definition of all programs related to cleaner production promotion and still continues to be a valid definition.

meetings played an important role in communicating the benefits of CP and, more importantly, influence national policies.

A program to facilitate the setting of the National Cleaner Production Centres (NCPC) program was launched in January 1995. Initially, NCPCs were established in partnership with national governments in China, India, Mexico, Tanzania, Brazil, Czech Republic, Slovakia, and Zimbabwe.

The NCPCs trained consultants on CP audits (later referred to as CP opportunity assessments) and identified and implemented CP demonstration projects. The outcomes of these projects showed the environmental, economic, and social benefits of CP. A database of case studies was developed, called International Cleaner Production Information Clearinghouse (ICPIC). ICPIC contained around 400 successful examples on implementation of CP cutting across various sectors, making a convincing argument that CP does pay.

Inspired by the CP approach, several national governments, bi-lateral, and multi-lateral development agencies set up programs. These programs focused on building capacities, supporting demonstration projects, and providing finance, especially for SMEs. In this process, numerous guidance manuals describing methodologies and tools were evolved. A compilation called Cleaner Production Companion was developed by UNEP to serve as a one-stop resource for CP community. A Global Status report on CP was published in 2002[44] summarizing 72 countries' achievements.

Post 20-year operation of the CP program, UNIDO & UNEP jointly formulated a program on Resource Efficient Cleaner Production (RECP) in developing and transition countries[45]. RECP entailed the continuous application of preventive environmental strategies to processes, products, and services to increase resource efficiency and reduce risks to humans and the environment. Resource efficiency was thus the focus.

RECP addresses the three sustainability dimensions[46] individually and synergistically:

- Heightened economic performance through improved productive use of resources.
- Environmental protection by conserving resources and minimizing industry's impact on the natural environment.
- Social enhancement by providing jobs and protecting the wellbeing of workers and local communities.

In 2009, the NCPC program's name was changed to "UNIDO-UNEP Joint Global Resource Efficient and Cleaner Production (RECP) Program for developing and transition countries". At the end of 2014, the UNIDO-UNEP RECP Program had worked with 58 NCPCs in 56 countries. The centres' regional distribution was as follows: 13 in Africa, 11 in Asia, 19 in Europe, 10 in Latin America, and 5 in the Middle East and North Africa[47].

To ensure knowledge sharing, UNIDO established Network for Resource Efficient and Cleaner Production (RECP*net*)[48] that brought together more than 70 providers of RECP related services on a global level.

2.2.2 Green Productivity Program

Green Productivity (GP) is a concept which evolved to address the growing concern of consumers and stakeholders of business communities toward products, productivity, and the environment. To address this demand, Asian Productivity Organization (APO) developed the concept of GP as a strategy for enhancing productivity while improving environmental performance. GP encourages the application of productivity tools along with environmental management tools. While the concept of GP is similar to CP, GP focuses more on productivity improvement and social development[49]. Box 2.14 presents definition of GP.

GP program focused on capacity building by training GP professionals and other stakeholder institutions in the APO member counties to institutionalize GP's promotion and adoption. Supporting GP demonstration projects was also included in the GP program. APO published a compendium of case studies and training manuals on how to implement GP.

The GP program of APO gradually expanded to address more areas, especially the greening of supply chains. Under this program two flagship projects were undertaken viz., holding eco-products international fair and development of the eco-products database. These projects were implemented to support leading Japanese corporations with years of practice experience in greening supply chains.

BOX 2.14 DEFINITION OF GREEN PRODUCTIVITY[50]

GP is a holistic, proven approach for strengthening competitiveness, protecting the environment, achieving sustainable low-carbon growth to combat the adverse impacts of climate change, and alleviating poverty. It should be adopted by all stakeholders including governments and the public and private sectors.

2.2.3 3R Forum

The Regional 3R Forum in Asia was launched in November 2009 by UN Centre for Regional Development, Nagoya (UNCRD) with the Government of Japan's support. The concept of 3R, i.e., Reduce, Reuse, and Recycle, existed long before the creation of the 3R forum. Japan was the first country to adopt the 3R approach in its policy formation[51].

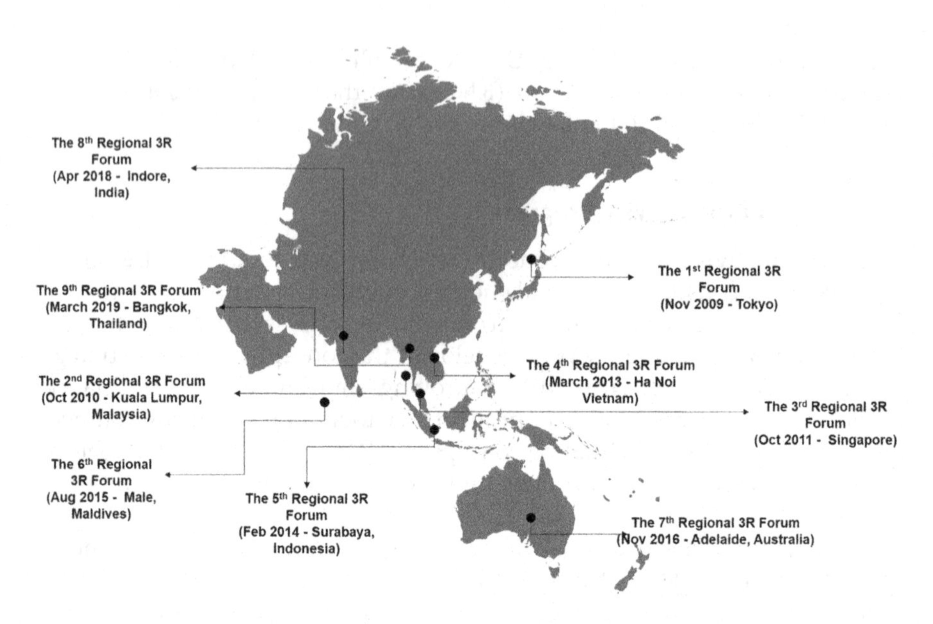

FIGURE 2.6
Locations of the annual meetings of the 3R forum.

The Forum had an objective to provide strategic policy advice to national governments in Asia-Pacific Region in mainstreaming 3Rs in the overall policy, planning, and development. The Forum addressed topics such as resource efficiency, sustainable consumption, and production, integrated solid waste management, sound material-cycle, low-carbon, and zero waste society. Later, the Forum focused on circular economy. The website of 3R Forum[52] holds a significant repository of knowledge.

Figure 2.6 shows the locations of the 3R Forum events. In the 4th 3R Forum held in Hanoi in Vietnam, 3R Goals (2013–2023) were adopted to guide and provide a framework for reporting progress on 3R policies. Taking inspiration from the 3R Forum, Government of Bangladesh prepared a 3R Action plan[53].

2.2.4 Sustainable Consumption and Production (10 YFP)

Sustainable consumption and production (SCP) was included as a standalone goal (SDG 12) of the 2030 Sustainable Development agenda. SDG 12 plays a critical role in decoupling. In 2012, under the Target 12.1, a 10-Year Framework Program (10 YFP) was launched for implementation. The 10 YFP is a global commitment to accelerate the shift toward SCP in both developed and developing countries[54].

The One Planet Network was formed to implement the commitment of the 10 YFP. It is a multi-stakeholder partnership for sustainable development, generating collective impact through its six programs: Public Procurement, Buildings and Construction, Tourism, Food Systems, Consumer Information, and Lifestyles and Education[49].

In 2017, the 10-year framework completed half its term. It served as a good opportunity to take stock of the progress, achievements and lessons learned during its first five years and further define priorities for the next five years. The progress report released for the period 2012–2017 identifies a variety of solutions within the network that support the shift to SCP, such as knowledge resources, technical tools, policies, and policy instruments. While cost effective and high-impact solutions exist, reporting results indicate that maintainable changes in consumption and production remain a challenge[55].

2.2.4.1 SWITCH Asia[56]

SWITCH-Asia (2007–2020) is the largest SCP program supported by the European Union in Southeast Asia, South Asia, Central Asia, Mongolia, and China. Under this program, nearly EUR 300 million have been invested toward promoting SCP in Asia and Central Asia.

The program funds pilot projects through its grants scheme to help consumers and companies act more responsibly in their daily choices and lifestyles and adopt cleaner technologies and more sustainable industrial practices. Adopting sustainable practices will ultimately support local companies to easily access the supply chains of multinational companies seeking to establish green and fair conduct within their value chains.

Over the last 14 years, 130 projects have been funded in the region; of which 37 projects are ongoing and 22 are new projects as of 2020. Over 500 Asian and European non-for-profit partners, about 100 private sector associates and 80,000 Asian Micro, Small, and Medium-sized Enterprises (MSMEs) supported SWITCH-Asia.

The experience and evidence provided by these projects is expected to flow into policy and regulatory discussions with national governments. The program provides technical assistance to governments, supporting them in improving their national strategies and action plans regarding sustainable production and consumption practices. It also carries out advocacy activities at regional level.

2.2.4.2 SWITCH Africa Green Program – Framework on Circular Economy[57]

SWITCH Africa Green program was launched in 2019. The implementation of the program is guided by both global and regional settings and priorities. During the 17th Ministerial session of the African Ministerial Conference on the Environment in South Africa in 2019, a regional framework on the circular economy was proposed to focus on contextualizing the circular economy to African countries' needs. SWITCH Africa Green program works with Burkina Faso, Ethiopia, Ghana, Kenya, Mauritius, South Africa, and Uganda to achieve sustainable development by engaging in transition toward an inclusive green economy, based on SCP patterns.

SWITCH Africa Green is developed and funded by the European Commission and implemented by UNEP. Project partners include United Nations (UN) agencies, notably the United Nations Development Program (UNDP) and the United Nations Office for Project Services, One Planet Network, and the African Roundtable on SCP. The main governmental partners are the relevant ministries (environment, industry, agriculture, tourism, economy, and finance) as well as national environmental protection or management agencies and authorities.

2.2.5 The Life Cycle Initiative

The Life Cycle Initiative was established in 2002 by the UNEP. It is a public-private, multi-stakeholder partnership helping the private and public decision makers access credible life cycle knowledge[58]. The initiative facilitates the application of life cycle knowledge in the global sustainable development agenda in order to achieve global goals faster and more efficiently.

The Life Cycle Initiative brings together users and experts of life cycle approaches. To support decisions and policies related to the shared vision of sustainability as public good, it provides a global forum of knowledge which is science based. Additionally, it delivers authoritative opinions on comprehensive tools and approaches by engaging its multi-stakeholder partnership (governments, businesses, scientific, and civil society organizations).

For more information, visit the Interactive road map of LCI - https://www.lifecycleinitiative.org/?da_image=5203

According to the progress report published by UNEP[59], 2019 was a landmark year for the Life Cycle Initiative. It received unprecedented attention to mainstream LCT at the global policy level. The initiative successfully passed the midterm point for the strategy 2017–2022 and showcased significant successes in three main program areas: technical and policy advice, capacity development, and knowledge.

The Initiative has since engaged in technical support to the plastics, tourism and chemicals sectors, as well as delivered the Hotspot Analysis Tool for Sustainable Consumption and Production (SCP-HAT) which is currently being used to guide national action plans for SCP[53].

2.2.6 Resource Efficiency Initiative

Resource efficiency, which draws on the concept of Factor 4 and Factor 10, has been a key component of "Europe 2020", the EU's strategy for economic growth. The strategy's seven flagship objectives included a resource-efficient Europe initiative which provides a long-term framework for action, supporting policy agendas for environment and climate change, energy, transport, industry, agriculture, fisheries, and regional development. The initiative aims

BOX 2.15 DEFINITION OF RESOURCE EFFICIENCY BY EU[61]

Resource efficiency means using the Earth's limited resources in a sustainable manner while minimizing environmental impacts. It allows us to create more with less and to deliver greater value with less input.

to enhance certainty for investment and innovation and create opportunities for sustainable economic growth by ensuring that all relevant policy areas factor in resource efficiency in a consistent manner[60]. Box 2.15 shows the definition of Resource Efficiency by the EU.

Several countries of the EU took actions toward Resource Efficiency. For instance, Germany undertook Resource Efficiency Program (ProgRess) in February 2012. ProgRess describes measures for increasing energy efficiency along the entire value chain – from raw material extraction and product design to production, use, and circular economy[62]. Case studies are presented in Chapter 10 where resource efficiency has been anchored in the circular economy's policies and strategies.

Resource efficiency is vulnerable to rebound effects because monetary savings can lead to increased consumption – savings from the use of peer-to-peer lodging (e.g., AirBnb) can lead to more travel GHG emissions. Appropriate economic instruments need to be introduced to address such rebound effects.

2.2.7 Green Growth

The concept of green growth originated in the Asia and Pacific Region. At the 5th Ministerial Conference on Environment and Development held in Seoul in 2005, 52 Governments and other stakeholders from Asia and the Pacific agreed to move beyond the sustainable development rhetoric and follow a path of "green growth". The governments adopted a Ministerial declaration (the Seoul Initiative Network on Green Growth) and a regional implementation plan for sustainable development. This began a broader vision of green growth as a regional initiative of the United Nations Economic and Social Commission for Asia and the Pacific (UNESCAP), where it is viewed as a key strategy for achieving sustainable development as well as the Millennium Development Goals (in particular two and seven relating to poverty reduction and environmental sustainability).[63] Box 2.16 gives the definition of green growth.

In 2008, partly in response to the global financial crisis, the Republic of Korea (RoK) adopted "low carbon green growth" as the country's new development vision. In 2009, the National Strategy for Green Growth and Five-Year Plan for Green Growth (which were accompanied by the enactment of a Framework Act on Low Carbon Green Growth) were released. The RoK has

BOX 2.16 DEFINITION OF GREEN GROWTH[64]

Green growth is the pursuit of economic development in an environmentally sustainable manner. Green growth seeks to spur investment and innovation in ways that give rise to new, more sustainable sources of economic activity and jobs. For green growth to succeed as a transformative development strategy, it must be supportive of good governance, transparency, and equity.

since been instrumental in promoting the concept more broadly, including through the OECD countries[65].

At the OECD Ministerial Council Meeting in June 2009, 30 members and 5 prospective members (comprising approximately 80% of the global economy) approved a declaration acknowledging that green and growth can go hand-in-hand and asked the OECD to develop a green growth strategy bringing together economic, environmental, technological, financial, and development aspects into a comprehensive framework. Since then, the OECD has become a major proponent of green growth and supports efforts of countries to implement green growth[56].

The Global Green Growth Institute (GGGI) was formed to support developing countries and emerging economies to achieve sustainable inclusive economic growth. Since this initial founding, the organization has made significant progress and has been transformed into one of the leading international organizations supporting green growth policy and green investments. As of 2020, GGGI has 37 Members and delivers programs for more than 30 Members and partners – in Africa, Asia, the Caribbean, Europe, Latin America, the Middle East, and the Pacific – with technical support, capacity building, policy planning and implementation, and by helping to build a pipeline of bankable green investment projects[60].

2.2.8 Green Economy

The term "green economy" was first coined in a pioneering report for the Government of the United Kingdom by a group of leading environmental economists, entitled Blueprint for a Green Economy[66].

The financial crash of 2008 led to the need to look at Green Economy as a stimulus for sustainable growth. Green Economy was looked at one of the leading strategies to revive the economy. UNEP took the leadership. Box 2.17 provides definition of green economy.

A series of publications by UNEP, UNCTAD, UNDESA, and the UNCSD Secretariat elaborate on the concept and outline guiding principles, benefits, risks, and emerging international experience. In December 2011, the UN Environment Management Group (a system-wide coordination body of more

BOX 2.17 DEFINITION OF GREEN ECONOMY[67]

A green economy is defined as low carbon, resource efficient, and socially inclusive. In a green economy, growth in employment and income are driven by public and private investment into such economic activities, infrastructure, and assets that allow reduced carbon emissions and pollution, enhanced energy and resource efficiency, and prevention of the loss of biodiversity and ecosystem services.

than 40 specialized agencies, programs, and organs of the United Nations) also released its system-wide perspective on green economy – *Working Toward a Balanced and Inclusive Green Economy* – which identifies and clarifies the use of green economy and other related terms. This report adopts the definition provided by UNEP in its 2011 Green Economy Report. Several non-government organizations and partnerships have also developed in recent years, promoting green economy as a concept, and undertaking research, analysis and outreach[68].

The Green Economy Initiative aims to demonstrate that investing in green sectors – such as energy efficient technologies, renewable energy, public transport, sustainable agriculture, environment friendly tourism, and sustainable management of natural resources including ecosystems and biodiversity – has a better chance to bring about recovery and sustainable growth, increase competitiveness, save and create jobs, improve the quality and decency of jobs, and reduce poverty, while tackling acute environmental problems[69] .

A Green Economy Coalition (GEC) was formed with 50+ members representing different interests: environment, labor, socio-economic status, business, and government. Its members came from different parts of the world but were united by the same vision. Its mission was prosperity for all within one planet limits. It set the following five themes consisting of valuing nature, tackling inequality, greening economic sectors, reforming financial systems, and measuring and governing. The GEC is governed by a Steering Group, which meets every two months to decide on strategic and management issues and is coordinated by a small Secretariat hosted by the International Institute of Environment and Development in London, and by Finance Watch in Brussels[70].

The Partnership for Action on Green Economy (PAGE) was launched in 2013 as a response to the call at Rio+20 (the United Nations Conference on Sustainable Development) in 2012 to support those countries wishing to embark on greener and more inclusive growth trajectories[71]. PAGE brought together five UN agencies – UNEP, International Labour Organization (ILO), UNDP, UNIDO, and UN Institute for Training and Research (UNITAR) – whose mandates, expertise, and networks combined can offer integrated and holistic support to countries on inclusive green economy, ensuring coherence and avoiding duplication. Although the five agencies have previously undertaken joint green initiatives, this program was the first time that all five partners collaborated to coordinate their support, expertise, and resources at a national level.

PAGE focuses on SDG 8: "Promote sustained, inclusive and sustainable economic growth, full and productive employment and decent work for all." PAGE partner countries receive services for policy development and implementation, capacity building, and financing for inclusive green economy transitions, tailored to their individual needs and circumstances. It allows each country to develop its own pathway to an economy that is low-carbon, resource-efficient, and equitable. PAGE has partnered with 20 countries in their efforts for deeper economic reframing around sustainability. More than 90 countries have benefited from the capacity building and knowledge sharing services of PAGE[66].

With the increase in ocean pollution, in 2010 Blue Economy Program was launched focusing on marine-based economy[72]. World Wide Fund (WWF) described Blue Economy as an economy that[73]

- Provides social and economic benefits for current and future generations.
- Restores, protects, and maintains the diversity, productivity, resilience, core functions, and intrinsic value of marine ecosystems.
- Uses clean technologies, renewable energy, and circular material flows.

According to Mark Spalding, President of The Ocean Foundation, "...the core of the new Blue Economy concept is the de-coupling of socioeconomic development from environmental degradation... a subset of the entire ocean economy that has regenerative and restorative activities that lead to enhanced human health and well-being, including food security and creation of sustainable livelihoods[74]."

2.3 Concluding Remarks

The key concepts described in this chapter help to appreciate the canvas and potential of circular economy. In "designing circular economy" for a region or for a corporate ecosystem, it is important that we build on the essential concepts and familiarize ourselves with some of the leading programs, take their advantage and importantly draw on the lessons learnt.

All the programs described above were formed to institutionalize the key concepts related to sustainability and decoupling between economic growth and generation of wastes and emissions to the environment. Although

ACTIVITY 2.2 DISTINGUISHING BETWEEN KEY CONCEPTS

Attempt distinguishing the following concepts by stating the key principles, scope, and intended outcomes. Also, highlight the overlaps between the same.

1. Green Economy and Circular Economy.
2. Waste Minimization and Pollution Prevention.
3. Cleaner Production, Green Productivity and Sustainable Consumption and Production.

organizations define and named the programs differently, they share the same core concepts and have differences in the emphasis.

Over the last five decades, many of these programs gained popularity and enjoyed adoption across sectors and over national, regional, and global scales. The programs involved key stakeholders such as policy makers, regulators, business, investors, academia, and communities. Collectively, these programs have considerably influenced the national policy frameworks, investment focus as well as building of capacities.

Most of the programs followed a partnership approach by drawing synergy and taking a collaborative approach to setting the targets and implementation. Still, there have been overlaps and competition that could have been avoided. Due to limited funding, poor capacities, and lack of national governments' ownership, some programs were not very effective and could not be sustained. Unless the national governments lead efforts, the transition to a circular economy will continue to remain a challenge.

The concepts and programs described in this chapter led to the development of several useful methodologies, tools, and knowledge resources. Chapter 3 presents an overview of some of these resources.

2.4 Key Takeaways

- Even though the term "Circular Economy" has recently come into play, its idea has evolved from several concepts, strategies, and programs.
- The instruments (concepts, strategies, and programs) to practice a circular economy can be divided into three levels according to their implementation – Facility/corporate focused, regional/across life cycle/across supply chains, policy/strategy.

- Many of these instruments have commonalities, overlaps, and interlinkages and should not be considered in silo.
- Although organizations define and named the programs differently, they share the same core concepts and have differences in the emphasis.
- Waste minimization was one of the early concepts that evolved from facility and corporate levels to the regional and neighborhood scale.
- Even though "pollution prevention" is a term used interchangeably with "waste minimization", there is a difference between the two terms.
- Industrial ecology promoted a "systems thinking" approach and the concept was implemented in the design and implementation of EIPs.
- Eco-efficiency was a management philosophy that encourages businesses to search for environmental improvements that yield parallel economic benefits.
- LCT has become a key tool in policy and decision making in circular economy. It has also inspired innovation, especially in product design and influenced consumer behavior.
- Cradle to cradle concept was associated with "closing the loop" and provided understanding of a circular economy.
- Eco-design considers environmental aspects at all stages of the product development process.
- EPR has influenced waste management globally and plays an important role in a circular economy.

ADDITIONAL READING

1. Waste minimization:
   A guide to establishing and managing waste minimisation clubs in South Africa – The Manual is aimed at a person/organisation that wishes to initiate a waste minimisation club and requires guidelines for undertaking such a project. It addresses aspects such as how to form a club, call meetings, determine the level of contributions from companies, identify some of the problems that can occur, and explains the various roles of the people involved. It also provides sample letters and presentations and provides sources of information.

Source: Barclay, S. and Buckley, C., 2006. *A Guide to Establishing and Managing Waste Minimization Clubs in South Africa.* Water Research

Commission. <http://www.wrc.org.za/wp-content/uploads/mdocs/TT283%20Web.pdf> [Accessed 26 October 2020].

2. Eco-design:
   D4S sustainability manual – A global guide for designers and industry, it provides support to ecodesign novices and those looking to further their understanding of the field. The manual focuses on three different design approaches: redesign of existing products, radical sustainable product innovation, and new product development. An additional section acts as a comprehensive "how-to" guide for first time users.

Source: Crul, M. and Diehl, J., 2006. *Design for Sustainability: A Practical Approach for Developing Economies.* [online] UNEP, Delft University of Technology. <https://d306pr3pise04h.cloudfront.net/docs/issues_doc%2FEnvironment%2Fclimate%2Fdesign_for_sustainability.pdf> [Accessed 7 October 2020].

3. LCT:
   Why take a life-cycle approach? – The purpose of this brochure is to introduce a life cycle approach as one means to help us recognize opportunities, balance opportunities with risks and make choices that contribute value to our economies, our natural environments, and our communities.

Source: UNEP. 2004. *Why Take a Life Cycle Approach.* <http://www.unep.fr/shared/publications/pdf/DTIx0585xPA-WhyLifeCycleEN.pdf> [Accessed 7 October 2020].

4. CP program:
   a. **Global status report on sustainable consumption and CP (2002)** – The Global Status Reports on sustainable consumption and CP take stock of what has been achieved so far, what lessons have been learnt and propose future courses of action. This publication contains the executive summary of each Global Status Report.

Source: UNEP. 2002. *Global Status 2002: Sustainable Consumption and Cleaner Production.* <http://www.unep.fr/shared/publications/pdf/3211-GlobalStatus02.pdf > [Accessed 23 October 2020].

b. **National Cleaner Production Centres (NCPCs) & Networks** – A UNIDO sponsored knowledge page for NCPCs and associated programs.

Source: UNIDO. n.d. *National Cleaner Production Centres (NCPCs) & Networks.* <https://www.unido.org/our-focus/cross-cutting-services/partnerships-prosperity/networks-centres-forums-and-platforms/national-cleaner-production-centres-ncpcs-networks> [Accessed 27 October 2020].

Notes

1. NEPIS | US EPA. 1966. *Report To Congress: Minimization of Hazardous Waste Appendices.* https://bit.ly/33nDJnM [Accessed 3 October 2020].
2. Fandrich, V., 2011. Business Waste Prevention Evidence Review. http://randd.defra.gov.uk/Document.aspx?Document=WR1403-L2-m4-7-Waste-Minimisation-Clubs.pdf [Accessed 3 October 2020].
3. Apo-Tokyo. 2013. *Waste Minimization Circles Under The NPC.* https://www.apo-tokyo.org/wp-content/uploads/sites/5/2013_Mar-Apr_p8.pdf [Accessed 3 October 2020].
4. Zero Waste. 2020. Who Started The Zero Waste Movement? <https://www.zerowaste.com/blog/what-is-it-who-started-the-zero-waste-movement/> [Accessed 21 October 2020].
5. Jennings, R., 2019. *The Zero-Waste Movement Is Coming For Your Garbage.* https://www.vox.com/the-goods/2019/1/28/18196057/zero-waste-plastic-pollution [Accessed 3 October 2020].
6. Seldman, N., 2016. *Zero Waste: A Short History And Program Description – Institute For Local Self-Reliance.* Institute for Local Self-Reliance. https://ilsr.org/zero-waste-a-short-history-and-program-description/> [Accessed 7 October 2020].
7. Zero Waste. 2020. *Who Started The Zero Waste Movement? – Zero Waste.* https://zerowaste.com/blog/what-is-it-who-started-the-zero-waste-movement/ [Accessed 5 October 2020].
8. Zero Waste USA. https://www.zerowasteusa.com/
9. GRRN. https://archive.grrn.org/
10. Unmaking Waste. n.d. *Keynotes.* https://www.unmakingwaste.org/speakers-2/ [Accessed 7 October 2020].
11. Unmaking Waste. n.d. *Zero Waste SA.* https://www.unmakingwaste.org/zero-waste-sa/ [Accessed 7 October 2020].
12. UN Habitat. 2010. Solid Waste Management in the World's Cities https://thecitywasteproject.files.wordpress.com/2013/03/solid_waste_management_in_the_worlds-cities.pdf [Accessed 7 October 2020].
13. South Australian Government Data Directory. n.d. Zero Waste SA- Inactive https://data.sa.gov.au/data/organization/about/zero-waste-sa [Accessed 7 October 2020].

14. Green Industries SA. n.d. About us. https://www.greenindustries.sa.gov.au/about-us. [Accessed 7 October 2020].
15. Date Source: Fitzgerald, S., National Geographic. 2018. *25 Places That Have Committed To Going Zero-Waste*. https://www.nationalgeographic.com/travel/lists/zero-waste-eliminate-sustainable-travel-destination-plastic/ [Accessed 3 October 2020].
16. US EPA. n.d. *Pollution Prevention Law And Policies*. https://www.epa.gov/p2/pollution-prevention-law-and-policies#define [Accessed 3 October 2020].
17. Anastas, P. and Warner, J., 1998. *Green Chemistry: Theory and Practice*. New York: Oxford University Press.
18. A concept to measure the efficiency of a synthetic process, in terms of the number of atoms required in all the starting materials and reactants versus how many of these atoms are wasted (i.e., atoms that do not become part of the final product). A process which uses many atoms not found in the final product is said to have low atom economy, whereas a process in which most atoms of the starting material and reactants end up in the final product is said to have high atom economy.
Source: https://www.chem.ucla.edu/~harding/IGOC/A/atom_economy.html#:~:text=Atom%20economy%3A%20A%20concept%20to,part%20of%20the%20final%20product).
19. Frosch, R. and Gallopoulos, N., 1989. Strategies for Manufacturing. *Scientific American*, 261(3), pp.144–152.
20. GDRC. n.d. *Sustainability Concepts: Industrial Ecology*. [online] Available at: <https://www.gdrc.org/sustdev/concepts/16-l-eco.html> [Accessed 26 October 2020].
21. Li, X., 2018. Industrial Ecology and Industrial Symbiosis – Definitions and Development Histories. Industrial Ecology and Industry Symbiosis for Environmental Sustainability. Cham: Palgrave Pivot. https://doi.org/10.1007/978-3-319-67501-5_2
22. UNIDO. n.d. *Eco-Industrial Parks*. [online] Available at: <https://www.unido.org/our-focus-safeguarding-environment-resource-efficient-and-low-carbon-industrial-production/eco-industrial-parks> [Accessed 5 October 2020].
23. UN ESCAO. 2009. Eco-efficiency Indicators: Measuring Resource-use Efficiency and the Impact of Economic Activities on the Environment. https://sustainabledevelopment.un.org/content/documents/785eco.pdf> [Accessed 5 October 2020].
24. Environmental Justice Organisations, Liabilities and Trade. n.d. *Eco-Efficiency*. <http://www.ejolt.org/2015/02/eco-efficiency/#:~:text=According%20to%20the%20WBCSD%2C%20%E2%80%9Ceco,line%20with%20the%20Earth's%20estimated> [Accessed 21 October 2020].
25. World Business Council for Sustainable Development (WBCSD). n.d. *Our History*. <https://www.wbcsd.org/Overview/Our-history> [Accessed 7 October 2020].
26. International Institute for Sustainable Development. n.d. *Responsible Business*. <https://www.iisd.org/business/tools/principles_factor.aspx> [Accessed 21 October 2020].
27. Schmidt-Bleek, F., 2008. *Factor 10 Institute*. http://www.factor10-institute.org/pages/factor_10_institute_2008.html [Accessed 5 October 2020].

28. GDRC. n.d. *Sustainability Concepts: Factor 10.* https://www.gdrc.org/sustdev/concepts/11-f10.html [Accessed 5 October 2020].
29. World Business Council for Sustainable Development (WBCSD). n.d. *Factor 10.* <https://www.wbcsd.org/Programs/Circular-Economy/Factor-10> [Accessed 5 October 2020].
30. Life Cycle Initiative. n.d. *What Is Life Cycle Thinking?* https://www.lifecycleinitiative.org/starting-life-cycle-thinking/what-is-life-cycle-thinking/ [Accessed 5 October 2020].
31. European Union. 2010. *LCT-Making Sustainable Consumption and Production A Reality.* <https://eplca.jrc.ec.europa.eu/uploads/LCT-Making-sustainable-consumption-and-production-a-reality-A-guide-for-business-and-policy-makers-to-Life-Cycle-Thinking-and-Assessment.pdf> [Accessed 7 October 2020].
32. Koroneos, C., Nanaki, E., Rovas, D. and Krokida, M., 2013, November. Life cycle assessment: A strategic tool for sustainable development decisions. In The Third World Sustainability Forum, Greece.
33. UNEP. 2004. *Why Take A Life Cycle Approach.* <http://www.unep.fr/shared/publications/pdf/DTIx0585xPA-WhyLifeCycleEN.pdf> [Accessed 7 October 2020].
34. Wautelet, T., 2018. The Concept of Circular Economy: Its Origins and its Evolution. https://www.researchgate.net/publication/322555840_The_Concept_of_Circular_Economy_its_Origins_and_its_Evolution [Accessed 7 October 2020].
35. European Commission. n.d. *Sustainable Product Policy & Ecodesign – Internal Market, Industry, Entrepreneurship And Smes – European Commission.* <https://ec.europa.eu/growth/industry/sustainability/product-policy-and-ecodesign_en> [Accessed 7 October 2020].
36. Crul, M. and Diehl, J., 2006. *Design For Sustainability: A Practical Approach For Developing Economies.* [online] UNEP, Delft University of Technology. <https://d306pr3pise04h.cloudfront.net/docs/issues_doc%2FEnvironment%2Fclimate%2Fdesign_for_sustainability.pdf> [Accessed 7 October 2020].
37. Ellen Macarthur Foundation. n.d. *Circular Economy Schools of Thought.* <https://www.ellenmacarthurfoundation.org/circular-economy/concept/schools-of-thought#:~:text=The%20circular%20economy%20concept%20has,%2C%20thought%2Dleaders%20and%20businesses.> [Accessed 7 October 2020].
38. Cyr, S., 2015. *Social Innovation… Nature's Way.* Philanthropy Journal. <https://pj.news.chass.ncsu.edu/2015/06/08/social-innovation-natures-way/> [Accessed 21 October 2020].
39. OECD. n.d. *Extended Producer Responsibility.* <https://www.oecd.org/env/tools-evaluation/extendedproducerresponsibility.htm> [Accessed 21 October 2020].
40. Multi-Material Stewardship Western. n.d. *History of EPR.* <https://www.mmsk.ca/residents/history-epr/> [Accessed 9 October 2020].
41. Mule, J.Y., 2012. Design for disassembly approaches on product development. *International Journal of Scientific Engineering and Research*, 3(6), pp. 996–1000.
42. UNEP. n.d. *Resource Efficient and Cleaner Production.* <http://www.unep.fr/scp/cp/> [Accessed 23 October 2020].
43. World Business Council for Sustainable Development. n.d. *Cleaner Production and Eco-Efficiency: Complementary Approaches to Sustainable Development* <http://www.gcpcenvis.nic.in/PDF/eco%20effiency%20and%20CP.pdf> [Accessed 23 October 2020].

44. UNEP. 2002. *Global Status 2002: Sustainable Consumption and Cleaner Production.* <https://wedocs.unep.org/bitstream/handle/20.500.11822/8436/-Sustainable%20Consumption%20and%20Cleaner%20Production_%20Global%20Status%202002%20-%20A%20Contribution%20to%20Sustainable%20Development-2002256.pdf?sequence=2&%3BisAllowed=> [Accessed 23 October 2020].
45. UNIDO. 2017. *Joint UNIDO-UNEP Programme on Resource Efficient and Cleaner Production (RECP) In Developing and Transition Countries.* <https://www.unido.org/sites/default/files/files/2018-03/100050-RECPpercent20Indpercent20Evalpercent20Report.pdf> [Accessed 23 October 2020].
46. UNIDO. n.d. *Resource Efficient and Cleaner Production (RECP).* <https://www.unido.org/our-focus/safeguarding-environment/resource-efficient-and-low-carbon-industrial-production/resource-efficient-and-cleaner-production-recp> [Accessed 23 October 2020].
47. UNIDO. n.d. *NCPC 20 Years.* <https://www.unido.org/our-focus/cross-cutting-services/partnerships-prosperity/networks-centres-forums-and-platforms/national-cleaner-production-centres-ncpcs-networks/ncpc-20-years> [Accessed 23 October 2020].
48. RECPnet. https://www.recpnet.org/
49. Pathak, P. Green Productivity: A Better Way to Sustainable Industrial Economy <https://www.academia.edu/10345730/Green_Productivity_A_Better_Way_to_Sustainable_Industrial_Economy>[Accessed 23 October 2020].
50. United Nations Partnerships for SDGs platform. n.d. Asian Productivity Organization's Green Productivity Initiative <https://sustainabledevelopment.un.org/partnership/?p=2242> [Accessed 21 October 2020].
51. Ministry of Environment. 2005. *Japan's Experience in the Promotion of the 3Rs.* [online] Available at: <https://www.env.go.jp/recycle/3r/en/approach/02.pdf> [Accessed 14 December 2020].
52. 3 R Forum in Asia and the Pacific. https://www.env.go.jp/recycle/3r/en/
53. Department of Environment. 2010. *National 3R Strategy For Waste Management.* [online] Available at: <http://old.doe.gov.bd/publication_images/4_national_3r_strategy.pdf> [Accessed 14 December 2020].
54. UN Environment Programme. n.d. *One Planet Network.* <https://www.unenvironment.org/explore-topics/resource-efficiency/what-we-do/one-planet-network> [Accessed 21 October 2020].
55. United Nations Social and Economic Council. 2018. Progress report on the 10-year framework of programmes on sustainable consumption and production patterns <https://digitallibrary.un.org/record/1627351?ln=en> [Accessed 22 October 2020].
56. SWITCH Asia. n.d. *About Us.* <https://www.switch-asia.eu/switch-asia/about-us/> [Accessed 27 October 2020].
57. UNEP. 2020. *Promoting African Green Business and Circular Economy for Better Policies.* <https://www.unenvironment.org/news-and-stories/story/promoting-african-green-business-and-circular-economy-better-policies> [Accessed 14 December 2020].
58. Life Cycle Initiative. n.d. *About The Life Cycle Initiative.* <https://www.lifecycleinitiative.org/about/about-lci/> [Accessed 21 October 2020].
59. Life Cycle Initiative. 2019. Life Cycle Initiative Progress Report 2019. <https://www.lifecycleinitiative.org/wp-content/uploads/2020/03/2019-LC-Initiative-progress-report-5.3.20.pdf> [Accessed 21 October 2020].

60. European Commission. 2011. *Resource Efficiency – A Business Imperative*. [online] Available at: <https://ec.europa.eu/environment/resource_efficiency/documents/factsheet_en.pdf> [Accessed 21 October 2020].
61. European Commission. n.d. *Resource Efficiency*. <https://ec.europa.eu/environment/resource_efficiency/#:~:text=Ongoing%20work-,Resource%20Efficiency,greater%20value%20with%20less%20input.> [Accessed 21 October 2020].
62. Federal Ministry for the Environment, Nature Conservation, and Nuclear Safety. n.d. German Resource Efficiency Programme (Progress) – An Overview. <https://www.bmu.de/en/topics/economy-products-resources-tourism/resource-efficiency/overview-of-german-resource-efficiency-programme-progress/> [Accessed 21 October 2020].
63. Sustainable Development Knowledge Platform. n.d. *Green Growth*. <https://sustainabledevelopment.un.org/index.php?menu=1447> [Accessed 21 October 2020].
64. Green Growth Knowledge Platform. n.d. *Explore Green Growth*. <https://www.greengrowthknowledge.org/page/explore-green-growth> [Accessed 21 October 2020].
65. Sustainable Development Knowledge Platform. n.d. *Green Growth*. <https://sustainabledevelopment.un.org/index.php?menu=1447> [Accessed 30 October 2020].
66. Sustainable Development Knowledge Platform. n.d. *Green Economy* <https://sustainabledevelopment.un.org/index.php?menu=1446#:~:text=The%20term%20green%20economy%20was,Markandya%20and%20Barbier%2C%201989).> [Accessed 21 October 2020].
67. UN Environment Programme. n.d. *Green Economy*. [online] Available at: <https://www.unenvironment.org/regions/asia-and-pacific/regional-initiatives/supporting-resource-efficiency/green-economy> [Accessed 21 October 2020].
68. Sustainable Development Knowledge Platform. n.d. *Green Economy*. <https://sustainabledevelopment.un.org/index.php?menu=1446> [Accessed 23 October 2020].
69. United Nations System. n.d. *Green Economy Initiative*. <https://www.unsystem.org/content/green-economy-initiative-gei> [Accessed 27 October 2020].
70. Green Economy Coalition. n.d. *Our Members*. <https://www.greeneconomycoalition.org/members> [Accessed 30 October 2020].
71. United Nations Partnerships for SDGs platform. n.d. *Partnership For Action On Green Economy*. <https://sustainabledevelopment.un.org/partnership/?p=7468> [Accessed 30 October 2020].
72. Drishti IAS. 2018. *Blue Economy*. [online] Available at: <https://www.drishtiias.com/to-the-points/paper3/blue-economy#:~:text=Thepercent20conceptpercent20waspercent20introducedpercent20by,jobspercent2Cpercent20andpercent20oceanpercent20ecosystempercent20health.> [Accessed 21 October 2020].
73. World Wide Fund. n.d. *Principles For a Sustainable Blue Economy*. <http://d2ouvy59p0dg6k.cloudfront.net/downloads/15_1471_blue_economy_6_pages_final.pdf> [Accessed 21 October 2020].
74. The Ocean Foundation. n.d. *Blue Economy* <https://oceanfdn.org/blue-economy/> [Accessed 21 October 2020].

3

Circular Economy Toolbox

In Chapter 2, some of the key concepts that influenced development of several programs and initiatives are reviewed. These programs helped in setting the stage for a Circular Economy. The practice experience then led to development of resources including methodologies, tools, and knowledge bases. While it was not be possible to cover the entire gamut of such developments, some of these resources are introduced as an essential "tool-box" at the end of this chapter.

3.1 Doing More with Less

3.1.1 Waste Minimization

United States Environment Protection Agency's (US EPA) Hazardous Waste Engineering Research Laboratory published "Waste Minimization Opportunity Assessment Manual" in 1988[1]. This manual provided a step-by-step guidance on how to reduce waste at source and how to ensure safe treatment and disposal. Figures 3.1 and 3.2 illustrate the Waste Minimization Assessment Procedure and Waste Minimization Techniques, respectively.

Later, in 1991, United Nations Industrial Development Organization and United Nations Environment Programme (UNIDO-UNEP) jointly published a methodology for waste audit with three case studies. Figure 3.3 shows the Waste Minimization Audit Procedure.

The terms "Opportunity Assessment" and "Audits" are interchangeable. The term "Audit" however, has a negative connotation as generally audits lead to reports that identify faults in a process. Hence, the term "Opportunity Assessment" is a better term as it focuses on improvements.

It may be observed that the concept of reduction in resource consumption was embedded in waste minimization techniques before the term "Resource Efficiency" was popularized. The implementation was proposed in phases to help in scoping and optimize data collection related efforts. Forming a team and getting top management commitment was stressed upon. Segregation

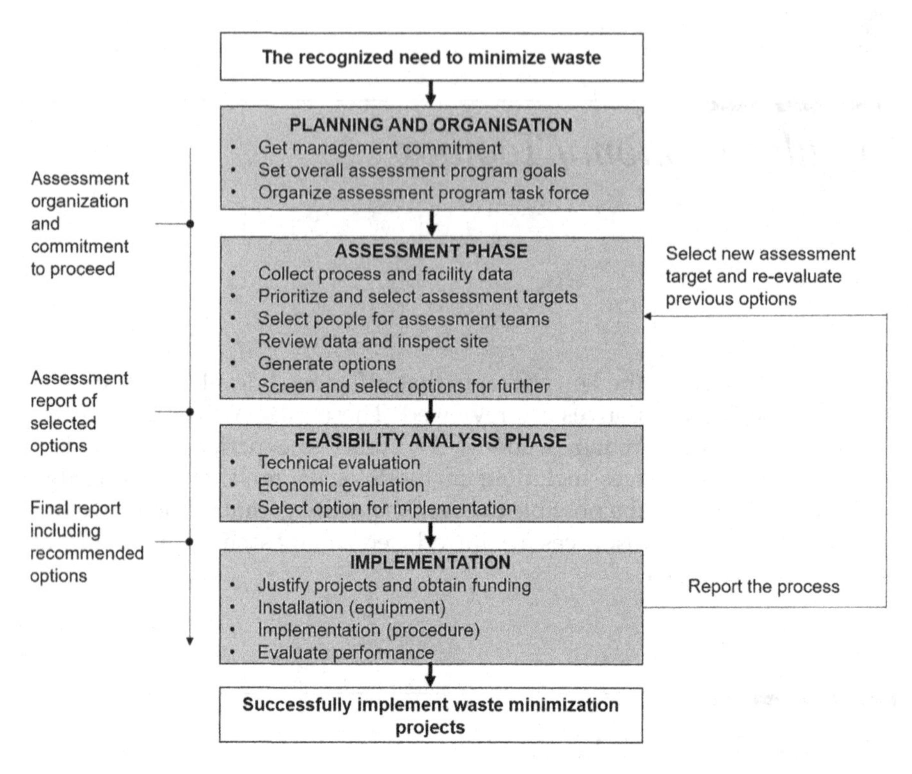

FIGURE 3.1
US EPA's waste minimization assessment procedure[1].

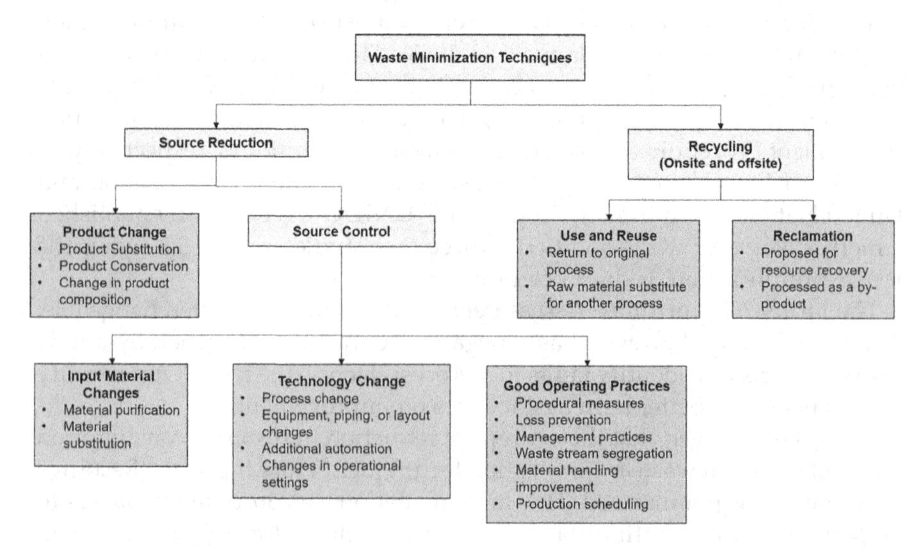

FIGURE 3.2
Waste minimization techniques suggested by US EPA[1].

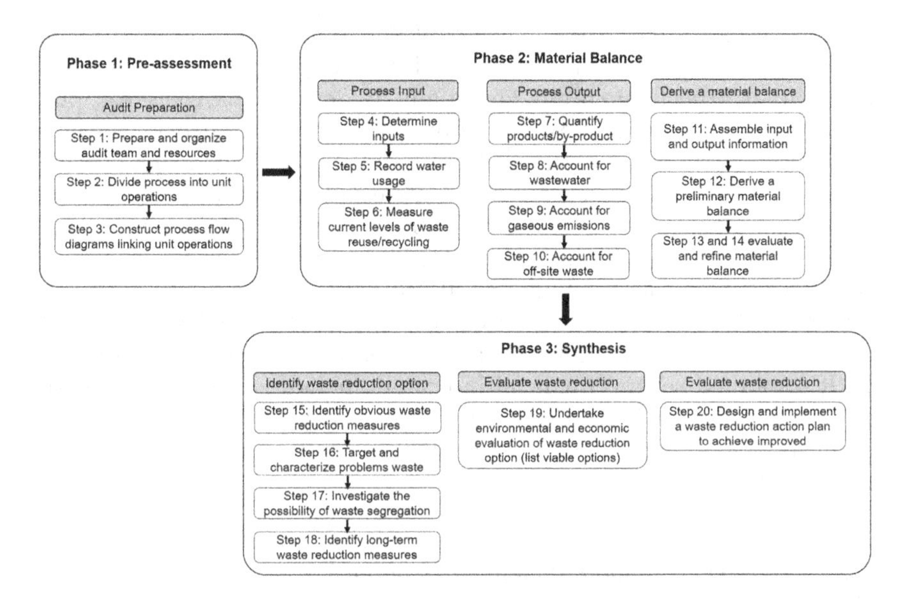

FIGURE 3.3
UNIDO-UNEP's waste minimization audit methodology[2].

of waste streams was a step recommended to explore onsite recycling and recovery.

Software tools were developed to assist in finding and assessing waste minimization related options. One such tool was – The Strategic Waste Minimization Initiative (SWAMI) Software. SWAMI required user-supplied information for process definition, as well as material inputs and products for each unit operation and outputs associated with waste streams[3]. SWAMI provided a scheme for identifying and prioritizing (on a cost or volume basis) waste reduction opportunities in process units and treatment operations, perform mass balance calculations, draw process flow diagrams, and guide in selection of potential waste minimization strategies. This software system was developed in conjunction with the US EPA publication "Waste Minimization Opportunity Assessment Manual"[3]. SWAMI software is not used anymore as its technicalities are outdated. However, it is recommended to explore its scope.

3.1.2 Pollution Prevention

P2/Finance[4] was one of the most popularly used software tool for financial analysis of pollution prevention measures. It was developed by the Tellus Institute for US EPA in 1992. E2/Finance, a companion to P2/Finance was also made available focusing on energy efficiency.

P2/Finance accounted for conventional and hidden costs of inputs and outputs. Liability costs could be included by specifying the year in which the liability is expected to occur. Potentially hidden tangible costs (e.g., public image) could be entered as part of the "other costs" category. However, external (social) costs were not included in the analysis. The major cost elements included purchased equipment, materials, utility connections and new utility systems, site preparation, construction/installation, engineering/contractor fees, start-up/training, contingency, permitting initial charge for catalysts and chemicals, working capital, and salvage value.

P2/Finance provided a mechanism to compare current and alternative practices. Its use was not limited to the analysis of pollution prevention projects. It computed several financial indicators including NPV (Net Present Value), IRR (Internal Return Rate), and simple payback over user-defined time horizon. Both P2/Finance and E2/Finance software are now outdated and not available for use.

3.1.2.1 Pollution Prevention Abatement Handbook

One of the early knowledge resources on pollution prevention was the *"Pollution Prevention Abatement Handbook: Towards Cleaner Production"* published by the World Bank in 1999. The handbook was specifically designed to be used in the context of the World Bank Group's environmental policies, as set out in Operational Policy 4.01, "Environmental Assessment"[5]. It consists of the detailed guidelines representing state-of-the-art thinking on how to reduce pollution emissions from the production process covering almost 40 industrial sectors. In many cases, the guidelines provided information on the benchmarks which could be achieved through cleaner production.

3.1.3 EHS Guidelines of the World Bank Group

Subsequently, in 2008, the sector specific best practice guidelines were updated and broadened by International Finance Corporation, working in close cooperation with the World Bank. These guidelines are now referred to as the World Bank Group Environment, Health, and Safety (EHS) Guidelines that apply to all World Bank Group projects[6]. These EHS guidelines cover 64 sectors providing information on benchmarks on resource consumption.

3.1.4 Cleaner Production Audit

Based on the waste minimization related methodologies, several methodologies with variants were developed for cleaner production audit. Figure 3.4 shows a methodology based on manual developed by IVAM in 2008[7].

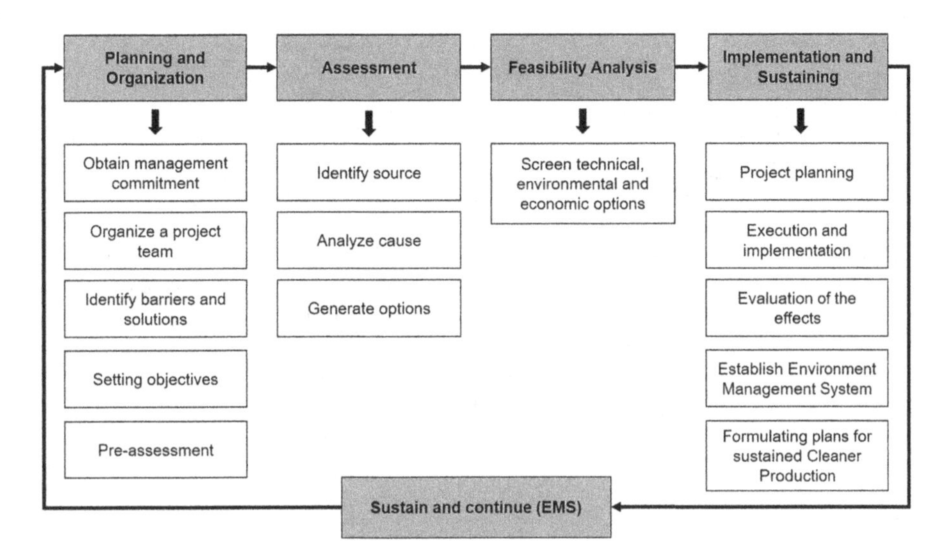

FIGURE 3.4
Cleaner production audit.

It is observed that this methodology of cleaner production auditing recommends establishment of Environmental Management System (EMS). This is to ensure that there is a continual improvement process, and the audit does not become a one-time exercise. This approach is also in line with the focus on pollution prevention, as earlier cited in the EMS. However, there are few instances, where integration of EMS is done during cleaner production audits.

Methodologies followed for Resource Efficient Cleaner Production (RECP) are similar. In RECP, the focus of the audit is to improve resource efficiency in the process operations to improve productivity, reduce operating costs and increase profitability while improving environmental performance.

The cleaner production and RECP audits help in identification of several actions such as better housekeeping (where tools like 5S can be used), process rationalization, process optimization (that can include material substitution, equipment retrofits, reuse, recycle, and recovery), process/equipment change with more resource efficient technologies, using renewable energy to the extent possible and considering product redesign. In this process, there are co-benefits such as reduction of Greenhouse Gases (GHG) emissions, improved health and safety of the workers and meeting environmental compliance.

3.1.5 APO's GP Methodology

To substantiate the Green Productivity (GP) concept, the Asian Productivity Organization (APO) adopted a multi-dimensional micro-to-macro approach

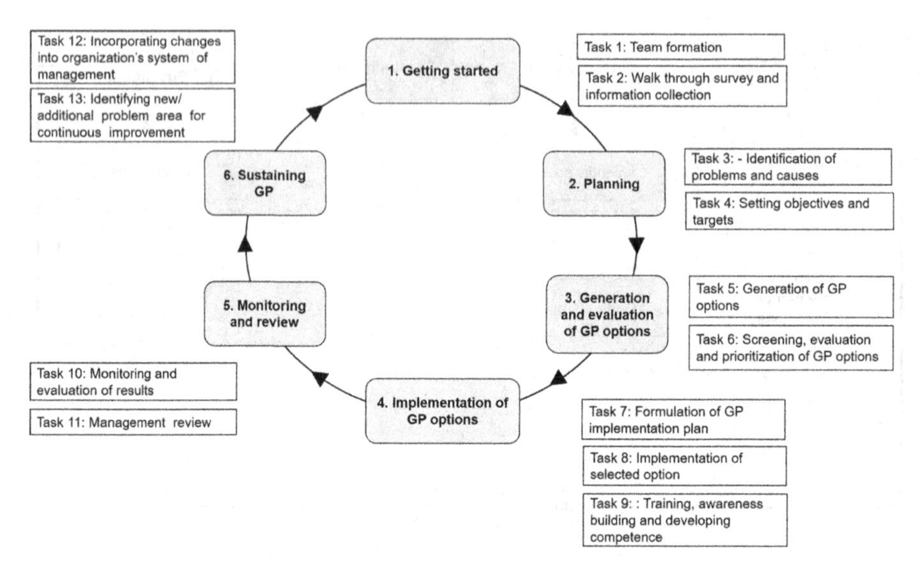

FIGURE 3.5
Methodology for implementing green productivity.

to promote GP practices. It focuses on the enterprise level through the applications of productivity and management tools (such as Total Quality management, 5S, Total Productive Maintenance, etc.) that go in tandem with waste and emission prevention, energy conservation, pollution control, and Environmental Management Systems (EMS).[8]

To operationalize GP at practical levels, the APO has developed a six-step, 13-task methodology following Deming's plan-do-check-act (PDCA) cycle that is used in the EMS. This is illustrated in Figure 3.5. This methodology has been successfully applied in the past few years throughout the APO region in various GP demonstration programs and has been found to be very effective and productive. While the PDCA framework provides the basic structure of GP implementation, the distinctive part of GP methodology is its ever-expanding set of tools and techniques to complement the PDCA framework. This methodology has been field tested and is being disseminated through various training programs in APO member countries. Figure 3.5 shows the GP methodology. Initially, GP methodology was applied more to the manufacturing sector, but now GP is increasingly applied to agriculture, service industry, and even communities.

GP Guidance and Trainer's Resource Manuals are the culmination of the efforts of many GP experts from the region over the past few years. These manuals have so far been used in many of the APO's multi-country workshops and by the National Productivity Organizations in their in-country GP programs.

3.1.6 Design for Sustainability

The manual on Design for Sustainability (D4S) published by UNEP of was cited in Chapter 2 under additional reading. Another important resource to refer is the D4S website[9] jointly developed by TU Delft with UNEP. This website has several important resources such as D4S manual structured in modules, case studies, and worksheets. Importantly, the D4S steps cover topics such as redesign, new product development, and Product as a Service (PaaS). PaaS are business models that provide for joint delivery of products and services. PaaS models are emerging as a means to enable collaborative consumption of both products and services. PaaS will be discussed further in Chapter 8. Figure 3.6 illustrates step wise methodology for D4S.

In the interest of extending life of products, concepts such as Design for Disassembly (D4D) and Design for Repairability (D4R) emerged. Two useful sources on D4D and D4R are shown in Box 3.1.

GHG emissions have become a recent global concern for manufacturing. As product design has a profound effect on a product's carbon footprint in its life cycle. It is important to estimate the carbon footprint of a product at its conceptual design stage[14]. Combining the concepts of D4S, D4D, and D4R, with Life Cycle Assessment (LCA), helps in to innovate designs that are less carbon intensive.

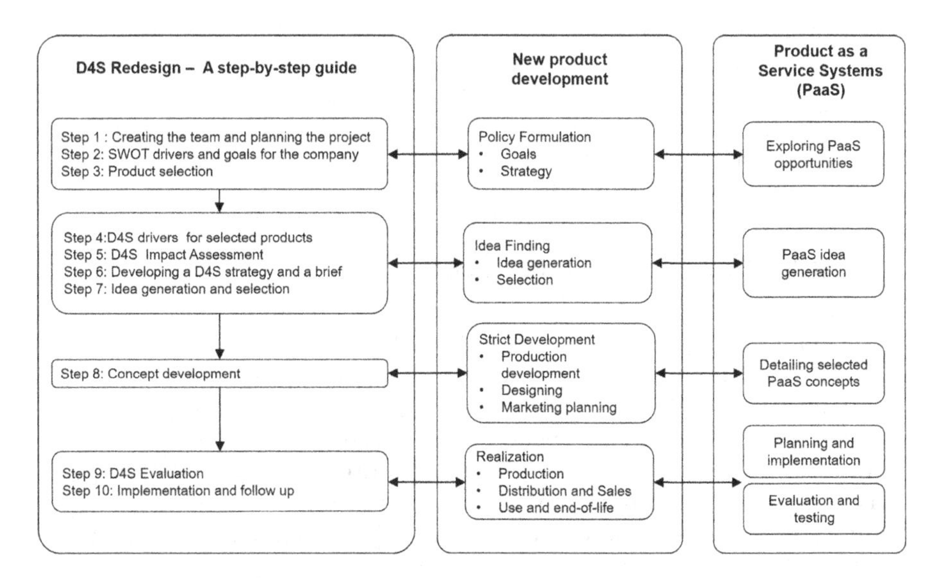

FIGURE 3.6
Stepwise methodology for D4S[10].

BOX 3.1 USEFUL RESOURCES FOR DESIGN FOR DISASSEMBLY AND DESIGN FOR REPAIRABILITY

DESIGN FOR DISASSEMBLY (D4D)

The report Art of D4D[11], released in 2013 by the Ellen McArthur Foundation, discusses D4D in lines with the teaching from Dr. Joe Chiodo, who has been working in the field of disassembly for more than three decades. The examples conversed are:

1. Bloom Laptop: The laptop is built on a modular basis and can be disassembled by anyone in 10 easy steps, without tools, in less than two minutes. The Bloom was never commercially launched, but perhaps its design inspired Hewlett Packard. They released the HP Z1 in 2012, calling it "the world's first all-in-one workstation with a 27" (diagonal) display that snaps open to let you swap out parts and make upgrades. No tools required.[12]
2. Think Chair: The chair is designed to be disassembled by the final user into its component pieces so it can be used again in a new process. The disassembly process takes about five minutes and requires only standard tools. Furthermore, the materials used in the seat are non-toxic, 99% recyclable, and carry the lightest environmental impact.
3. Yorkon Modular Construction: Yorkon is one of the UK's leading modular building manufacturers. This off-site manufacturer of buildings pieces together the final building in situ, having done most of the work elsewhere. This can be a money and time saver, as there is less disruption than would be the case for an on-site build.

DESIGN FOR REPAIRABILITY (D4R)

Research Council for Automobile Repairs Design Guide[13]: This report is a manufacturers' guide to ensure good design practice for repairability and limitation of damage. Improving vehicle damageability/repairability while maintaining safety design standards is critical to the automobile buying consumer, as it relates to personal safety, the affordability of the vehicle being purchased, and the costs associated with maintaining and insuring the vehicle. This guide shows a range of some of the good and poor examples of vehicle repairability.

3.2 GHG Accounting

An important aspect to consider while developing low carbon strategies is calculation of GHG emissions. This is done through GHG accounting. A methodology was released by an institution known as GHG Protocol[15] that was set by the World Resources Institute (WRI) and World Business Council for Sustainable Development (WBCSD) in the late 1990s. They created a global standardized framework to measure and manage GHG emissions from private and public sector operations, value chains and came up with mitigation actions.

GHG Protocol also developed a suite of calculation tools[16] to assist companies in calculating their GHG emissions and measure the benefits of climate change mitigation projects. GHG accounting is intimately related to energy audits as energy usage and its source are often the main contributor of GHG emissions. Suite of tools supported by Government of Canada on energy conservation, energy efficiency and renewable could be useful[17].

3.3 Management Systems

Management systems play an important role in ensuring that the interventions identified for waste minimization, improving resource efficiency and reduction in GHG emissions are systematically implemented and are continuously improved with a tracking or monitoring mechanism. Several management systems are prescribed by the International Standards Organization (ISO). Section 3.3.1 describes EMS and its variants and the management standard on circular economy that has been recently introduced.

3.3.1 Environmental Management System

An EMS helps an organization to understand its impact on the environment and increase its environmental performance. ISO 14001:2015 is an international voluntary standard that specifies these requirements in a structured manner and provides a step-by-step approach to establish an effective EMS in an organization[18]. There are numerous economic, social, and legal benefits of an EMS. Economic benefits include resources savings, reduced cost of operations and access to new global markets. Social benefits include improved morale and productivity of workers/staff, development of skilled workers and safer working conditions on the shop floor. Some legal benefits include improved compliance and reduction of environmental risks associated with operations.

ISO 14001 EMS primarily focuses on process issues, which are determined and attended or managed by the organization itself (refer to ISO 14001:2015 standard under section A. "Context of the Organization" and A.4.3 "Determining the Scope of the EMS"). While determining the scope, an organization is free to consider its products and services. As the adoption of the standards spread, several organizations began using EMS to focus on their product-related issues[19]. This concept of linking product design to EMSs became the foundation of a concept termed as Product-Oriented Environmental Management System (POEMS). Some motivations to consider a product's characteristics stem from concerns relating to the environmental impact of the products (direct and indirect), policies related to specific products such as the extended producer responsibility (European directive on Waste Electrical and Electronic Equipment), role of original equipment manufacturers to dominate development process because of access to technology and resources in comparison to other agents in the market[20]. This led organizations to take responsibility to decrease the environment impact of their product throughout its life cycle.

3.3.2 British Standard 8001: Circular Economy

BS 8001:2017 was developed in 2017 as a framework for implementing the principles of a circular economy in organizations. It is the first practical framework and guidance for organizations to implement the principles of a circular economy. The framework is intended to apply to any organization, regardless of location, size, sector, and type and regardless of varying levels of knowledge and understanding of the circular economy. It provides practical ways to achieve "quick-wins", right through to helping organizations re-think holistically how their resources could be managed to enhance financial, environmental, and social benefits[21]. BS 8001 will evolve further as there is more practice experience. Figure 3.7 shows the structure of BS 8001 along with its limitations.

3.4 Material Flow Analysis

A Material Flow Analysis (MFA) is a systematic reconstruction of the way a material takes through the natural and/or economic cycle. An MFA is generally based on the principle of mass balance.

Objectives of an MFA are:

- Trace the flow of raw materials through the company to establish connections within the process.

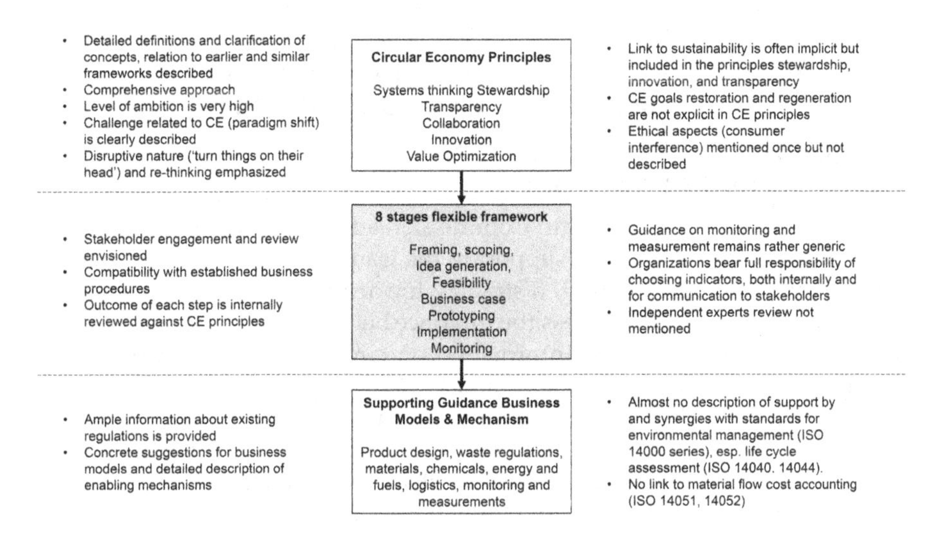

Detailed definitions and clarification of concepts, relation to earlier and similar frameworks described
Comprehensive approach
Level of ambition is very high
Challenge related to CE (paradigm shift) is clearly described
Disruptive nature ('turn things on their head') and re-thinking emphasized
Circular Economy Principles
Systems thinking Stewardship
Transparency
Collaboration
Innovation
Value Optimization
Link to sustainability is often implicit but included in the principles stewardship, innovation, and transparency
CE goals restoration and regeneration are not explicit in CE principles
Ethical aspects (consumer interference) mentioned once but not described
Stakeholder engagement and review envisioned
Compatibility with established business procedures
Outcome of each step is internally reviewed against CE principles
8 stages flexible framework
Framing, scoping,
Idea generation,
Feasibility
Business case
Prototyping
Implementation
Monitoring
Guidance on monitoring and measurement remains rather generic
Organizations bear full responsibility of choosing indicators, both internally and for communication to stakeholders
Independent experts review not mentioned
Ample information about existing regulations is provided
Concrete suggestions for business models and detailed description of enabling mechanisms
Supporting Guidance Business Models & Mechanism
Product design, waste regulations, materials, chemicals, energy and fuels, logistics, monitoring and measurements
Almost no description of support by and synergies with standards for environmental management (ISO 14000 series), esp. life cycle assessment (ISO 14040. 14044).
No link to material flow cost accounting (ISO 14051, 14052)

FIGURE 3.7
Structure of BS 8001 along with its limitations[22].

- Retrace waste and emissions to the point where they are generated.
- Identify weak points (inefficiencies) or areas of concern (hot spots).
- Set priorities for appropriate measures to minimize waste and emissions.

Although not as popular as LCA, MFA is now used as a tool to formulate circular economy strategies based on quantitative considerations and allow building of scenarios of material flows. MFA is used to understand urban metabolism and studies related to industrial ecology to guide policy formulation and develop strategies.

MFA's standards are based on ISO 14051:2011 that provides a general framework for Material Flow Cost Accounting (MFCA). The MFCA framework includes common terminologies, objective and principles, fundamental elements, and implementation steps. However, detailed calculation procedures or information on techniques for improving material or energy efficiency are outside the scope of ISO 14051:2011. ISO 14051:2011 is not intended for the purpose of third-party certification. Further, although an organization can choose to include external costs in an MFCA analysis, dealing with external costs are outside the scope of ISO 14051:2011[23].

Box 3.2 provides two examples that show the strategic use of MFA for policies on circular economy.

BOX 3.2 STRATEGIC USE OF MFA – CASE STUDIES IN EUROPEAN UNION AND SOUTH KOREA

MATERIAL FLOWS IN EUROPEAN UNION'S CIRCULAR ECONOMY[24]

In January 2018, the European Commission adopted a new set of measures as part of its ongoing support to the transition to a more circular economy. These included: (i) a strategy toward a more circular use of plastics; (ii) options to address the interface between chemical products and waste legislation; (iii) information on circular use of critical raw materials; and (iv) a monitoring framework toward a circular economy.

MFA showed that a large part of the European Union's (EU) mass material use consists of construction materials, many of which are accumulated in long-living in-use stocks. In-use stocks for products made from metals, biomass, and fossil fuels are also growing. The level of circularity varied by type of material and was found to be the highest for metals. Even with increasing end-of-life re-use and recycling rates, primary resource extraction would still be needed to meet the EU's materials demand. This is because it will take at least decades for materials contained in some growing in-use stocks to become available for recycling. Therefore, the MFA concluded that sustainable materials extraction and efficient use of resources will continue to be of paramount importance.

USE OF MFA FOR DECIDING RESOURCE RECOVERY AND RECIRCULATION PLANS IN SOUTH KOREA[25]

This study carried out in Korea highlights upcycling as the key concept for improving the value of waste by redefining the concept as "the recycling of waste materials and discarded products in ways that enhance their value". Four upcycling strategies were linked to material flow analyses conducted on waste electronic and electrical equipment, specifically waste refrigerators, and waste computers. The objective was to examine the technologies available for implementation and suggest guidelines for the promotion of upcycling.

It was found that the amount of waste refrigerators collected by the formal sector was 121,642 tons per year and the informal sector was 63,823 tons per year. The current recycling ratio of waste refrigerators was estimated as 88.53%. A total of 7585 tons per year of waste computers were collected by the formal sector and 3807 tons per year by the informal sector after discharge. The current recycling ratio of waste computers was estimated as 77.43%. It was found that it is possible to introduce 28 upcycling technologies in the case of refrigerators, and 15 technologies are available to promote upcycling in the case of computers. By refining the broad concept of upcycling and looking at the stages of material flow, the study presented universally applicable directions for incorporating upcycling in resource recovery and recirculation plans.

3.5 Life Cycle Assessment

LCA is a technique to assess the environmental and social impacts associated with the various stages of a product's life, from raw material through use until final disposal. LCA is the most popular tool for analysis when it comes to circular economy and sustainability. It has been applied to various products and organizations across the world and the application experience has helped to guide research, development databases, and computational tools.

According to the ISO 14040 series, LCA is structured in four phases, regardless of the type of LCA, the process of the assessment remains the same. Types of LCA include environmental LCA (e-LCA), social LCA (s-LCA), life cycle sustainability assessment (LCSA) and more recently organizational LCA (o-LCA)[26]. Figure 3.8 shows the LCA framework.

LCA is used not just to understand the impacts across the life cycle or identify the "hot spots" but also help in decision making. Figure 3.9 describes how such an application of LCA can be made.

While LCA has been attractive tool for decision making, it does face few limitations. Some of these limitations are listed below[28]

- LCA thoroughness and accuracy depends on the availability of data; gathering of data can be problematic; hence, a clear understanding of the uncertainty and assumptions is important.

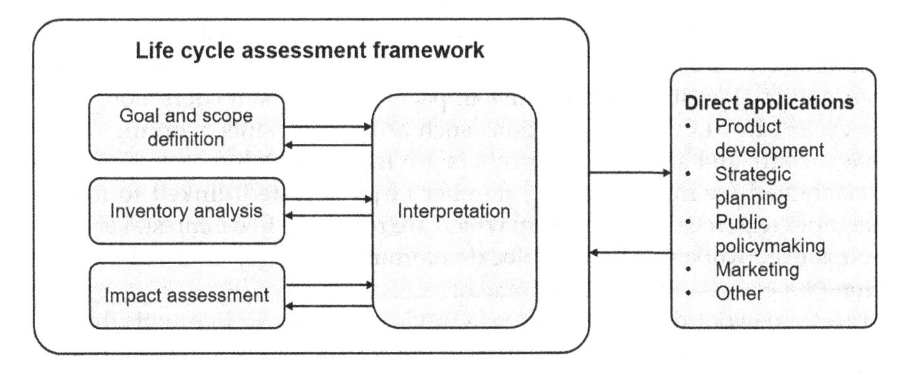

FIGURE 3.8
The LCA framework[27].

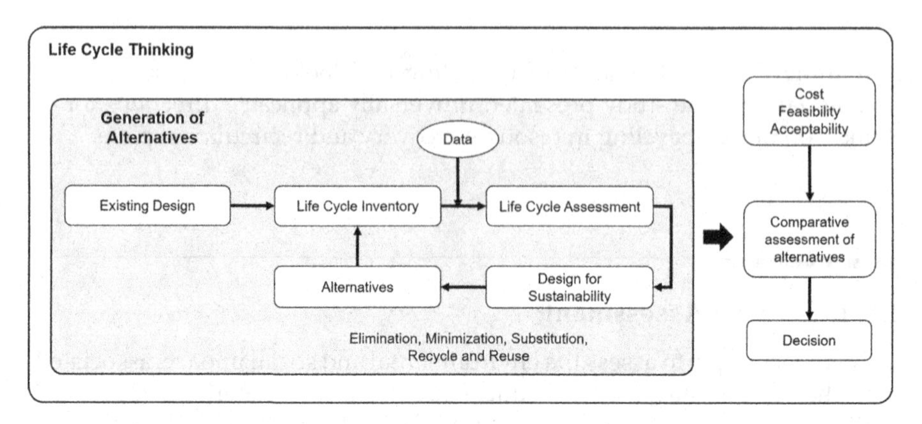

FIGURE 3.9
Application of LCA.

- Unlike traditional risk assessment, LCA does not necessarily attempt to quantify any specific actual impacts. While seeking to establish a linkage between a system and potential impacts, LCA models are suitable for relative comparisons, but may be not sufficient for absolute prediction and assessment of impacts/risks.
- The system boundaries determine which unit processes to be included in the LCA study. Defining system boundaries is partly based on a subjective choice, made during the scope phase when the boundaries are initially set. The results of LCA can change the outcomes or the decisions depending on the boundaries as there are spillover effects.

Recognizing the need for the integration of social criteria into environmental or e-LCA, in 2009 the UNEP/The Society of Environmental Toxicology and Chemistry (SETAC) Life Cycle Initiative published the Guidelines for s-LCA of Products. A s-LCA was described as a social impact assessment technique that aims to assess the social and socio-economic aspects of products and their potential positive and negative impacts on the stakeholders along their life cycle[29]. In s-LCA, considerations such as human rights, working conditions, health, and safety, among others are included. A life cycle inventory is elaborated for indicators (e.g., number of jobs created) linked to impact categories (e.g., local employment) which are related to five main stakeholder groups (e.g., worker, consumer, local community, society, and value chain actors)[30].

The framework detailed in the s-LCA Guidelines is in line with the ISO 14040 and 14044 standards for LCA. Adaptations for the consideration of social and socio-economic issues are described in the framework. It proposes a two-fold classification of social impacts: by stakeholder categories and impact categories. A set of subcategories, which are social and socio-economic issues of concerns, to be used in s-LCA are presented[28].

Life Cycle Costing (LCC) is the oldest of the three life cycle techniques. Developed originally from a conventional financial cost accounting perspective, in recent years LCC has gained importance. "Total" LCC considers external environmental and social costs and benefits anticipated. LCC is more relevant to the full costing of long-life goods (i.e., building objects, infrastructure, railways, trains, and aviation projects) that have a long-term maintenance and use phase as well as high costs.

As they have similar perspectives and aims and because they are all based on the ISO 14040:2006 stages (Phases 1, 2, 3, and 4), it is possible to combine these techniques into an overarching LCSA. Walter Klöpffer put this idea into a conceptual formula[31,32], when he suggested the combining of the three techniques. LCSA refers to the evaluation of all environmental, social, and economic negative impacts and benefits in decision-making processes toward more sustainable products throughout their life cycle.

$$\text{LCSA} = \text{e-LCA} + (\text{full})\,\text{LCC} + \text{s-LCA}$$

The overall objective of an LCSA is to provide a combined assessment of a product system. The results of an LCSA will show not only the negative impacts but also the benefits. When evaluating the results, it is recommended that data quality is considered. Interpreting the results in a combined fashion can be a key challenge to communicate the results to the non-expert audience of real-world decision-makers in public and private organizations. While LCA was originally developed for products, the benefits of the life cycle approach may be extended to the more complex prospect of organizational assessment. Recognizing this prospect, the UNEP/SETAC Life Cycle Initiative launched the flagship project "LCA of organizations" to further explore the capabilities and applicability of organizational life cycle assessment (o-LCA). The Guidance document developed as an outcome of this project, builds on key existing works and initiatives, like the Greenhouse Gas Protocol initiative, and especially strives to align with ISO/TS 14072, and with ISO 14040 and ISO 14044[33].

O-LCA analyzes the whole organization (i.e., organizational approach), including not only the facilities of the organization but also upstream and downstream activities (i.e., life cycle approach) and considers a set of relevant environmental aspects (i.e., multi-impact approach). In this way, it provides organizations with environmental understanding at the level at which most of the decisions are made, thus supporting them effectively to improve their environmental performance[34].

Product and o-LCA are complementary in more than one way as they provide insights of an organization at different levels. Methodology for o-LCA is set out in the ISO/TS 14072:2014 and is designed to assess the entire collection of goods and services of an organization. This methodology can serve multiple goals at the same time (e.g., identifying environmental hotspots throughout the value chain, tracking environmental performance over time, supporting strategic decisions, and informing corporate sustainability reporting)[28].

Presented below are some of the software tools and databases used for LCA:

Ecoinvent database – Ecoinvent is a not-for-profit association. The Ecoinvent database provides well documented process data for thousands of products, helping consumers make truly informed choices about their environmental impact. It builds on more than 20 years of experience in LCA methodology development and Life Cycle Initiative data compilation for different industrial sectors. Ecoinvent is used in a broad range of environmental studies including LCA, Environmental Product Declaration (EPD), Design for Environmental or Carbon Foot printing and allows one to conduct studies on different levels of detail: from screenings for basic, initial answers to extensive studies such as peer-reviewed, ISO compliant studies.

Visit https://www.ecoinvent.org/

SimaPro – SimaPro is one of the leading LCA software solution, with a 30-year reputation in industry and academia in more than 80 countries. It is a professional tool to collect, analyse and monitor the sustainability performance data of company's products and services. The software can be used for a variety of applications, such as sustainability reporting, carbon and water foot printing, product design, generating EPDs and determining key performance indicators.

Visit https://simapro.com/

GaBi – GaBi like SimaPro models every element of a product or system from a life cycle perspective, equipping businesses to make the best-informed decisions on the manufacture and life cycle of any product. It provides an easily accessible and constantly refreshed content database that details the costs, energy and environmental impact of sourcing and refining every raw material or processed component of a manufactured item. In addition, it looks at the impact on the environment presenting alternative options for manufacturing, distribution, recyclability, and sustainability.

Visit http://www.gabi-software.com/international/index/

Umberto LCA+ – Umberto LCA+ is a versatile LCA software that allows you to calculate LCA. You can also use this LCA tool to calculate EPD and Product Environmental Footprints (PEF), if your business has advanced to following these standards. It can also be used to find carbon footprints. Using Umberto LCA+, you start by generating a material flow network. Integrated databases such as ecoinvent or GaBi are useful in realizing the system's boundaries. These databases already contain a large number of data sets, allowing you to easily visualize the background systems. The next two phases of LCA, the life cycle inventory analysis and the life cycle impact assessment, are completed by the software. You then interpret and evaluate the results. The LCA software has functions to make it easy to evaluate the results in a way that they can actually be applied, for example, in the reduction of waste emissions. One of these functions is representing results as a Sankey diagram, which is helpful in securing an intuitive understanding of the outcome.

Visit: https://www.ifu.com/en/umberto/lca-software/

Global LCA Data network (GLAD) – The "Global LCA Data Access" network (GLAD) aims to achieve better data accessibility and interoperability. The network is comprised of independently operated LCA databases (nodes), providing users with an interface to find and access life cycle inventory datasets from different providers. GLAD thus supports LCA through easier access to data sources around the world. UN Environment serves as the Secretariat of the GLAD network, with representatives from 14 governments in the Steering Committee.

Visit https://www.globallcadataaccess.org/

The Social Hotspots Database (SHDB) – The SHDB provides solutions for continuous improvements to all forward-thinking companies who aim at having a socially responsible supply chain. It empowers buyers and suppliers by giving them the tools to assess sector-specific risks and opportunities. It has two important tools for Risk Mapping and for assessing how ethical is the supply chain. The Risk Mapping Tool provides the ability to visualize global sector risks on a map, analyze risks for multiple issues and compare the social hotspots index for country-specific sectors identifying contributing themes within a category. The ethical supply chain tool provides the ability to model supply chains, calculate a social footprint by impact category and subcategory and identify hotspots using a web-based interface. Many LCA software like Sima pro have integrated SHDB to allow conduct of both e-LCA and s-LCA.

Visit: http://www.socialhotspot.org/

3.6 System Dynamics Modeling

In understanding impact of policy related decisions on the environment, economy, and society, it is important that a "systems" approach used. Although framework such as LCSA is useful to track the economic, environmental, and social impacts, it is not able to address the "connections" or "interlinkages" and the "feedback mechanisms" in the consumption and production systems. System dynamics modeling helps to address these issues to reveal complex interconnections, dependencies, and causal relationships between the sustainability indicators.

Attempts have been made however to research and develop an integrated approach. A tool called System Dynamics Life Cycle Modeling was applied for taking policy decisions on container glass recycling in Sydney in Australia[35]. Another application of an integrated approach reported to understand impact of policies on electric vehicles in the United States of America[36].

System dynamic based modeling was developed by Jay Forester at MIT in mid 1950s. Later he wrote the classics such as World Dynamics, Industrial Dynamics, and Urban Dynamics. The report *Limits to Growth* was prepared by Club of Rome[37] 1972 based on a system dynamic model. System dynamics is particularly useful when it helps us understand the impact of time delays and nonlinearities in the system. Diagrams of loops of information feedback and circular causality are tools for conceptualizing the structure of a complex system and for communicating model-based insights[38].

UNEP while preparing the Green Economy report commissioned the Millennium Institute to apply Threshold 21 (T21) model to recommend policy decisions that will revive global economy at minimum investments while ensuring path toward sustainability. The T21 model based on system dynamics could generate scenarios tracking the dynamic interactions between key social, economic, and environmental drivers of development. During this course, T21 model was applied integrated planning tool by more than 40 nations, regional groups, and multilateral development institutions, including the governments of Kenya, Senegal, Swaziland, Peru, Venezuela, China, The Philippines, as well as the ECOWAS Commission and UNEP. Some projects include: The Green Economy Report published by UNEP; Changing Course in Global Agriculture; the Africa Adaptation Programme of the UNDP, and more[39].

Under the Partnership for Action on Green Economy (PAGE), UNEP collaborated with modeling experts from around the globe to develop the Integrated Green Economy Modeling (IGEM) framework[40]. This framework was intended to better respond to countries' needs in terms of analysing the cross-sectoral impacts of Green Economy policies, so as to incorporate some of the lessons learned from the application of existent modeling tools, such as the T21 model. The IGEM framework was designed to serve three purposes:

- Build on UNEP's past country experience with modeling green economy policies to answer increasingly complex requests from governments.
- Support the endowment of countries with solid quantitative tools to inform the design and implementation of green economy policies.
- Advance the process of implementing and monitoring some of the Sustainable Development Goals (SDGs), adopted in September 2015.

The IGEM framework was based on the linkages between system dynamics model and a computable general equilibrium (CGE) model, building on input–output and social accounting matrix (IO-SAM) models. The goal of IGEM framework was to test its effectiveness for green economy policy assessment as well as evaluate potential impacts.

Later, after the agreement on SDGs, Millennium Institute integrated Threshold21 with SDG (iSDG) – that simulated the trends for the SDGs until 2030 for each of the SDG indicators under a business-as-usual scenario and supports analysis of alternative policy scenarios.

The iSDG Model gives policymakers and planning officials the capacity to[29]:

- Visualize progress toward each of the SDGs, highlighting specific areas requiring more attention or resources.
- Evaluate the likely benefits of proposed policies and strategies and reduce undesired long-term impacts (up to 2050).
- Ensure policy coherence across areas of interventions and facilitate the alignment of SDG strategies with other national development plans.
- Define an efficient policy implementation schedule that facilitates high-impact results and monitors progress toward achieving policy objectives.

System dynamics certainly offers an opportunity to understand implications of the policy decisions taken to set the stage on circular economy. Some of the recent research has shown its potential to analyze the systemic effects of combining multiple product design and business model strategies for slowing and closing resource loops in a circular economy. Results from such research provide insights into the relationship between design considerations at the beginning of a product's life and their implications for the product's take-back stage when a new manufacturing cycle begins[41].

3.7 Epilogue

Several methodologies, tools, and knowledge bases have been evolved over past five decades that have potential applications in circular economy. There is no one single methodology and tool that may be considered as the silver bullet. Depending on the context and scale of application, a suit of methodologies and tools would need to be applied and some adaptation may be necessary. Table 3.1 presents a Toolbox that summarizes the potential methodology and/or tool, related key concepts, outcomes, and the target stakeholders.

TABLE 3.1

Toolbox

Potential Methodologies/Tools	Key Concepts	Outcomes	Target Stakeholders
Waste Minimization Audits, Cleaner/ Resource, Efficient Opportunity Assessments, Energy Audits, GP Methodology, GHG Inventorization, Chemical audits, Health & Safety Audits, Compliance Audits	Waste Minimization, Pollution Prevention, Eco-efficiency, Cleaner Production, GP.	Resource Efficiency, Sustainable Energy Management, GHG emission reduction, Improvement in Health & Safety, Compliance by improving competitiveness	Facility, Corporate, Supply Chains, Investors
MFA, Integrated MFA with LCA	Resource Efficiency, Resource re-circulation	Scenario Building Policies and Strategies toward circular economy	Policy Makers, Regions, Cities, Industrial Parks/ Clusters, Corporates
Management Systems and ISO Standards (14040, 14044), BS 8001 on circular economy	Continual improvement	Management commitment, Process Standardization, Continual improvement toward the targets	Facility, Corporate, Cities, Industrial Parks/Clusters
Life Cycle Thinking, environmental-LCA, LCC, Social-LCA, Organizational-LCA, LCSA	Eco-efficiency, Design for sustainability, D4D, Design for Recyclability, Extended Producer Responsibility, Cradle to cradle, Biomimicry	Life Cycle Management, Scenario building, Comparative assessment of options/alternatives, Branding	Corporates, Brand owners, Product designers, Material and Process Innovators, Reverse logistics operators, Informal sector
System Dynamic Modeling integrated with LCSA	Systems thinking with Causal loops, Feedback and delays	Scenario building with spill-over effects, Cross-sectoral considerations policy decisions, Assessment of the effectiveness of economic, market-based and information related instruments	International bodies, National governments, Regional authorities, Involvement of Multiple stakeholders

As regards to audits, the trend now is to expand the scope, e.g., reflecting chemical audits in resource efficiency audits to prepare a strategy and action plan to secure an eco-label. Including resource efficiency in conventional compliance audits that focus on EHS will add a value, turning audits into opportunity assessments. Combining energy audits with GHG accounting (inventorization) will be useful as measures taken on energy efficiency, energy conservation, and renewable energy greatly help in reducing GHG emissions.

EMS 14001 could be very useful to ensure that interventions identified in audits lead to an Environmental Management Plan (EMP) that gets implemented with the support of top management and its implementation is tracked on a regular basis by third party. Revisiting concepts like POEMS may be worth to address life cycle impacts of the products and introduce D4S as an element in the EMP. EMS over supply chains integrated with social, ethical, and environmental codes of conduct could be very effective to address circularity. BS 8001 on circular economy has been only recently introduced and it will be useful to assess its on ground experience.

MFA should be used as a strategic tool to develop strategies to foster circular economy. Application of MFA at industrial clusters/parks to come up with plans for industrial symbiosis is one example. MFA can also be very useful in taking decisions on options related to resource recovery and resource circulation.

e-LCA has been generally used for comparative assessment of products or during D4S. Emergence of s-LCA and subsequent integration of LCC, e-LCA, and s-LCA into LCSA has been a welcome step. Communicating results of LCSA to decision makers has however been a challenge.

Combination of System Dynamics with LCSA is recommended to prepare regional scale and national level circular economy action plans. The modeling tools based on system dynamics are getting prominence in international fora with leadership taken by UNEP.

It should be remembered however that as one uses elevated application of these methodologies and tools, the data requirements increase and these can become a challenge, especially in developing economies where data is of poor quality and not well organized. Application of s-LCA and LCSA can also be difficult in situations where the informal sector plays a major role in the material flows. Scenario building, considering the data uncertainties difficulties in the definition of system boundaries is perhaps one way to address these challenges.

Moving ahead from the methodologies and tools described in this chapter, certain basic or foundational strategies that greatly help to implement circular economy will be discussed ahead. These strategies are based on the key concepts introduced in Chapter 2 and have been evolved following practice experience of the methodologies and tools described in this chapter. In Chapter 4, these foundational strategies are described as "12Rs" of Circular Economy.

3.8 Key Takeaways

- Several methodologies exist for implementing waste minimization, cleaner production, and implementing GP. These methodologies share overlaps. There are considerable resources available that can be readily used by the practioners.
- Focusing on resource efficiency in audits will greatly help in the context of circular economy and the audits will become opportunity assessments.
- Integration of aspects such as EHS compliance, chemical screening, mitigation of GHG emissions, and energy management will lead to more value.
- Environmental management plans developed from EMS should expand to include D4S considerations like in the POEMS approach.
- MFA should be used as a powerful tool for taking policy decisions. Combination of MFA with o-LCA will help Corporates in managing circularity in the supply chains.
- Data can be a major issue while conducting LCA, especially when dealing with developing economies and thus informal sector plays a crucial role in circular economy.
- Systems Dynamic modeling in combination with LCSA is a growing area of interest and is particularly useful in drawing national and regional plans on circular economy.
- Finally, there is no one single methodology or tool that is suitable to develop and implement a plan for circular economy. One must make use of a judicial combination of the "toolbox" presented in this chapter depending on the context, scale and the interest of the key stakeholders.

ADDITIONAL READING

1. **Guide to design strategies** – Sustainable design strategies are best known as starting off with Victor Papanek in the 1970's and have been contributed to over the years by many different people and approaches. This article summarizes sustainable design strategies and the environmental, social, and economic impacts of the strategies from the initial phase through to the end of life. It also argues that Eco-Design is a core tool in the matrix of approaches that enables the Circular Economy.

Source: Acaroglu, L., 2020. *Quick Guide to Sustainable Design Strategies.* Medium. <https://medium.com/disruptive-design/quick-guide-to-sustainable-design-strategies-641765a86fb8> [Accessed 29 October 2020].

2. **The Handbook of design for sustainability** – *The Handbook* presents the first systematic overview of the subject that, in addition to methods and examples, includes historical perspectives, philosophical approaches, business analyses, educational insights, and emerging thinking. It is an invaluable resource for design researchers and students as well as design practitioners and private and public sector organizations wishing to develop more sustainable directions.

Source: Walker, S., 2013. *The Handbook of Design for Sustainability.* A&C Black.

3. **Design guidelines to develop circular products: Action research on Nordic industry** – This paper presents a set of generic design guidelines for different circular strategies. The guidelines are then used to map companies' circular product design initiatives in the early stages of product design and development. The guidelines have proved to support decision-making and enhance the circularity of products. The guidelines were developed, validated, and tested at four companies within the Nordic countries through an action research approach.

Source: Shahbazi, S. and Jönbrink, A. K., 2020. Design guidelines to develop circular products: Action research on Nordic industry. *Sustainability,* 12(9), p. 3679.

4. **How circular is the global economy? An assessment of material flows, waste production, and recycling in the European Union and the World in 2005** – This article applies a socio-metabolic approach to assess the circularity of global material flows. All societal material flows globally and in the European Union (EU-27) are traced from extraction to disposal and presented for main material groups for 2005. The results indicate that strategies targeting the output side (end of pipe) are limited given present proportions of flows, whereas a shift to renewable energy, a significant reduction of societal stock growth, and decisive eco-design are required to advance toward a CE.

Source: Haas, W., Krausmann, F., Wiedenhofer, D. and Heinz, M., 2015. How circular is the global economy? An assessment of material flows, waste production, and recycling in the European Union and the world in 2005. *Journal of Industrial Ecology*, 19(5), pp. 765–777.

5. **Low-carbon conceptual design based on product LCA** – This paper presents a carbon footprint model and a low-carbon conceptual design framework where the environmental impacts throughout the life cycle of a product can be assessed. In the carbon footprint model, the amount of carbon emission is estimated at the five stages of the entire product life cycle. The carbon footprint analysis is based on product LCA. Sensitivity analysis for design parameters is also performed to measure the effects of design parameters on the estimation of product carbon footprint quantitatively. The conceptual design of a cold heading machine is used to demonstrate the proposed methodology.

Source: He, B., Tang, W., Wang, J., *et al.*, 2015. Low-carbon conceptual design based on product life cycle assessment. *International Journal of Advanced Manufacturing Technology*, 81, pp. 863–874. https://doi.org/10.1007/s00170-015-7253-5

6. **Social Life Cycle Assessment (s-LCA) of product value chains under a circular economy approach: A case study in the plastic packaging sector** – This paper describes the theoretical framework and impact assessment approach for the s-LCA of product value chains under a circular economy approach by applying a scoring system in different subcategories and indicators, considering the plastic packaging sector as a case study.

Source: Reinales, D., Zambrana-Vasquez, D. and Saez-De-Guinoa, A., 2020. Social life cycle assessment of product value chains under a circular economy approach: A case study in the plastic packaging sector. *Sustainability*, 12(16), p. 6671.

7. **Application of life cycle thinking toward sustainable cities: A review** – This paper reviews LCT studies related to urban issues to identify the main research gaps in the evaluation of these improvement strategies. The review identifies the main sustainability strategies associated with each urban issue and compiles articles that deal with these strategies through LCT, including environmental LCA, LCC, social LCA (s-LCA), and LCSA, as well as integrated analyses with combined tools.

Source: Petit-Boix, A., Llorach-Massana, P., Sanjuan-Delmás, D., Sierra-Pérez, J., Vinyes, E., Gabarrell, X., Rieradevall, J. and Sanyé-Mengual, E., 2017. Application of life cycle thinking towards sustainable cities: A review. *Journal of Cleaner Production*, 166, pp. 939–951.

8. **UNEP-SETAC guideline on global land use impact assessment on biodiversity and ecosystem services in LCA** – This paper addresses the calculation of land use interventions and land use impacts, the issue of impact reversibility, the spatial and temporal distribution of such impacts and the assessment of absolute or relative ecosystem quality changes. Based on this, the authors propose a guideline to build methods for land use impact assessment in LCA.

Source: Koellner, T., De Baan, L., Beck, T., Brandão, M., Civit, B., Margni, M., Milà i Canals, L., Saad, R., De Souza, D. M. and Müller-Wenk, R., 2013. UNEP-SETAC guideline on global land use impact assessment on biodiversity and ecosystem services in LCA. *The International Journal of Life Cycle Assessment*, 18(6), pp. 1188–1202.

9. **Toward a LCSA** – This publication can be used as a springboard for stakeholders to engage in a holistic and balanced assessment of product life cycles and to consider the three pillars of sustainability in a unique and instructive approach. In this way, this publication will provide further guidance on the road toward the consolidation of LCSAs. The publication includes eight case studies to illustrate how current and emerging LCA techniques are being implemented worldwide from Asia through Europe and Latin America

Source: UNEP. 2011. *Towards A Life Cycle Sustainability Assessment*. <https://www.lifecycleinitiative.org/wp-content/uploads/2012/12/2011%20-%20Towards%20LCSA.pdf> [Accessed 29 October 2020].

10. **Organizational LCA: The new member of the LCA family – Introducing the UNEP/SETAC Life Cycle Initiative guidance document** – This editorial is the official announcement of the publication of this new UNEP/SETAC Life Cycle Initiative product and shortly presents it to the LCA community. Organizational LCA is a relevant and promising new member of the LCA family. Just like social LCA was promoted and supported by the publication of the Guidelines for Social

LCA (UNEP/SETAC 2009), the authors expect that the guidance document will help promote organizational LCA.

Source: Martínez-Blanco, J., Inaba, A., Quiros, A., *et al.*, 2015. Organizational LCA: The new member of the LCA family – Introducing the UNEP/SETAC Life Cycle Initiative guidance document. *International Journal of Life Cycle Assessment*, 20, pp. 1045–1047. https://doi.org/10.1007/s11367-015-0912-9

11. **Guidelines for s-LCA of products and organizations 2020 –** In 2009, UNEP's Life Cycle Initiative launched the first Guidelines for s-LCA. This 2020 edition also looks at how to link the social impacts of a product's production and consumption to the larger impacts associated with an organization's influence across the life cycle of a product. Social organizational LCA (also known as so-LCA) strengthens s-LCA by providing an organizational perspective that guides many organizational decisions. so-LCA also complements Organizational LCA guidance, another tool developed by the Life Cycle Initiative.

Source: UNEP. 2020. *Guidelines for Social Life Cycle Assessment of Products and Organizations 2020*. Benoît Norris, C., Traverso, M., Neugebauer, S., Ekener, E., Schaubroeck, T., Russo Garrido, S., Berger, M., Valdivia, S., Lehmann, A., Finkbeiner, M., Arcese, G. (Eds.). United Nations Environment Programme (UNEP).

12. **Addressing climate-sustainable development linkages in long-term low-carbon strategies: The role of Millennium Institute's iSDG Model –** A cohesive, integrated planning, and implementation framework that ensures that climate actions contribute to the achievement of the SDGs and vice versa is essential if these synergies are to be efficiently leveraged for global sustainable development. To support this type of integrated planning at the national level, the Millennium Institute has recently developed the Integrated Sustainable Development Goal (iSDG) model, a simulation tool that enables the analysis of individual policies and complex strategies designed to achieve the SDGs, as well as the testing of their likely impacts before they are adopted. This model could be of great use to countries in conceiving long-term low-emissions development strategies.

Source: World Resources Institute. n.d. *Addressing Climate-Sustainable Development Linkages in Long-Term Low-Carbon Strategies: The Role of Millennium Institute's iSDG Model.* <https://www.wri.org/climate/expert-perspective/addressing-climate-sustainable-development-linkages-long-term-low-carbon> [Accessed 29 October 2020].

Notes

1. US EPA. 1988. *Waste Minimization Opportunity Assessment Manual.* <https://bit.ly/2E3wKX7> [Accessed 17 August 2020].
2. UNIDO and UNEP. 1991. *Audit and Reduction Manual for Industrial Emissions and Wastes.* <https://open.unido.org/api/documents/4990579/download/AUDIT%20AND%20REDUCTION%20MANUAL%20FOR%20INDUSTRIAL%20EMISSIONS%20AND%20WASTES%20(19593.en)> [Accessed 27 October 2020].
3. United States Environmental Protection Agency. n.d. *USER's GUIDE: Strategic Waste Minimization Initiative (SWAMI) Version 2.0.* <https://cfpub.epa.gov/si/si_public_record_Report.cfm?Lab=NRMRL&dirEntryID=124892> [Accessed 27 October 2020].
4. Tellus Institute. n.d. *P2/FINANCE: Pollution Prevention Financial Analysis and Cost Evaluation System.* <https://p2infohouse.org/ref/01/00047/6-03.htm> [Accessed 27 October 2020].
5. International Finance Corporation. 1999. *Pollution Prevention and Abatement Handbook.* <https://www.ifc.org/wps/wcm/connect/topics_ext_content/ifc_external_corporate_site/sustainability-at-ifc/publications/publications_handbook_ppah__wci__1319577543003> [Accessed 27 October 2020].
6. International Finance Corporation. n.d. *Environmental, Health, and Safety Guidelines.* <https://www.ifc.org/wps/wcm/connect/topics_ext_content/ifc_external_corporate_site/sustainability-at-ifc/policies-standards/ehs-guidelines> [Accessed 27 October 2020].
7. IVAM. 2008. *Cleaner Production Manual.* <http://www.gcpcenvis.nic.in/Manuals_Guideline/CP_Manual_Improving_Living_and_Working_Condition_of_People_around_Industries.pdf>[Accessed 27 October 2020].
8. APO. 2006. *Handbook on Green Productivity.* <https://www.apo-tokyo.org/publications/ebooks/apo-handbook-on-green-productivity-pdf-7-6mb/> [Accessed 27 October 2020].
9. D4S website. http://www.d4s-sbs.org/
10. Source: Crul, M. and Diehl, J., 2006. *Design For Sustainability: A Practical Approach For Developing Economies.* UNEP, Delft University of Technology. <https://d306pr3pise04h.cloudfront.net/docs/issues_doc%2FEnvironment%2Fclimate%2Fdesign_for_sustainability.pdf> [Accessed 7 October 2020].
11. Ellen Macarthur Foundation. 2013. *Engineering the Circular Economy.* <https://www.ellenmacarthurfoundation.org/assets/downloads/news/EMF_Engineering-the-Circular-Economy_300913.pdf> [Accessed 29 October 2020].

12. HP. n.d. *HP Z1*. <https://www8.hp.com/ca/en/campaigns/workstations/z1.html> [Accessed 31 October 2020].
13. Research Council for Automobile Repairs. n.d. *RCAR Design Guide – A Manufacturers' Guide to Ensure Good Design Practice for Reparability and Limitation of Damage*. <https://www.rcar.org/Papers/DesignGuides/DesignGuide_v1_1.pdf> [Accessed 29 October 2020].
14. He, B., Tang, W., Wang, J., *et al.*, 2015. Low-carbon conceptual design based on product life cycle assessment. *International Journal of Advanced Manufacturing Technology* 81, 863–874. https://doi.org/10.1007/s00170-015-7253-5
15. Greenhouse Gas Protocol. n.d. *About Us*. <https://ghgprotocol.org/about-us> [Accessed 27 October 2020].
16. Greenhouse Gas Protocol. n.d. *Calculation tools*. <https://ghgprotocol.org/calculation-tools><https://ghgprotocol.org/about-us> [Accessed 27 October 2020].
17. Government of Canada. n.d. *Data Analysis Software and Modelling Tools*. <https://www.nrcan.gc.ca/maps-tools-publications/tools/modelling-tools/7417> [Accessed 29 October 2020].
18. ISO. n.d. *ISO 14001:2015*. <https://www.iso.org/obp/ui/#iso:std:iso:14001:ed-3:v1:en> [Accessed 31 October 2020].
19. Charter, M., Tischner, U., & Tischener, U. (Eds.), 2001. Sustainable Solutions: Developing Products and Services for the Future. Sheffield: Taylor & Francis Group.
20. Karna, A., 1999. Managing Environmental Issues From Design to Disposal—A Chain Reaction? Licentiate Thesis. Helsinki: Federation of Finnish Electrical and Electronics Industry.
21. BSI Group. n.d. *BS 8001 Circular Economy*. <https://www.bsigroup.com/en-GB/standards/benefits-of-using-standards/becoming-more-sustainable-with-standards/BS8001-Circular-Economy/> [Accessed 29 October 2020].
22. Pauliuk, S., 2018. Critical appraisal of the circular economy standard BS 8001:2017 and a dashboard of quantitative system indicators for its implementation in organizations. *Resources, Conservation and Recycling*, 129, pp. 81–92.
23. ISO. n.d. *ISO 14051:2011*. <https://www.iso.org/standard/50986.html> [Accessed 31 October 2020].
24. European Innovation Partnership on Raw Materials. n.d. *Material Flows In The Circular Economy*. <https://rmis.jrc.ec.europa.eu/uploads/scoreboard2018/indicators/15._Material_flows_in_the_circular_economy.pdf> [Accessed 29 October 2020].
25. Yi, S., Lee, H., Lee, J. and Kim, W., 2019. Upcycling strategies for waste electronic and electrical equipment based on material flow analysis. *Environmental Engineering Research*, 24(1), pp. 74–81.
26. Life Cycle Initiative. n.d. *Environmental LCA*. <https://www.lifecycleinitiative.org/starting-life-cycle-thinking/life-cycle-approaches/environmental-lca/> [Accessed 31 October 2020].
27. Nieuwlaar, E., 2013. Life cycle assessment and energy systems.
28. Penn State University. n.d. *Technologies for Sustainability Systems*. <https://www.e-education.psu.edu/eme807/node/690> [Accessed 31 October 2020].
29. UNEP. 2009. *Guidelines for Social Life Cycle Assessment of Products*. <http://www.unep.fr/shared/publications/pdf/DTIx1164xPA-guidelines_sLCA.pdf> [Accessed 29 October 2020].

30. Life Cycle Initiative. n.d. *Social Life Cycle Assessment (S-LCA).* <https://www.lifecycleinitiative.org/starting-life-cycle-thinking/life-cycle-approaches/social-lca/> [Accessed 29 October 2020].
31. Klöpffer, W., 2008. Life cycle sustainability assessment of products. *International Journal of Life Cycle Assessment,* 13(2), pp. 89–95.
32. Finkbeiner, M., Schau, E., Lehmann, A. and Traverso, M., 2010. Towards life cycle sustainability assessment. *Sustainability,* 2(10), pp. 3309–3322. Open access doi:10.3390/ su2103309.
33. UNEP. 2015. *Guidance On Organizational Life Cycle Assessment.* <https://www.lifecycleinitiative.org/wp-content/uploads/2015/04/o-lca_24.4.15-web.pdf> [Accessed 29 October 2020].
34. Martínez-Blanco, J., Inaba, A., Quiros, A., *et al.,* 2015. Organizational LCA: The new member of the LCA family—Introducing the UNEP/SETAC Life Cycle Initiative guidance document. *International Journal of Life Cycle Assessment,* 20, pp. 1045–1047. https://doi.org/10.1007/s11367-015-0912-9
35. Changsirivathanathamrong, A., Moore, S., Linard, K., n.d. *Integrating Systems Dynamic With Life Cycle Assessment.* <https://www.mssanz.org.au/MODSIM01/Vol%203/Changsirivathanathamrong.pdf> [Accessed 29 October 2020].
36. Onat, N., Kucukvar, M., Tatari, O. and Egilmez, G., 2016. Integration of system dynamics approach toward deepening and broadening the life cycle sustainability assessment framework: A case for electric vehicles. *The International Journal of Life Cycle Assessment,* 21(7), pp. 1009–1034.
37. Meadows, D. H., Meadows, D. L., Randers, J. and Behrens III, W. W., 1972. "The limits to growth." *New York* 102, no. 1972, 27.
38. System Dynamics Society - https://www.systemdynamics.org/
39. UNECE. n.d. *Integrated Simulation Tool.* <https://www.unece.org/fileadmin/DAM/env/documents/2016/wat/12Dec_06-08_Nexus_WS_TF_Geneva/WS_docs/Marketplace/S1.Zuellich_iSDG_FAQ_brochure_FINAL-EN.pdf> [Accessed 29 October 2020].
40. UN PAGE. 2017. *The Integrated Green Economy Modelling Framework – Technical Document.* <https://www.un-page.org/files/public/gep-modelling-final_jh_amend.pdf> [Accessed 29 October 2020].
41. Franco, M. A., 2019. A system dynamics approach to product design and business model strategies for the circular economy. *Journal of Cleaner Production,* 241, 118327.

4

12 Rs of Circular Economy

Chapter 2 looked at some of the key concepts and programs that have influenced the linear economy to move toward circular. Chapter 3 presented some of methodologies, tools, and knowledge bases for putting these concepts into practice. The application experience has led to 12 strategic principles or the "12 Rs" as shown in Figure 4.1.

Any one of the 12 Rs is not effective in isolation. It would be best to apply them in combination or collective and strategically to move the wheel of a circular economy. No one R is a silver bullet to help close the materials loop or reach a regenerative economy.

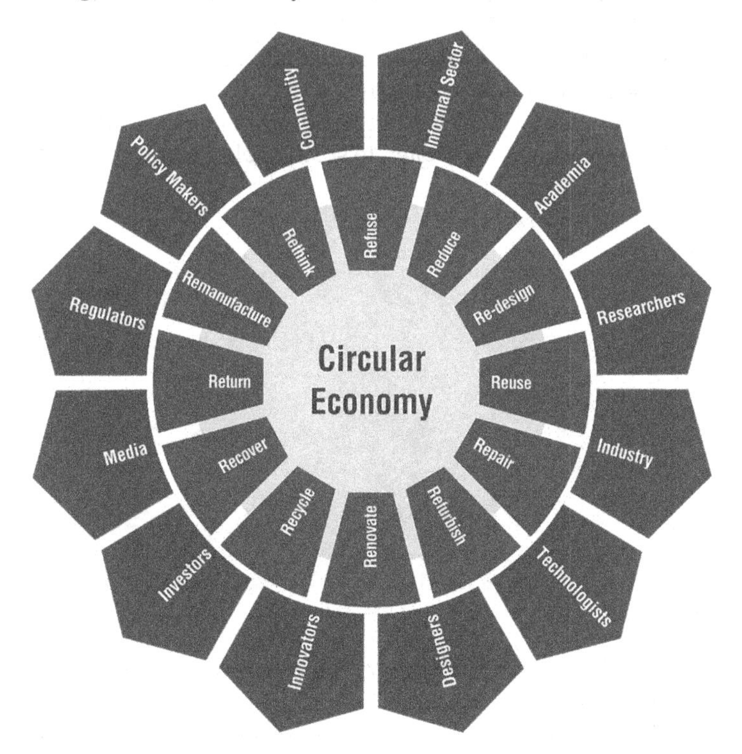

FIGURE 4.1
The 12 Rs of the wheel of circular economy.

While each of the Rs has their role, at times, they overlap in application or practice. For instance, Repair, Refurbish, and Renovate overlap when we think about restoring an old product into a usable state and extending product life.

Figure 4.1 also shows the 12 key stakeholders that need to get involved in putting the various Rs into practice depending on the context. All these stakeholders have to work together to ensure benefits across the value chain and align and adopt one or more Rs. For example, companies in the electronics industry may *reduce* the content of the product's hazardous substances and *re-design* for disassembly and reparability through innovation. Policymakers and regulators may then play a role to introducing relevant regulations and standards to push eco- designs. Companies may develop reverse logistics plans to take back used products for *repair, refurbish,* and *remanufacturing.* Finally, the community will need to show a preference for the refurbished products understanding how this preference will result into reduced extraction of virgin resources and extension of product life. Media and academia may play a role in removing misconceptions about the quality of refurbished goods and about the risks of their usage. It is also crucial to recognize the role that the informal sector plays in application of these strategies. The formal sector, especially in developing economies, has successfully built businesses in partnership with the informal sector to collect, sort, clean, and *recycle* the waste streams.

Figure 4.2 shows how the 12 Rs can be visualized in four circles to structure interventions in a circular economy.

The innermost or core circle should be the priority or the "bulls-eye". *Refuse, Reduce,* and *Reuse* are intimately related to sustainable consumption. This circle

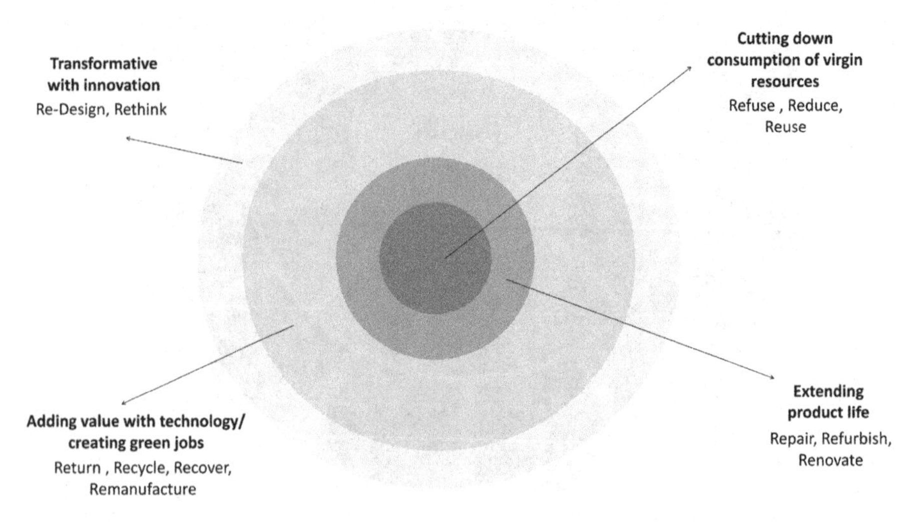

FIGURE 4.2
Four circles of the 12 Rs.

stresses the importance of moving toward sustainable lifestyles. Progress on a circular economy heavily depends on the adoption of this core circle.

The second circle gives importance to product life extension through *Repair, Refurbish,* and *Renovate*. In most cases, this circle promotes community-level initiatives and small businesses. These 3 Rs harbor considerable product related innovation that is generally not tapped. Businesses following these 3 Rs have a great potential to create green jobs, especially in repair, refurbish, and renovate related services. Establishing market linkages and introducing a certification system for refurbished products are necessary to scale up economic activities in this circle.

The third circle signifies the importance of adding value to the economy through "greener and more resource efficient" technologies and materials. These 4 Rs consisting *Return, Recycling, Recovery,* and *Remanufacturing* attract investments and are more suitable for medium to large-scale businesses. These Rs help in not only reducing the extraction and transportation of virgin resources but also provide opportunities to make significant profits through innovative business models. For example, the R such as *Return* requires setting of a reverse logistic chain and deployment of Reverse Vending Machines (RVM). Lately, it has been seen that digital technologies are playing an essential role in implementing return operations.

Recycling and recovery are possible both on-site and off-site and attract engagement models based on Public Private Partnership (PPP) with support from Government when on a scale. *Remanufacturing* is generally a high investment and high-tech option, often requiring a policy intervention in defining what is the End of Life (EoL).

The outermost or the fourth circle is as critical as the core or inner circles. It aims to transform society by innovation in materials, technologies, and product design and bring in a behavior change. The Rs of *Re-Design* and *Rethink* thus cut across core as well as the inner circles.

When we assess the worth of the Rs in economic terms, we should not limit only to the monetary value (i.e., return on investment or profitability). Instead, we should also include other social and environmental benefits, especially to the informal sector, such as the generation of livelihoods, improvement in health and safety, and imparting of skills, i.e., building capacities.

This chapter illustrates each of the 12 Rs with examples.

4.1 Refuse

To refuse is to decline, deny, reject, or resist[1].

Frugal lifestyles and sustainable living help to teach the idea of refusal and consume only for the basic needs and not wants. In behavioral science, frugality has been defined as "the tendency to acquire goods and services in a restrained manner, and resourceful use of already owned economic goods and services, to achieve a longer-term goal"[2].

Refuse essentially means living within your basic needs and not consume more than needed.

ACTIVITY 4.1 CALCULATING PERSONAL SUSTAINABILITY SCORE

Read my blog – 30 resolutions for sustainable living in 2020

https://prasadmodakblog.com/2019/12/30/30-resolutions-for-sustainable-living-in-2020/

Follow the scoring scheme suggested in the blog and compute your personal sustainability score. Compare the score with your colleagues and members in the family and discuss how you can improve the sustainability score both individually and collectively.

Box 4.1 includes examples of the strategy of Refuse.

BOX 4.1 EXAMPLES OF REFUSE

NO STRAWS ATTACHED[3]

Location: New Delhi, India

The "No Straws Attached" campaign was started in Delhi (India) in 2018 by three friends – Priyanka Kalra, Umang Manchanda, and Mallika Arya. They wanted to tackle single-use plastic pollution by introducing the concept of "refuse". They started with single-use plastic straws. They focused on plastic straws for three reasons:

- Plastic straws cannot be recycled as they are too light to go through the recycling process.
- It is simple to avoid plastic straws altogether, as one can easily drink directly from the glass.
- Giving up plastic straw is one little step toward a behavior change on minimizing plastic waste.

To achieve results through this campaign, a two-step strategy was used. First, the advantage of social media was taken to spread the message among various stakeholders. Second, conversations regarding the benefits of plastic straw alternatives (such as straws made of steel or bamboo) were launched with various business owners. The reason behind involving restaurants and cafes was that if these restaurants and cafes do not serve plastic straws in the first place, the customers will not use.

ACTIVITY 4.2 APPLYING ENVIRONMENTAL-LIFE CYCLE ASSESSMENT AND LIFE CYCLE COSTING TO STRAW ALTERNATIVES

How "effective", "safe", and "environmentally friendly" are the alternatives suggested for replacing the plastic straws? Can you apply environmental-Life Cycle Assessment and Life Cycle Costing to identify which one should be the preferred option?

4.2 Reduce

To reduce is to bring down consumption. The idea is also to reduce the size, quantity, toxicity, or intensity of material and energy use in the product.

According to research conducted in the European Union (EU), more than 80% of all product-related environmental impacts across life cycle can be influenced during the design phase[4]. Design for Sustainability (D4S) discussed in Chapter 2 is used to implement the R of reduce. Box 4.2 provides some examples of the Reduce strategy.

BOX 4.2 EXAMPLES OF REDUCE

TIDE ECOBOX[5]

Location: United States of America

Packaging waste in the consumer sector has been on the rise. To tackle this issue, organizations need to actively reduce their packaging and make it eco-friendly. On these lines, the liquid detergent company Tide, released its "sustainable detergent" called Tide Eco-Box. The company claims to produce the detergent with the help of 100% renewable energy.

Tide Eco-box is supplied in a smart package compressed into a shipping-safe form, made from 60% less plastic. In addition, it features a No-Drip twist tap to minimize any leaks and spills.

REPLENISH[6]

Location: United States of America

Typically, a bottle of liquid cleaning detergents is 90% water and the active ingredient make up is only 10%. Most of the product volume is thus just plastic packaging and water. Company Replenish produces reusable, durable spray bottles that have an attachment for pods at the base containing liquid concentrates ranging from cleaning to personal care. One pod has enough concentrate to make up for 6 bottles of cleaning liquid by adding water. This is claimed to reduce the volume of plastic packaging requirement by 90%, thereby reducing transport costs and associated emissions. In addition, it reduces generation of plastic waste.

The company offers various brands this solution to adapt their product offerings in a Replenish 3.0 bottle. Downstream companies can white label these bottles under their own brand without any development costs. This enables Replenish to get to the market faster with a completely reusable, concentrate-based solution.

Note that this case study illustrates application of 2 Rs – *redesign* and *reduce*, at manufacturer's level and 2 Rs – *reuse* and *reduce* at the consumer level.

FlyBird Farm Innovations[7]

Location: Bengaluru, India

Farming is a resource intensive industry. To reduce consumption of natural resources, we need to create solutions with modern technology that help control resource consumption. FlyBird Farm innovation is a social enterprise focused on helping farmers conserve water and control fertigation. The technology helps farmers improve crop yield and increase production from their fields while reducing water and electricity consumption. This is done by making the most affordable and cutting-edge irrigation control systems.

This innovation claims up to 30% water savings, while increasing the crop yield by 10–15% with 15–20% reduction in fertilizer input.

ECO-DOME MARCO[8]

Location: Morocco, North Africa

In developing regions where the population is high and economic growth is slow, there is an urgent need to develop sustainable means of resource consumption. In 2016, the company Eco-Dome, designed dome structures using a mixture made from 90% earth and 10% cement. These domes offer several benefits. They are soundproof and offer thermal insulation as well as natural cooling. They reduce CO_2 emissions up to 64% and provide extra strength in natural disasters like earthquakes, reducing the need for artificial reinforcement. Construction time is reduced as well as 45% cost reduction per m^2 is achieved when switching from reinforced concrete.

ACTIVITY 4.3 NATIONAL DESIGN STANDARDS RELATED TO "REDUCE"

"Reduce" has significant environmental, social, and economic benefits. Have you come across examples, where the innovations and experience of Reduce have resulted into change in the national design standards?

4.3 Re-design

Re-design is a plan for making changes to the structure and functions of an artefact, building or system so as to better serve the purpose of the original design, or to serve purposes different from those set forth in the original design. The idea is also to make durable products that are not toxic and can be easily repaired and disassembled. Box 4.3 includes some examples of "Re-design".

BOX 4.3 EXAMPLES OF RE-DESIGN

COMPANY ECOVATIVE[9]

Location: New York, United States of America

Company Ecovative Design makes packaging and building materials using mycelium as a natural binding agent. Mycelium is a fast-growing vegetative part of a fungus, which is safe and inactive and grows in a mass of branched fibers. It mainly originates from biological and agricultural wastes. The self-assembling bonds formed by mycelium grow quickly and produce miles of tiny white fibers that envelop and digest the seed husks, binding them into a strong and biodegradable material[10].

Mycelium based materials are cheaper and have the potential to become an alternative to polystyrene for a wide variety of applications.

In this example, the concept of biomimicry is used.

KUNGSBACKA

Location: Multi-country

Furniture retail company, IKEA, launched a product Kungsbacka. It consists of kitchen fronts which is made from recycled plastic bottles and reclaimed industrial wood. The units were designed by Swedish studio, Form Us With Love. Twenty-five plastic bottles are used for every 40 × 80 cm unit. The plywood on the inside is made of recycled industrial wood, making the kitchen unit 100% recycled[11].

This case study shows application of re-design with recycling as the strategy.

Mitticool[12]

Location: Gujarat, India

Earthen clay has been used since ancient times for making ornaments, tools, utensils, and clay pots. Products made from earthen clay are biodegradable and can also be crushed and repurposed to be used in other places such as in the construction industry.

In villages, due to a lack of resources and infrastructure, it is crucial to re-design basic amenities. After 2001 Gujarat Earthquake, from the headlines of newspaper, Mr. Mansukhbhai Prajapti came up with an idea of a clay refrigerator running without electricity. He named it Mitticool ("mitti" is the word for clay in the Hindi language). By 2004, the company Mitticool designed a refrigerator, completely made from earthen clay, to store drinking water, food, vegetables, and milk. The company claims that the refrigerator can store food for 3 to 5 days without staling it.

Over the years, the company has steadily grown and has its presence all over India as well as in international countries. Today, the company offers various earthen cookware, earthen pots, and other tableware as a part of its product line-up.

ABWEZI[13]

Location: Blantyr, Malawi

The company Abwezi has come up with a unique design for a school bag which not only reduces plastic waste but also provides illumination to off-grid rural households in Malawi. The company has distributed school bags that are made by upcycling disposed plastic bags. The school bag is integrated with a solar panel. This solar panel inserted at the back of the bag, gets charged as the child walks to school. The bag provides families with a clean and free source of energy for lighting after-school-learning.

Upcycling bags locally enabled Abwezi to reduce costs to produce the school bags. The company targeted NGOs (Non-Governmental Organization), companies, and well-wishers to buy bags by the bulk and distribute them for free to less privileged communities.

This case study highlights several benefits. It reduces the disposal costs of used plastic and damage to the environment. It also reduces consumption of virgin materials used to make plastic. Integration with a solar panel with the school bag provides a source of energy to the families for better quality of life.

Hemp as an Alternative

The industrial hemp plant can be used in the most diverse forms. It not only helps with bioremediation but also can be used as a sustainable alternative in a variety of industries.

One hectare of industrial hemp can absorb 22 tons of CO_2 per hectare. Hemp's rapid growth (grows to 4 m in 100 days) makes it one of the fastest CO_2-to-biomass conversion tools available, more efficient than agro-forestry. In addition, the CO_2 is permanently bonded within the fiber that can be used for anything from textiles to paper and as a building material[14].

It is estimated that the global market for hemp consists of more than 25,000 products in nine submarkets: agriculture, textiles, recycling, automotive, furniture, food/nutrition/beverages, paper, construction materials, and personal care. For construction materials, hempcrete (a mixture of hemp hurds and lime products) can be used as a lightweight insulating material[15].

The fashion industry has been consuming hemp textile since several decades. The brand Giorgio Armani started using hemp in an Emporio Armani collection in 1995, while the company Ralph Lauren makes hemp floor rugs. In India, the company Boheco that was launched in 2013, has been a leading advocate for not only hemp textiles but various other products made out of different parts of the plant.

ACTIVITY 4.4 RISKS AND DRIVERS OF "RE-DESIGN"

Re-design examples provided in Box 3.3 focus on material change or substitution. Do you see any risks in such approach? How can these risks be addressed? What could be the drivers for re-design?

4.4 Reuse

Reusing refers to using an object as it is once again. Reused items include anything that was bought second hand, often furniture and clothing.

Re-use is the process of re-using a product for the same purpose without conducting any significant repair to the product[16]. Box 4.4 presents some examples of Reuse.

BOX 4.4 EXAMPLES OF REUSE

RePack[17]

Location: Finland

RePack is a sustainable packaging service for e-commerce based in Finland. Reusable packaging is shared by a pool of webstores. The packaging is made out of durable and recycled polypropylene and comes in three adjustable sizes.

Consumers can choose the RePack shipping option while shopping from partner companies. At the time of the purchase, the customer can choose this delivery option for an extra fee. The packaging can be returned by users for free, from anywhere in the world. Users retrieve economic value when returning the packaging, by being rewarded with a voucher for their next purchase.

RePack packaging can be reused 40 times on average. Such reuse has a smaller carbon footprint than single-use packaging by a big margin, even accounting for the emissions resulting from the collection or reverse logistics.

This example shows how reuse is supported by redesign and economic instruments such as Deposit Refund Scheme (DRS).

CLOTHES DONATION

Today, the average consumer buys 60% more items of clothing than in 2000, but each garment is kept for half as long since consumers discard items more quickly[18]. These buying habits contribute to the 39 million tons of post-consumer textile waste that is generated (at a minimum) worldwide each year – primarily in the form of garments.

Donating clothes that are in good condition can increase their life and reduce textile waste at landfills.

An NGO in India, Goonj[19], collects all forms of discarded clothes and textiles to convert it into blankets and sanitary pads to distribute in rural India.

LIBRARY OF THINGS[20]

Location: United Kingdom

Library of Things is a social enterprise, rooted in the simple ideology of "Why buy when you can borrow". The enterprise allows citizens of the United Kingdom (UK) to borrow items for a variety of applications like gardening, cooking, adventures, cleaning, etc. They operate across 5 sites within the UK and borrowers from just one of the sites have saved 15,900 kg of waste from going to the landfill by borrowing rather than buying.

This is a regenerative business model, creating affordable options for borrowers and revenue-generating streams for community space partners. It helps reduce the purchase of new material and gives products a longer life through multiple uses.

KHILONEWALA[21]

Location: India

The company Khilonewala is an online toy library that provides rental services for toys, games, and books. The service is catered for kids in the age group of 1–12 years.

In less than 3 years, the model is successfully operating in 35 cities with 45 centers across India. Their catalogue offers about 500 products to choose from.

The examples cited for reuse show how consumption can be collaborative. In many ways this is how a shared economy will operate.

4.5 Repair

Repair means to put something damaged, broken, or not working correctly back into good condition or make it work again.

For a given fault within a product, if an operation has been conducted to correct the fault, then the product is said to have been repaired. Almost certainly all repaired products are not restored to original standard and any guarantee issued will generally only cover the corrected fault. This process involves less work than remanufacture and refurbishing[18]. Box 4.5 illustrates some examples of Repair.

BOX 4.5 EXAMPLES OF REPAIR

REPAIR CAFES[22]

Location: Amsterdam

Amsterdam was the first city in the world to introduce repair cafés. These community-driven cafes regularly organize local meetups where residents are encouraged to fix their broken belongings over a cup of coffee. Help is provided by experienced professionals and local enthusiasts. Everyone is welcome at these meetings, regardless of skill-level.

They can bring along broken household appliances, ripped clothing, impaired bicycles, or worn-out possessions.

After the first repair café launched in 2009, the idea quickly spread throughout Europe and eventually travelled well beyond the continent's borders. In 2017, there were more than 600 in the Netherlands alone.

iFixit[23]

Location: United States of America

iFixit is a website that teaches people how to fix almost any electronic device. Along with using the resources available on the website, anyone can create a repair manual for a device as well as edit the existing set of manuals to improve them. The site empowers individuals to share their technical knowledge on repairing with the rest of the world thus encourage repair instead of discarding broken products.

FAIRPHONE[24]

Location: Netherlands

In 2010, the company Fairphone had a vision of creating a modular phone that made repairing possible at the users end. As of today, the company has released three generations of smartphones which have adopted this modular design.

This design makes repair easy for the customers by adding or removing broken or malfunctioning parts, only by using a single screwdriver. This prevents the option of unnecessarily buying a new product thus making the phone last longer and reducing the environmental impact.

Fairphone is not only focused on developing products with minimal impact to the environment but also sources components for the phone which are greener than their counterparts such as recycled plastics and conflict-free minerals. The company also ensures its social commitment by providing its workers living wage bonuses. Fairphone is the only smartphone company to be Fairtrade gold certified.

This product which is based on a "redesign" in comparison to its contemporaries, may encourage customers to "rethink" about what goes into making a smartphone and how ethically the phone is produced. It is also an example of D4D and D4R (see Chapter 3) and a demonstration of how cleverly a product could be designed for longer life and reduced generation of e-waste.

ACTIVITY 4.5 DISCUSSION ON "REPARABILITY"

Giving access and encouraging repairing may violate the intellectual property rights (IPR) of the original equipment manufacturer (OEM). In the United States of America, "right to repair" has been a subject of discussion. In France, a repair bill is under discussion. You may like to read more about this debate. "Reparability" (the concept of D4R) is now considered as an important element of D4S of a product. We discuss some of these developments in Chapter 5.

4.6 Refurbish

Refurbish improves older or damaged equipment to bring it to a workable or better-looking condition. See Box 4.6 for examples. Refurbish is often treated equivalent to reconditioning.

Refurbishing involves taking a product and restoring/replacing all component parts that have failed or are on the verge of failure. This results in

BOX 4.6 EXAMPLES OF REFURBISH

GOOGLE DATA CENTRES[25]

In partnership with the Ellen MacArthur Foundation, Google has made provisions to custom build and remanufacture its own servers so that those can be refurbished at the end of their life.

After ensuring safety of the data on servers, the withdrawn servers get dismantled into separate components (motherboard, CPU, hard drives, etc.), inspected, and prepared for use as refurbished inventory. The refurbished parts are used to build remanufactured servers with performance equivalent to brand new machines. In 2016, 36% of servers Google deployed were remanufactured machines.

MICROSOFT-UGANDA REFURBISHMENT PARTNERSHIP[26]

Location: Uganda

A partnership was signed in 2007 between the United Nations Industrial Development Organization (UNIDO) and Microsoft Corp to support business opportunities and entrepreneurship among local small and medium enterprises (SMEs) in Uganda. The idea was to address the challenge of shortage of high-quality, affordable hardware and software.

Funded primarily by local private and public sector investors, a business model was designed with a focus on commercial and environmental sustainability. Based on a survey of the market for refurbished PCs in Uganda, a target to refurbish 10,000 quality-brand PCs a year was set. A retail price was estimated to start at USD 175, one-third of the price of a new business PC. The PCs included a one-year warranty and genuine Windows software from Microsoft.

This example shows how refurbishing can be set up as a business and on a scale. While there are benefits of reduced consumption of virgin materials and reduced liability on waste management, the model also shows the advantage of providing access to legal software like the Microsoft operating system. Such an ethical perspective of refurbishing computer hardware systems should not be overlooked.

GREENSOLE[27]

Location: Multiple Countries

Greensole was founded in India with an objective to contribute to social good, by creating a self-sustaining infrastructure that facilitates the provision of the necessity of footwear to everyone. To meet this objective, Greensole refurbished discarded shoes thus diverting waste from landfill.

Greensole appealed to organizations and communities to either donate their old shoes or money to Greensole for refurbishment of old shoes. The minimum amount to be donated is INR (Indian Rupees) 199 for the refurbishment of one pair of shoes. The company donates a particular number of pairs to be distributed to the people in need in a particular village or any specific place based on the requirement.

the product being returned to an acceptable standard (typically less than virgin standard). Generally, any warranties issued for the refurbished product are typically less than a warranty given to a virgin product. Reconditioning involves less work than remanufacture but more than repair[18].

4.7 Renovate

Renovate is to restore something to a newer or better state. Generally, renovation is seen more applicable to existing built structures such as housing, commercial buildings, and public places. Box 4.7 illustrates a case study of CleanBC Better Homes.

BOX 4.7 EXAMPLES OF RENOVATE

CleanBC BETTER HOMES[28]

Location: Canada

CleanBC Better Homes is British Columbia's online hub for homeowners and businesses to access information, rebates, and support to reduce energy use and greenhouse gas emissions in new and existing homes and buildings. It is funded by the Province of British Columbia and the Government of Canada under the Low Carbon Economy Leadership Fund. CleanBC Better Homes rebates are administered by BC Hydro, FortisBC, and BC Housing.

Some of the resources offered by this platform include – rebate search tools, free energy coaching services, energy advisor consulting services, and contractor directories.The platform offers a range of financial services:

- CleanBC home options – offers low-interest financing for fuel switching to a heat pump.
- Bank or local credit union options for energy efficiency upgrades. FortisBC loans for upgrading electric furnace or baseboard heaters which can be availed up to CAD 6500 at 1.9% fixed interest rates.
- Penticton home energy loan program (HELP) – loan available for residents of Penticton for energy efficiency upgraded. Maximum loan amount available is CAD 10,000 to be repaid in 10 years.

ACTIVITY 4.6 FINANCING FOR RENOVATION

Financing for renovation is the strategy in most case studies on housing and office renovations in cities as well as in the national programs. Renovation not only extends the life of the asset but lowers energy consumption leading to reduction of GHG emissions. How should the renovation loans be structured for its uptake, considering the economic and environmental benefits? Read more on this at the website of CleanBC Better homes and look for more such case studies.

4.8 Recycle

In recycling, we turn an item into raw materials which can be used once again, usually for a completely new process or product. There are numerous examples of recycling and they dominate in examples to illustrate opportunities and benefits in circular economy. Unfortunately, many consider circular economy thinking as equivalent to widespread practice of recycling.

Recycling is a series of processes where waste products/materials are collected, processed, and returned to the pool of raw materials. The products with recycled content are recognized as recycled products as in the case of a plastic bottle for instance, a recycled bottle[18]. Box 4.8 provides illustrative examples.

BOX 4.8 EXAMPLES OF RECYCLE

GREEN TOYS[29]

Location: The United States of America

Green Toys is a California-based company that sells children's toys made from recycled plastic milk jugs along with other forms of recycled plastic. Since its inception more than 10 years ago, it has already recycled more than 55 million milk jugs. Having safety as a priority, the toys are produced with adherence to domestic and international safety and environmental regulations.

The company is also committed to sustainable shipping and hence the boxes carrying toys are printed with soy ink, which biodegrades four times faster than petroleum-based inks.

ALLBIRDS[30]

Location: Multiple countries

Shoe manufacturing company Allbirds, produces footwear from 100% recycled materials, including sheep's wool, recycled cardboard and plastic, and castor bean oil. The company also follows sustainable farming and animal welfare practices to ensure well-being of their livestock.

Compared to a company that sells synthetic shoes, Allbirds uses around 60% less energy during the manufacturing process. The standard sneaker emits 12.5 kg CO_2e while Allbirds' average shoe emits 7.6 kg CO_2e. Their goal is to become carbon neutral.

ROTHY'S[31]

Location: The United States of America

The company Rothy's uses recycled plastic to make fashionable shoes that are designed for working women. They blend their signature thread that is spun from water bottles with marine plastic collected within 30 miles of coastlines and marine environments. Rothy's website shows that more than 62,000,000 plastic water bottles have been recycled so far.

Their packaging boxes are also made to be sturdy and reusable for storage, shipping or as a recycle bin organizer. As of November 2020, they also produce and sell bags that are knit with 100% recycled materials.

Other interesting examples include furniture and other products made by discarded and shredded TetraPak containers that are essentially non-recyclable and upcycling discarded front-loader washing-machine drums into bookshelves and alike.

4.9 Recover

In recovery, waste is converted into resources (such as electricity, heat, compost, and fuel) through thermal, mechanical, chemical, and biological means. See Box 4.9 for examples.

BOX 4.9 EXAMPLES OF RECOVERY

COMPOST NOW[32]

Location: United States of America

When food scraps are sent to landfills, precious nutrients are wasted, starving the soil from the necessary organic material needed to be productive and healthy. Through composting, these nutrients are recovered and sent back to the soil, hence closing the nutrient cycle.

The company Compost Now's services are based on closing the loop by empowering community members and local businesses to divert their compostable waste from the landfill. They provide weekly pick-up services for collecting waste bins from client's doorstep, which are taken to local composting centers. Once the compost is ready, it is either given back to the client or shared with community gardens.

Since inception, the company has managed to divert about 11.8 million kgs of compost from landfills which has led to creation of 3.7 million kgs of nutrient-rich compost.

BIOGAS RECOVERY AT HOUSEHOLD LEVEL

In Côte d'Ivoire, sub-Saharan Africa, 60% of the local energy demand is met by the consumption of wood and coal. It is one of the major causes of deforestation in that area. With this regard, the start-up LONO has produced KubeKo Biogas that is an anaerobic micro-digester, which breaks down organic waste into biogas and fertilizer.

The micro-digester, accessible for rural populations, operates on a very simple principle: domestic and agricultural organic waste, whether liquid or solid, is placed in the digester which then produces biogas. It can be used for example for clean cooking in the home, as a biofertilizer, or even as electricity to cover a household's energy needs[33].

Similarly, in India, Dr. Anand Karve has developed a user-friendly compact biogas system that has a capacity of 1000 l and uses just 2 kg of starchy agro-waste, non-edible seeds, oilcake of non-edible seeds or leftover food. It produces just a couple of liters of watery effluent that is easy to dispose of. Dr. Anand Karve won the prestigious UK based Ashden Award for innovating this technology[34].

Both these technologies make the process for biogas production more efficient than other solutions, needing less space for the same outcome while still being affordable at the level of household.

FACEBOOK DATA CENTRE[35]

Facebook's data centre in Odense, Denmark began operations in September 2019. Like all of Facebook's data centres, this data centre is supported by 100% of renewable energy. Facebook partnered with local district heating company Fjernvarme Fyn, which operates Odense's district heating system. The heat generated by Facebook's servers is directed to copper coils filled with water instead of releasing to the atmosphere. This heat recovery system is facilitated by a heat pump system which runs on renewable energy. This recovered heat is then pumped directly into the district heating system. Facebook aims to ramp up its efforts to provide 100,000 MWh of energy per year from this project, which is enough to meet the requirements of nearly 6900 homes in Odense.

4.10 Return

In Return, we take back products and packaging after use as a responsible manufacturer. Box 4.10, talks about examples of businesses with Return model.

BOX 4.10 EXAMPLES OF RETURN

HEPI CIRCLE[36]

Location: Indonesia

Hepi Circle is Indonesia's first refill delivery network that offers everyday cleaning products in reusable bottles. Customers can buy a bottle of detergent at their local store, pay a deposit, and during their next visit return their empty bottles. The reuse habit is rewarded with a "hepi point", that can go toward purchasing food or reusable products at the store. The refill and distribution to local stores is powered by women on bikes. The project has demonstrated financial feasibility and long-term potential impact.

CUPABLE – REUSABLE BEVERAGE MUGS[37]

Location: India

Cupable is an Indian company that designs, manufactures and recycles drinkware. Cupable's cups, unlike single-use plastic cups, can be returned to the partnered restaurant or can be requested for pickup from Cupable. Post pick up, the cups get washed and sanitized for the next round of reuse. The cups are Internet of Things (IoT) enabled, which makes them easily trackable at each stage of life with instant information about each use and wash. Cupable works with several brands such as Starbucks, Burger King, The Thick Shake Factory, etc.

4.11 Remanufacture

In remanufacturing, we rebuild a product to specifications of the original manufactured product using a combination of reused, repaired, and new components. See Box 4.11 for examples of companies operating a refurbish business model.

Remanufacturing is the only end of life process where used products are brought at least to original equipment manufacturers' performance specification from the customer's perspective and at the same time, are given warranties that are equal to those of equivalent new products[18].

It may be observed that most operations of "return" require partnership with reverse logistics operators and the consumers. It is also questionable sometimes whether to remanufacture a product given the intensity of operations consisting collection, transportation, component disassembly and remanufacturing or processing. You may like to read literature on such assessment and comparisons carried out for products based on Life Cycle Assessment.

BOX 4.11 EXAMPLES OF REMANUFACTURING

CATERPILLAR[38]

Cat Reman is one of the leaders in remanufacturing services. At the end of serviceable life of its industrial equipment, it remanufactures them to same-as-new condition. This not only reduces operating costs by providing customers same-as-new quality at a fraction of the cost of a new part but also reduces waste, lowers greenhouse gas production, and minimizes the need for raw materials.

A complete Cat Certified Rebuild includes more than 350 tests and inspections, automatic replacement of approximately 7000 parts and a like-new machine warranty. In addition, trained dealer service professionals perform this work using genuine equipment and parts. Caterpillar provides information, data, training, and service tools to help dealers make the most appropriate decisions on which parts to reuse in order to achieve expected longevity of rebuilt components.

XEROX[39]

Xerox established its take back program in the early 1990s. Its process includes taking back "end of life" photocopiers, printers, scanners from customers, creating a remanufacture and parts reuse programme that forms the foundation of waste free Initiatives. Xerox takes advantage of remanufacturing through the design of their business model. Xerox's Production Systems Group works to strict corporate targets, e.g., in 2003 95% of all waste produced was recycled, which equates to 175,000 tons and increases year on year.

Milliken[41]

Milliken's has designed a remanufactured carpet range so that unwanted carpet tiles can be recovered, cleaned, retextured, and restyled in preparation for reuse, either within the same company or at another location. The cost is about half the cost of new carpet. Milliken has also developed a glue-free carpet installation system called TractionBack™, which is available on all modular carpet products manufactured by Milliken. This is a high-friction coating applied to the backing of the modular carpet tiles. The coating allows the tile to be installed without using wet adhesive or "peel and stick" adhesives. The absence of adhesive reduces contamination and makes reprocessing easier. Milliken's remanufacturing process allows the carpet to stay intact, eliminating carpet waste.

GUANGZHOU HUADU WORLDWIDE TRANSMISSION[40]

Huadu, a company in Guangzhou in China provides services around transmission boxes (power/automotive/gearboxes), such as testing, maintenance, remanufacturing, tech support, and training.

In 1996, the Company established an effective channel for collection & distribution of used and remanufactured transmission boxes.

Guangzhou Huadu has partnered with retail franchise 4S to collect used vehicle parts, remanufacturing them into as-new certified spare parts. The company collects 60% of its total transmission boxes through 4S stores while the remaining come from repair shops, consumers, and used auto parts dealerships. 4S model is a full-service approach that brings together sales, service, spare parts, and surveys (customer feedback). These remanufactured products have been authorized by more than 30 major auto companies like GM (General Motors), JATCO (Japan Automatic Transmission Company), & PSA (Groupe PSA) to provide maintenance service through 4S dealership system.

It has been estimated that if machinery is remanufactured rather than manufactured from virgin materials, components could be 50% cheaper to make, at the same time saving 70% in materials inputs and reducing emissions by 60%.

4.12 Rethink

Rethink means thinking out of the box to innovate. Innovations need not limit to materials or design but can include business models and policy instruments. Box 4.12 provides some examples of businesses following the Rethink strategies.

BOX 4.12 EXAMPLES OF RETHINK

ORIQA EDIBLE CUTLERY[41]

Oriqa Ltd. is an edible cutlery manufacturing company based in India. Their range of edible cutlery includes spoons, bowls, forks, plates, etc. The cutlery has different flavors, which increase their demand. It is made out of millets, rice, wheat and has no added preservatives. Food safety standards set by authorities are strictly followed during the production. They have an internal quality assessment mechanism in place to double sure the quality and safety of our products when consumed. The edible cutlery concept and products have been widely acclaimed due to its eco-friendly and 100% biodegradable nature. You may like to read more about similar companies, e.g.: Edible Pro: https://ediblepro.com/

ALTERNATIVE MEAT (PLANT/CELL-BASED MEAT)

Agriculture is one of the primary drivers of climate change, estimated globally at 14–15% of all GHG emissions, half of which is generated directly by livestock. Meat production will be unsustainable by 2050 at current and projected rates of consumption due to high resource intensity and destructive cost. This opens a large market for nutritious protein alternatives that can provide comparable taste, texture, and nutrition density[42]. You may like to read about companies such as:

- Beyond Meat: https://www.beyondmeat.com/
- Sweet Earth: https://www.sweetearthfoods.com/

AVANI[43]

Location: United Arab Emirates

As an alternative to plastic bags, Avani has created plastic-like bags using cassava, an edible root – also known as yuka, arrowroot, or tapioca. The cassava bag is 100% natural, biodegradable, sustainable, compostable, and recyclable. Each non-toxic bag will break down into carbon dioxide and water within six months and it dissolves in hot water in a few minutes. Similar companies exist all over the world offering 100% biodegradable bags made out of various types of vegetable starch.

EASTGATE CENTRE[44]

Location: Zimbabwe

The Eastgate Centre, a shopping center and office building in Harare, Zimbabwe, is designed to exploit more passive and energy-efficient mechanisms of climate control rather than using a traditional fuel-based air-conditioning system. The design of the building was inspired by the structure of termite mounds. The building's construction materials have a high thermal capacity. The ventilation fans operate on a cycle timed to enhance heat storage during the warmer daytime and heat release during the cooler night-time. Airflow within the building's large, internal open spaces are driven by the internal heat generated by the building's occupants and appliances as it rises from the lower floors toward the open rooftop chimneys. The building has various openings throughout, which further enable passive internal airflow driven by outside winds. All these design features reduce temperature changes within the building interior as temperatures outside fluctuate. The building had saved 10% on costs up-front by not purchasing an air-conditioning system.

Check out the Institute of Biomimicry for examples of designs inspired by nature: https://biomimicry.org/biomimicry-examples/

4.13 Epilogue

The 12 Rs and examples illustrated above show that there are overlaps in the definition and scope of each of the Rs. In practice multiple Rs are used as a strategy to arrive at a cost-effective and sustainable solutions to drive circular economy. The examples also show that partnership of stakeholders is essential for implementation. Without involvement of multiple stakeholders, the outcomes of 12 Rs are not going to be effective. Figure 4.3 shows a mapping

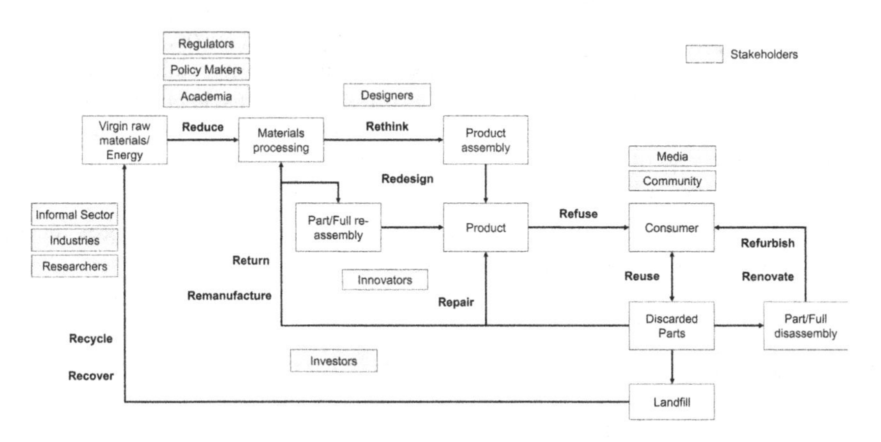

FIGURE 4.3
Manufacturing process with the 12 Rs and the key stakeholders.

of a manufacturing process with the 12 Rs and the key stakeholders. This figure emphasizes the point that any of the Rs and stakeholder should not be viewed in isolation or silo.

Highlighted below are some important observations in connection with the 12 Rs on extending product life and increase resource recirculation.

It is important that the costs of collection, sorting and reprocessing of the component or product post use or at end of life are lower than the total cost of the virgin products, reflecting the avoidance of end-of-life treatment costs. Only then setting up circular systems can make economic sense.

With increasing resource prices, challenges on resource availability and higher end-of-life treatment costs, circular systems are expected to be more attractive. Business models based on partnership, supported by innovation and financing can make this possible. Policy instruments (e.g., economic, market-based, and information driven) will make the circular options attractive and feasible.

Extending product life will substitute and differ virgin material inflows to counter the dissipation of material out of the economy. The 12 Rs and the stakeholders help to keep products, components, and materials in use longer within the circular economy. This is possible by integrating repair, refurbish and renovate services in the life cycle of the product. Devising business models based on Product as a Service (PaaS) systems will also help. Implementing extended producer responsibility will also play an important role.

Many of the examples illustrated in this chapter show a diversified re-use of materials across the value chain. During such a cross sector flow, the materials get cascaded leading to longer circulation.

Life Cycle Sustainability Assessment (LCSA) plays a very critical role in identifying 12R interventions that are technically and economically feasible as well as acceptable to the society. D4S will influence in upstream, downstream, and recirculation operations. Here, avoidance of contaminated materials or objectionable substances in the product design becomes crucial to ensure that operations such as collection, sorting, and reprocessing are safe to the workers and re-circulation is techno-commercially feasible. For example, when the company Desso decided to eliminate all toxic chemicals in its carpet tiles – in line with Cradle-to-Cradle principles – its business benefited from an uptake in the aviation market, where carpet off gassing can affect passenger health and comfort[45].

Chapter 5 further discusses the policies, strategies, and challenges faced for product life extension with case studies.

14.14 Key Takeaways

- The 12 Rs of circular economy help for resource use reduction, resource recovery, and resource recirculation.

- The 12 Rs are not to be applied in silo. They will need to be strategically deployed depending on the context, scope, and objectives. The four circles of 12 Rs described in this chapter provide the insight in formulating integrated strategies.
- Tools like cleaner production opportunity assessments can help in identifying the options, especially at the facility level. Material Flow Analysis (MFA) and LCSA have great potential to build scenarios and assess policy level options from economic, environmental, and social considerations. Techno-commercial evaluation is a must, and the assessment should include social acceptability.
- For putting 12 Rs in practice, stakeholders will have to work in partnership. Without forging partnerships, solutions toward circular economy will not be sustainable.
- Examples of 12 Rs illustrated in this chapter stress all of the above.
- Finally, product design and strategies to extend product life greatly influence the material circularity across the life cycle. These interventions lead to innovation, expand the value chain and reduce risks in the re-circulation of materials.

Notes

1. Vocabulary. n.d. *Refuse Definition*. <https://www.vocabulary.com/dictionary/refuse> [Accessed 24 October 2020].
2. Lastovicka, J., Bettencourt, L., Hughner, R. and Kuntze, R., 1999. Lifestyle of the tight and frugal: Theory and measurement. *Journal of Consumer Research*, 26(1), pp. 85–98.
3. Kanti, A., 2018. *Your Choice Matters: The Hidden Cost Of Plastic Straws*. Businessworld. <http://www.businessworld.in/article/Your-Choice-Matters-The-Hidden-Cost-Of-Plastic-Straws/02-02-2018-139322/> [Accessed 24 October 2020].
4. Murray, B., 2013. *Topic Guide: Embedding Environmental Sustainability In Product Design*. [online] Product Sustainability Forum.<http://www.wrap.org.uk/sites/files/wrap/Embedding%20sustainability%20in%20design%20%20-%20final%20v1.pdf> [Accessed 24 October 2020].
5. Tide. n.d. *Tide Original Liquid Laundry Detergent Eco-Box*. <https://tide.com/en-us/shop/type/liquid/tide-original-eco-box> [Accessed 24 October 2020].
6. Replenish. n.d. *Details*. <http://www.myreplenish.com/#details> [Accessed 24 October 2020].
7. FlyBird Innovations. n.d. *About Us*. <http://www.flybirdinnovations.com/> [Accessed 24 October 2020].
8. Jacob, S., 2018. *Eco-Dome: Building Ecological 'Houses Of Tomorrow' In Rural Morocco*. Business Line. <https://www.thehindubusinessline.com/circular-economy/ecodome-building-ecological-houses-of-tomorrow-in-rural-morocco/article9299727.ece> [Accessed 24 October 2020].

9. Ecovative Design. n.d. *About.* <https://ecovativedesign.com/> [Accessed 24 October 2020].
10. Abhijith, R., Ashok, A. and Rejeesh, C., 2018. Sustainable packaging applications from mycelium to substitute polystyrene: A review. *Materials Today: Proceedings,* 5(1), pp. 2139–2145.
11. Material District. 2017. *New IKEA Kitchen Units Kungsbacka Made From 100% Recycled Materials.* <https://materialdistrict.com/article/ikea-kungsbacka-recycled-materials/#:~:text=New%20IKEA%20kitchen%20units%20Kungsbacka%20made%20from%20100%25%20recycled%20materials,-Share&text=Each%20year%2C%20approximately%20100%20billion,fraction%20of%20those%20are%20reused.> [Accessed 25 October 2020].
12. Mitticool. n.d. *Clay Products Exporters.* <https://mitticool.com/> [Accessed 25 October 2020].
13. SEED. 2019. *Abwezi School Bags.* <https://www.seed.uno/enterprise-profiles/abwezi-school-bags> [Accessed 25 October 2020].
14. Vosper, J., n.d. *The Role of Industrial Hemp in Carbon Farming.* GoodEarth Resources PTY Ltd. <https://hemp-copenhagen.com/images/Hemp-cph-Carbon-sink.pdf> [Accessed 25 October 2020].
15. Datoo, S., 2013. *Hemp is Used in Over 25,000 Products, Now Including Bmws.* [online] Quartz. Available at: <https://qz.com/109268/hemp-is-used-in-over-25000-products-now-including-bmws/> [Accessed 25 October 2020].
16. Paterson, D., Ijomah, W. and Windmill, J., 2017. End-of-life decision tool with emphasis on remanufacturing. *Journal of Cleaner Production,* 148, pp. 653–664.
17. RePack. n.d. *Reusable Packaging Service For Ecommerce.* [online] Available at: <https://www.originalrepack.com/about/> [Accessed 25 October 2020].
18. Common Objective. n.d. *Fashion and Waste: An Uneasy Relationship.* <https://www.commonobjective.co/article/fashion-and-waste-an-uneasy-relationship> [Accessed 25 October 2020].
19. Goonj Organisation. n.d. <https://goonj.org/> [Accessed 25 October 2020].
20. Library of Things. n.d. *Library Of Things.* <https://www.libraryofthings.co.uk/> [Accessed 25 October 2020].
21. Khilonewala. n.d. *About Us.* <http://khilonewala.in/> [Accessed 25 October 2020].
22. Coggins, T., 2017. *In Amsterdam, Repair Cafés are a Thing.* Culture Trip. <https://theculturetrip.com/europe/the-netherlands/articles/in-amsterdam-repair-cafes-are-a-thing/> [Accessed 25 October 2020].
23. iFixit. n.d. <https://www.ifixit.com/> [Accessed 25 October 2020].
24. Fairphone. n.d. <https://www.fairphone.com/en/> [Accessed 25 October 2020].
25. Google. 2018. *Environment Projects: Once is Never Enough.* <https://sustainability.google/projects/circular-economy/> [Accessed 25 October 2020].
26. UNIDO. 2008. *Computer Refurbishment Centre Opens in Kampala.* <https://www.unido.org/news/computer-refurbishment-centre-opens-kampala> [Accessed 25 October 2020].
27. Greensole. n.d. About us <https://www.greensole.com/home> [Accessed 31 October 2020].
28. Better Homes BC. n.d. *Learn More About Us.* <https://betterhomesbc.ca/about-us/> [Accessed 25 October 2020].
29. Green Toys. n.d. *Our Story.* <https://www.greentoys.com/> [Accessed 25 October 2020].

30. Allbirds. n.d. *Our Sustainability.* <https://www.allbirds.com/pages/sustainability> [Accessed 25 October 2020].
31. Rothy's. n. d. *Sustainability.* <https://rothys.com/sustainability> [Accessed 25 October 2020].
32. Compost Now. n.d. *Our Services.* <https://compostnow.org/> [Accessed 26 October 2020].
33. LONO. n.d. *Products.* <https://www.lonoci.com/Services> [Accessed 26 October 2020].
34. ARTI – Appropriate Rural Technology Institue. n.d. *Compact Biogas System.* <http://www.arti-india.org/InsidePages/TechnologyDetails.aspx?TechnologyId=4&Title=Compact%20biogas%20system&Category=Energy%20Lighting%20And%20Fuel> [Accessed 26 October 2020].
35. Edelman, L., 2020. *Facebook'S Hyperscale Data Center Warms Odense.* Facebook Technology. <https://tech.fb.com/odense-data-center-2/> [Accessed 26 October 2020].
36. Hepi Circle. n.d. *About.* <http://www.hepicircle.org/#about> [Accessed 26 October 2020].
37. Cupable. n.d. <https://www.cupable.co/> [Accessed 26 October 2020].
38. Caterpillar. n.d. *Circular Economy.* <https://www.caterpillar.com/en/company/sustainability/remanufacturing.html> [Accessed 26 October 2020].
39. Gray, C. and Charter, M., n.d. *Remanufacturing and Product Design.* The Centre for Sustainable Design. <https://cfsd.org.uk/Remanufacturing%20and%20Product%20Design.pdf> [Accessed 26 October 2020].
40. Ellen Macarthur Foundation. n.d. *Case Studies.* <https://www.ellenmacarthurfoundation.org/case-studies/remanufacturing-at-scale> [Accessed 26 October 2020].
41. Oriqa Edible. n.d. *Edible Cutlery.* <https://www.oriqaedible.com/> [Accessed 26 October 2020].
42. Joshi,I.,Param,S.andGadre,M.,2015.*SavingthePlanet:TheMarketforSustainableMeat Alternatives.* Berkeley, University of California. <https://scet.berkeley.edu/wp-content/uploads/CopyofFINALSavingThePlanetSustainableMeatAlternatives.pdf> [Accessed 26 October 2020].
43. AVANI MIDDLE EAST. n.d. *Home.* <https://www.avanime.eco/> [Accessed 26 October 2020].
44. Ask Nature. 2016. *Eastgate Centre.* <https://asknature.org/idea/eastgate-centre/> [Accessed 26 October 2020].
45. Ellen Macarthur Foundation. n.d. *Cradle to Cradle Design of Carpets.* <https://www.ellenmacarthurfoundation.org/case-studies/cradle-to-cradle-design-of-carpets> [Accessed 31 October 2020].

5

Making Longer Lasting Products

Chapter 4 discussed repairing and refurbishing with some examples. Repairing is the restoration of a broken, damaged, or failed device to an acceptable operating or usable condition. A repair can involve replacement.

Refurbishing is refinishing and sanitization (beyond repair) to serve the product's original function with better aesthetics. Although in good condition, repaired and refurbished products may not be comparable with new or remanufactured products. Repair and refurbish are important strategies for product life extension.

Products are discarded if they do not function properly. In such cases, the consumer returns the malfunctioning product to the retailer requesting a replacement if it is still in the warranty.

Figure 5.1 illustrates a typical decision tree on the repair by a consumer, and Figure 5.2 shows the stakeholders involved in the repair business.

Post the warranty period if the product does not function properly due to any reason, such as mishandling, then such a product is often discarded. In cases where a repairing service is not available, the consumer will dispose of the product as waste that reaches the recycling market or for land disposal. If the opportunity exists to repair and refurbish at a reasonable price or the product is in warranty, then the consumer goes for this option and the product enters a second life cycle.

In 2018, a behavioral study was carried out by the European Commission on consumer engagement in a circular economy. The study focused on vacuum cleaners, televisions, dishwashers, smartphones, and clothes. The results showed that 64% of consumers always repaired broken products and 26% used a professional repair service. The reason for not repairing products was the high price of repair, followed by the preference to get a new product and the feeling that the old product was obsolete or out of fashion. Around 5–10% of consumers did not know how or where to repair the product, and 8–14% felt it would require too much effort. The study also indicated that a lack of trust in repair services was also an important factor. The study showed that consumers were ready to pay more for products with better reparability – around EUR 29–54 more for vacuum cleaners, EUR 83–105 for dishwashers, EUR 77–171 for TVs, EUR 48–98 for smartphones, and EUR 10–30 for coats[2].

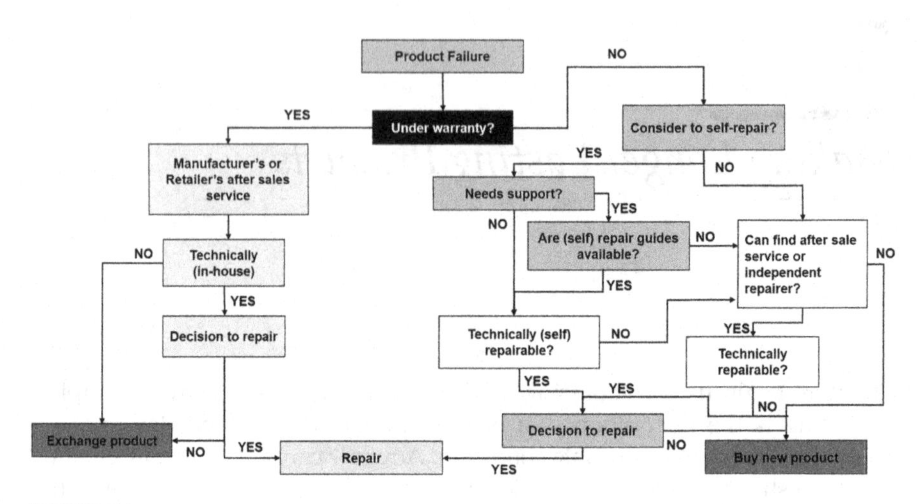

FIGURE 5.1
Decision tree on repair by a consumer[1].

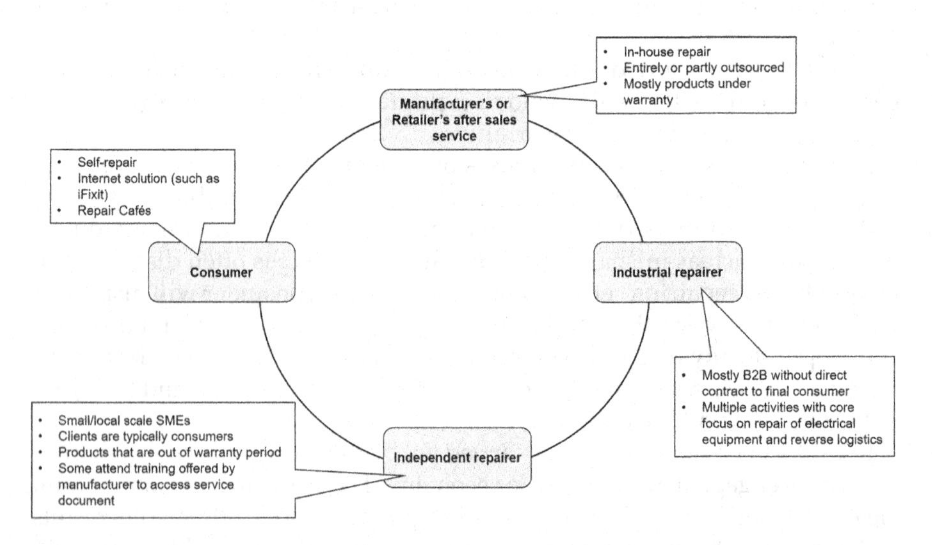

FIGURE 5.2
Stakeholders involved in the repair business[1].

5.1 Decision on Whether to or Not to Extend Product Life?

There are several important pre-requisites for product life extension. These include a repair and/or refurbish market, reverse logistics/take-back systems, research, development, and innovation that help re-make the product as close to the original and at low costs. Consumer understanding and cooperation and client relationship management in the post-warranty phase also play an important role.

Earlier, the product manufacturers were generally averse to product life extension and in some cases, even lobbied for planned obsolescence. The manufacturers also feared the loss of sales and/or market share if the products lasted longer. Later, allowing repair from the third parties became an issue, on account of the protection of intellectual property rights.

Perhaps, shifting to the business model of Product as a Service (PaaS) systems, where in products and services are blended helps to address these challenges. In the PaaS-based business models, repair and refurbishing considerations can be embedded. Concepts such as performance economy, collaborative consumption, shared economy, and servitization can become the drivers. In these business models, consumers are not just supplied with goods, but also supported with the repair and refurbish that extend across the product's life. Companies practicing such a strategy become the champions of a circular economy as they assume a lifelong ownership of the product for the consumers' interest and satisfaction. Table 5.1 shows typical expected product lifetimes of various products.

A longer lifetime for products than expected, however, does not mean a solution is sustainable. Shortening the lifetime could, in theory, provide the same benefits as a longer lifetime for products if a system of full recirculation is operational[3]. For example, reverse logistics and recycling infrastructure. This means that product's design, repair, recycling, and remanufacturing maximize the value of the material used and minimize the environmental burdens. In other words, all "components" of the circulating economy are in place and the stakeholders involved have business interest and commitment to work together. Tools like social Life Cycle Assessment (s-LCA) can help in deciding the "early exit" of the product before it reaches its end of life but remember that these computations are often fraught with data and could be questionable.

In theory, a product that is both easy to recycle and more durable would be the ideal case of design for sustainability (D4S). For instance, automobiles

TABLE 5.1

Average Expected Product Lifetimes[3]

1–2 Years	3–4 Years	5-6 Years	7–10 Years	> 10 Years
Small electrical appliances, (e.g., tooth-brushes, toys) mobile/smart phones, general clothing, shoes	Portable devices, personal computers, bed items, specific clothing (e.g., sports), bicycles, coats	Cameras, general kitchen-ware, lighting, power tools, vacuum cleaners, washing machines, curtains	Automotive, TVs, kitchen appliances, general furniture, carpets, beds, refrigerators	Appliances attached to house (boiler, sunroof, etc.), kitchen and bathroom specific furnishings

ACTIVITY 5.1 DISCUSSION ON "EXPECTED PRODUCT LIFETIME"

You may like to research the current trend in expected product lifetimes by the consumers, by age, gender, and income in different economies. Can the expected product lifetime be legislated in Extended Producer Responsibility (EPR)?

with thicker metal frames that last longer also have more recyclable materials. In such a scenario, EPR policies emphasizing durability and recyclability work hand in hand[4].

Beril Toktay, a professor at Georgia Tech's Scheller College, says that sometimes when you design for recyclability, you give up on durability, and when durability is the goal, recyclability is sacrificed. In the case of photovoltaic panels, Toktay and her team highlighted how thin-film panels are much more cost-effective to recycle than other panels because they contain precious metals. Meanwhile, crystalline silicon panels that are not as cost-effective to recycle, have much longer life spans because their components degrade much more slowly. More stringent EPR requirements can lead to a photovoltaic technology choice with lower recyclability and higher durability and, consequently, result in higher greenhouse gas emissions. These results call for a careful analysis of the benefits of EPR legislation in the context of durable goods. Atalay Atasu, a professor at the Scheller College of Business, points out that you must distinguish between different product categories to consider the recyclability and the durability implications and make sure that your policy is not conflicting with the objective[5].

Decisions on how long one should consider extending the product life are not easy to make. Figure 5.3 shows different scenarios on decisions taken on product life extension and possible implications on resource consumption.

- Scenario 1: Product is used till the expected lifetime and then discarded. A new product is bought that is more resource efficient.
- Scenario 2: Product life is extended further by repair and refurbish. Some parts are replaced as required. It is possible that the product in its extended life does not operate at the best possible resource efficiency.
- Scenario 3: Product is returned or sold to another consumer before the end of expected life and instead a new product that is more resource efficient is purchased.
- Scenario 4: Product that is returned to the manufacturer is disassembled and core components are used to remanufacture. Product

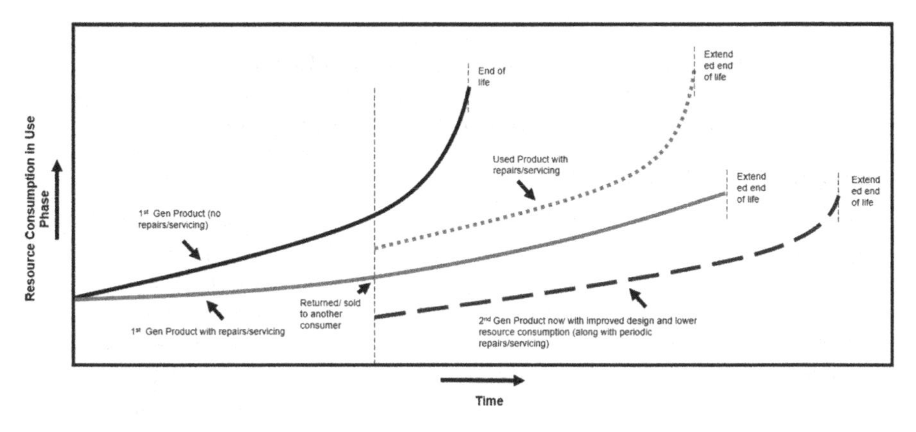

FIGURE 5.3
Scenarios on decisions taken on product life extension.

bought by the second customer is upgraded and maintained through repairs and replacement of parts as required. Attempt is thus made to achieve a reasonable resource efficiency in its second life cycle.

It is difficult to decide which of the above scenarios is more sustainable unless you do a quantitative assessment using techniques such as LCA. Remember that in such cases achieving circularity alone does not ensure sustainability.

Therefore, there is a need to identify the right balance in terms of environmental and social costs between continuing to use a product or replacing it with a better performing product. Suppose the existing product consumes more resources during its lifetime than a new one. In that case, it may be worth replacing it, provided the environmental and social costs to manufacture a new product do not offset the expected benefits linked to changing the product. Consumers generally welcome longer product lifetimes for some products. Still, in other cases, they are worried about the high costs of acquiring or maintaining products with long lifetimes or are worried about being locked into obsolescent products[6].

Box 5.1 illustrates two examples where LCA was used to assess the impact of product life extension and provide guidance on product life extension related policies.

A review of LCA studies was conducted on the optimal replacement moment of seven use-intensive product categories (washing machines, refrigerators, TVs, mobile phones, laptops, clothing, and vacuum cleaners). The results show that washing machines and refrigerators should be used for at least 10 years before they should be replaced with a more energy-efficient model. Vacuum cleaners, clothing, mobile phones, and laptops are usually replaced "before their time" and should be used and reused for longer.

BOX 5.1 CASE STUDIES THAT ILLUSTRATE THE APPLICATION OF LCA FOR ASSESSING IMPACT OF PRODUCT LIFE EXTENSION

PRAMS IN SINGAPORE[7]

In Singapore, prams for babies are used through a rental model rather than ownership. When product is used as a service, the product is an asset for the service provider rather than a mere consumable. This provides incentives for extending product life through maintenance and take-back options. A PaaS model thus saves resources and generates less waste in comparison to business-as-usual, i.e., purchasing the prams.

A comparative LCA study was carried out between Pram as a rental business model and a traditional pram ownership model for Singapore over the span of 5 years. The study used 15 impact indicators (climate change, fossil depletion, metal depletion, human toxicity, ozone depletion, etc.) to measure the impact of various pram business models on the environment.

Rental services call for continuous repair and cleaning. These processes are energy-intensive since they require electricity for drying, chemicals and water for washing, and fuel for transportation. Although the PaaS model's overall impact is less, indicators like ozone depletion were being impacted negatively due to the increased number of maintenance and cleaning cycles. Pram rental model could limit their impact by limiting their washing cycles to 4–5 times per year.

It was found that parents who wanted to retain complete ownership of prams, could potentially reduce their environmental impact by 50% by passing the prams to next users after completing its first usage-cycle (generally 3 years). This type of cascading ownership model causes the least negative impact. In order to leverage this type of model, stakeholders like pram manufacturers, distributers, and policy makers need to intervene to guarantee recovery, refurbishment, and redistribution of prams, or essentially put in place an operating system toward achieving circularity.

Pram distributers can create a network of pram users that will connect those who need it and those looking for subsequent users. Pram companies should also look at various methods of cleaning (light and heavy) based on the requirement to minimize the negative impact. A collaborative platform of stakeholders is thus necessary to make such a system function.

WASHING MACHINES IN THE UNITED KINGDOM[8]

A washing machine has varied levels of environmental impact at various stages of its life cycle. While the production phase creates impacts such as resource depletion, global warming, etc. the usage and End-of-Life phase generate much larger impacts. Similarly, while refurbishing is a good idea to extend product life, it also creates some negative impacts, but these impacts are less as compared to manufacturing a new washing machine. A study was carried out to evaluate refurbishing options and replacing washing machines based on their resource efficiency levels using LCA.

Washing machines of efficiency, ratings A and C were studied across a range of scenarios for replacement with higher-efficiency machines (rated A, A+, and A++). A total of 21 replacement and refurbishment scenarios were investigated, covering scenarios of machine upgradation from C to A and subsequently to A+ and A++. Based on 6 impact categories and resource efficiency levels of the washing machines, the following observations were made:

1. In case of A rated washing machines, refurbishing them will be a better option as compared to replacing them unless it is replaced with A++ rated.
2. For a C rated washing machine, decision for replacement and refurbishing will depend on the number of years, which can be added to the machine's life. In most cases, replacement is a feasible option.
3. A rated machine should be replaced with A+ rated machines, only if it leads to extension of machine's life by more than 3 years.

In the case of TVs, it makes sense to keep older LED (light-emitting diode) models in use, instead of replacing them with newer, less energy efficient models[9].

In the report published by United Nation Environment Program (UNEP)[7] it is cautioned that in general, policy makers should bear in mind that:

- Successive generations of electronic products are not always more energy efficient.
- Replaced products are not always taken out of circulation.
- LCA scenarios must be modeled as close to the "real, messy world" as possible, to do justice to the highly diverse user contexts and cultures. This requires large amounts of data, both technical (i.e., energy efficiency developments) and sociological (i.e., diverse use patterns).
- There is a bias toward European data. Data from developing economies is missing. Consumer use scenarios vary greatly across cultures, which impact the optimal replacement moment.

ACTIVITY 5.2 DISCUSSION ON "SECOND-HAND GOODS"

Products that are reasonably functional and get replaced, generally find a place in the market of second-hand goods. There are traders and aggregators who buy and sell such goods to the markets in rural areas or in developing economies. In many instances, the buyer gets the second-hand goods refurbished but generally the goods perform at low productivity or resource use efficiency. The second life cycle of the goods is therefore often sub-optimal. What could be the strategies to address this challenge? Is a regulation on the use of second-hand goods possible?

5.2 Obsolescence

Obsolescence of products can reach the end of their lifespan and become obsolete in various ways, sometimes also due to repair's impossibility.

Obsolescence can be[2]:

- Planned or built-in – Products are designed to deliberately fail after a certain time or a certain number of uses.
- Premature – Products last less than their normal lifespan compared to consumer expectations.
- Indirect – Components required for repair cannot be obtained, or it is not practical or cost-effective to repair the product.
- Due to incompatibility – Products no longer work properly once an operating system is updated.
- Style obsolescence – Influencing consumers to believe that their products are out of date although they are still perfectly functional.

Sometimes products, especially in consumer electronics like mobile phones are discarded even in good condition as the consumer wants to replace with something new that is more appealing, better in performance or has additional functions or features. The interest to change the product is often to remain contemporary or be a part of the fashion. Many manufacturing companies influence the consumers for buying the new product through advertising campaigns. This strategy is known as "planned obsolescence". While a "passion to change" or a "dynamic consumer behavior" is the trend of the day, some manufacturing industries have been practicing dubious strategies of "planned obsolescence". These manufacturers exploit planned obsolescence to make consumers buy new products, by purposefully making it difficult, or too costly, to make repairs, or by preventing backwards compatibility.

A classic case of planned obsolescence is nylon stockings. The original nylon was also used for parachutes in the military and is still used in climbing ropes enabling climbers to embark on daring climbs. Now it seems quite ridiculous that even after being such a strong polymer, a pair of "remanufactured" nylon stocking only lasts a few weeks or so.

In 1920's, the infamous "Phoebus cartel" was formed, wherein representatives from top light bulb manufacturers worldwide, such as Germany's Osram, the UK's Associated Electrical Industries, and General Electric in the United States (via a British subsidiary), colluded to artificially reduce bulbs' lifetimes to 1000 h. The original light bulbs made by Thomas Edison in 1879 are still going strong in some museums. Today, bulbs have to be changed at least once a year, sometimes more.

A glaring example of such design is the disposable cameras, where the customer must purchase an entire new product after using them a single time. Then there are design features meant to frustrate repairs, such as Apple's "tamper-resistant" Penta lobe screws that cannot easily be removed with common consumer tools.

Apparently, producers that make cheaper but not so durable products believe that the additional sales revenue they create offsets the additional costs of research and development, and further offsets the opportunity costs of repurposing an existing product line. A possible goal for such a design is to make the cost of repairs comparable to the replacement cost or prevent any form of servicing of the product to extend its life. The idea of product reparability is simply throttled in such designs.

The literature on planned obsolescence focuses on suppliers who intentionally supply products with a short lifetime in order to sell replacements to consumers. The degree to which this is the case is largely unknown – surprisingly, little is concretely known about producer preferences in terms of product lifetime. Whether economic incentives favor short product lifetimes might depend on the degree of global competition to which the product is subjected to. In any case, producer preferences about product lifetime may reflect legitimate trade-offs, just as is the case with consumer preferences. Irrespective of whether producers intentionally reduce their products' lifetime, it is likely that at least some producers place too little emphasis on long product lifetimes. According to a report by the Öko-Institut e.V. and Bonn University commissioned by the German Environment Agency, most electrical appliances' service life is becoming shorter and shorter[10].

While planned obsolescence and tampering of reparability, is appealing to producers to grab the market, it does a significant harm to the society in the form of negative externalities. Continuously replacing products rather than repairing them creates more waste and pollution, uses more natural resources, and results in more consumer spending – such a way of living challenges this planet's sustainability.

By making explicit linkages with mainstream innovation policies, the circular economy initiative opens synergies that capitalize on existing policy instruments that could support the extension of product lifetimes. Given the fast pace of product innovation, this also brings to the forefront a relative lack of coherence between current innovation policy and any policy package promoting a longer lifetime for products. A delicate balance must be stricken between policies enabling fast innovation that brings to the market better and more efficient products (in the context of resources and energy) and new initiatives aiming to optimize in the long term the embedded capital stock of products currently in use.

5.3 Policies and Regulations on Reparability of Products

Today, a consumer must look at the reparability of the product. The book "Products that last" published by TU Delft[11], contains innovative examples of reducing materials in products and orienting business models. The Austrian Standards for assessing the reparability of white and brown goods (Austrian Standards, 2014) and the criteria used by IFixit are examples of inspiring directions (IFixit, 2015)[12]. The European Consumer Organization (BEUC), has launched a new campaign for durable goods, more sustainable, and better consumer's rights called the BEUC campaign[13].

Repair needs to be affordable and accessible to consumers. A reduction of Value Added Tax on repair can further incentivize actions in this area. Box 5.2 shows various initiatives taken in France, Austria, and Sweden to extend product life through repair.

The European Commission has announced plans for new "right to repair"[16] rules that it hopes will cover phones, tablets, and laptops by 2021. From 2021, firms will have to make appliances longer lasting, and they will have to supply spare parts for machines for up to 10 years. These rules apply to lighting, washing machines, dishwashers, and fridges. If successful, these rules will mean these devices should remain useful for longer periods of time before needing to be recycled or ending up in landfills[17].

In addition to introducing new "right to repair" rules, the EU also wants products to be more sustainably designed in the first place. Under the new plan, products should be more durable, reusable, upgradeable, and constructed out of more recycled materials. The EU's hope is to reward manufacturers that achieve these goals. Finally, the EU is also considering introducing a new scheme to let consumers sell or return old phones, tablets, and chargers more easily.

The EU consumer legislation regulates consumers' right to have products repaired within the legal guarantee period, but not beyond its expiry or for defects not covered by the guarantee. Efforts to ensure access to repair are

BOX 5.2 POLICIES AND REGULATIONS RELATED TO EXTENDING PRODUCT LIFE IN FRANCE, AUSTRIA, AND SWEDEN

In 2015, as part of a larger movement against planned obsolescence across the European Union (EU), France passed legislation requiring that appliance manufacturers and vendors declare the intended product lifespans and inform consumers how long spare parts for a given product will be produced. Since 2016, appliance manufacturers in France are required to repair or replace, free of charge, any defective product within two years from its original purchase date. This will effectively create a mandatory two-year warranty for products such as stoves, washing machines, and mobile phones.

The Austrian standard establishes a guarantee of 5 years for brown goods and 10 years for white goods. It also considers the guarantee to reflect both the product's life span and the time for which spare parts should be available. One of the reasons that spare parts are often unavailable is rapid design changes or planned obsolescence. Spare parts should be available at least during the product's normalized lifespan and consumers should be informed about where to find or order them[14].

In 2017, Sweden decided to help its citizens become more sustainable by rewarding them for their repairing efforts, taking one more step toward a circular economy. The parliament adopted a 50% tax break for repairs on shoes, clothes, and bicycles, and allows its citizens to claim back from income tax half the labor cost of appliance repair[15].

also included in EU environmental and product legislation. The upcoming eco-design requirements for TV screens, refrigerators, lighting, household washing machines, and dishwashers are expected to ensure that independent repairers can access spare parts and repair information. The European Parliament has called for extending the eco-design requirements to non-energy related products, including the reparability of products, more systematically in eco-design legislation, and extending the duration of legal guarantees. Similar calls have come from a range of stakeholders.

In the USA, extensive discussions are taking place on the proposed "Right to Repair" legislation. Many states are "considering" the law and Massachusetts may be the first state to enshrine the right-to-repair in law[18]. The Right to Repair bill will make it easier for people to repair their broken electronic equipment – like cell phones, computers, appliances, cameras, and even tractors. The legislation would require manufacturers to release repair information to the public and sell spare parts to owners and independent repair shops. However, it is going to be a bumpy ride as giants like Apple and Microsoft are gearing up to oppose this legislation.

Promotion of the development of product buying/use guides or consumer awareness/marketplace campaigns, can increase the understanding of product durability, induce positive consumer attitude toward product maintenance and repair, and encourage consumers to hold companies to account. Advocacy on insisting that product designs should address reparability is now getting a traction. In Chapter 4, a case study of Repair Café was cited. Box 5.3 describes two similar initiatives.

BOX 5.3 THE RESTART PROJECT[19] AND RREUSE: A SOCIAL ENTERPRISE IN REUSE, REPAIR, AND RECYCLING[20]

RESTART PROJECT

Restart is a London-based social enterprise – encourages and empowers people to use their electronics longer to save money and reduce waste. Restart helps people learn to repair their own electronics in community events (parties) and in workplaces and speak publicly about repair and product resilience. Today, Restart is working with 54 people in 10 countries who are planning on replicating and adapting the Restart model.

RREUSE

RREUSE is a social enterprise in Belgium, active in reuse, repair, and recycling. RREUSE promotes more durable and repairable products by:

- Advocating policies to make repair and re-use activities more competitive in order to provide more repair choices for consumers
- Influencing product policy by including provisions with repair-friendly criteria including:
  1. Demand durable and easy to repair product design.
  2. Making spare parts available and guaranteed for longer periods.
  3. Providing Free access to repair service documentation and software for all re-use and repair centers.

To push these demands further, RREUSE takes part in a coalition of organizations fighting for a "Right to Repair" in the EU. This campaign advocates that consumers should be able to decide how their products should be repaired, by themselves or by professionals. Manufacturers should not be able to retain the right to repair their products by giving access to repair tools or information only to repairers that they authorized.

In 2017, UNEP produced a report[7] on product life extension and outlined several strategies for promoting the same.

For developing economies in which the informal second-hand and repair market is highly developed, the report recommends the following policy measures:

- Informal economic sectors that revolve around trading, repairing, and regaining material from redundant products currently lack access to investment capital and information to make repairs energy efficient, safe, and environmentally sound. It is recommended to recognize these professions and offer them social rights, official status, and training.
- The introduction of energy efficiency labeling, other eco-labels and awareness campaigns can stimulate the more affluent households to invest in high-quality, longer lasting, and/or energy efficient products.
- Stimulating the acceptance of alternative business models (the shift from "owning" to "leasing" products) in the Business-to-Consumer market, includes addressing privacy and other liability issues pro-actively.

Figure 5.4 summarizes few of the most viable strategies for product life extension.

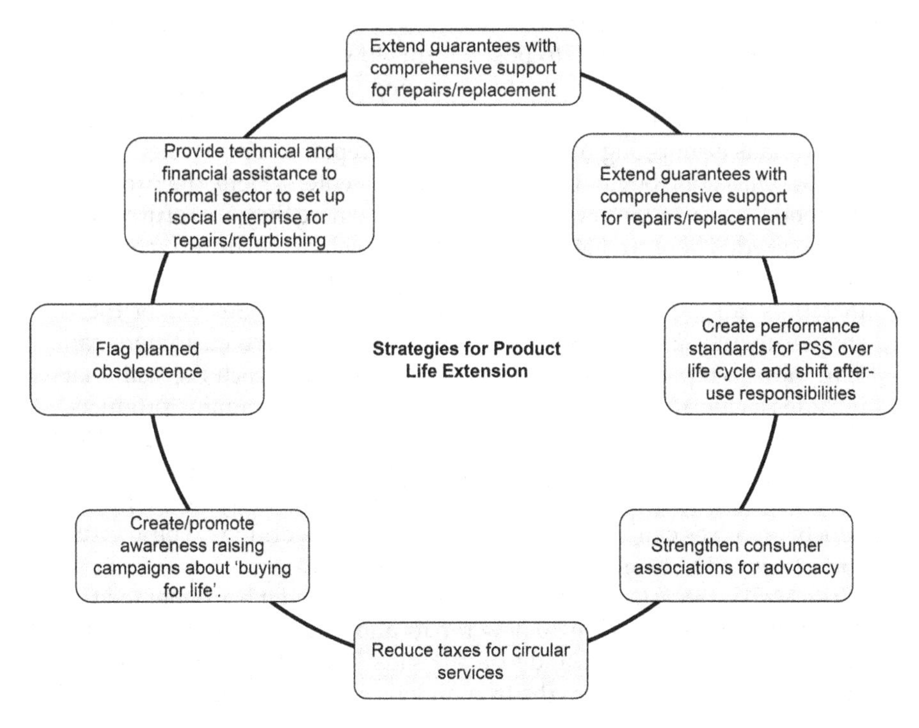

FIGURE 5.4
Strategies for product life extension.

There are a handful of reasons for the repair market still being underdeveloped despite the major benefits of repairing. Firstly, the attitude of people toward repaired goods is often perceived to be negative. Consumers prefer buying new products as they feel repaired/refurbished goods would not be as functional as a new product. Most of the times people do not repair products with manufacturers and instead upgrade to a new one due to high costs of repair and unavailability of repair centres near to their homes.

Box 5.4 presents an example of ReTuna Återbruksgalleria (ReTuna), the world's first shopping center that attempts to address this barrier.

There exist examples of malls where a corporation has taken the initiative to practice refurbishing. Box 5.5 shows an example of Walmart on the refurbishing of shopping carts.

BOX 5.4 RETUNA ÅTERBRUKSGALLERIA[21]

ReTuna is about 75 miles west of Stockholm. It opened in August 2015 in the city of Eskilstuna. ReTuna is not only the first shopping center to sell recycled and reclaimed goods in Sweden, but also in the world. Eskilstuna's 67,000 residents seem more and more open to the idea of buying repurposed and refurbished goods, which is inspiring ReTuna to set an even bigger goal: to position this town as a global destination that will showcase what sustainable living and the circular economy are all about. ReTuna is a living demonstration of the circular economy.

ReTuna is a shopping mall that sells only repaired or upcycled products and has gone beyond the local drop-off centers. Here, the dropped off goods are sorted into various workshops where they are refurbished or repaired. There are 14 workshops that include furniture, computers, audio equipment, clothes, toys, bikes, gardening, and building materials: all garnered from second-hand products. Several of these shops function as "do-it-yourself" showrooms, where customers can learn how to repair or refurbish. The products then reach the mall that includes a café and restaurant with a heavy focus on organic products. There is a conference and exhibition facility complete with a specialty school for studying recycling. Visitors can enroll in a one-year Design-Recycle-Reuse program. In addition, ReTuna offers study visits during which attendees can learn about the mall's inner workings. These visits cost about USD 136 and are held once a week.

The Mall is operated by the local municipality and it has benefited the local economy by creating 50 new repair and retail jobs and providing space for private start-ups and local artisans. The Swedish community's biggest bonus is relief from the tremendous burden and expense of disposing of unwanted goods.

BOX 5.5 REFURBISHING SHOPPING CARTS AT WALMART[22]

Walmart utilizes four refurbishing centers across the United States, where returned and damaged phones, tablets, TVs, computers, and game consoles are sent for refurbishing. In 2017, Walmart sent more than 7.5 million electronic items for refurbishing.

For the last 15 years, Walmart has worked with Unarco to refurbish aged and damaged shopping carts. This effort was launched to save virgin materials and energy used to produce new shopping carts.

Each shopping cart is refurbished up to four times. The carts are disassembled with all worn accessories stored for reuse. The painted finish is removed, and then extensive repairs are performed to return the cart's metal portion to "like new" condition. A new finish is then applied to complete the refurbishment process and new accessories as necessary are installed.

Refurbishment of shopping carts has resulted in the refurbishment of 791,000 Walmart shopping carts in 3 years, which has eliminated the need for 22,000 tons of metal that would have been needed to produce new carts.

Globally, the repair market has been poorly researched due to the complexity of the sector. The Indian repair market has often been criticized for being highly informal and unorganized. However, the recent bloom of the e-commerce sector and growing awareness of environmental issues are changing the repair market's functioning.

The maximum demand is driven by the "electronics manufacturing and communications" sector. Likewise, amongst the consumer-focused sectors in the repair market, the refurbished smartphone sector has been the largest growing sector. Box 5.6 describes the smartphone repair and refurbishing market in India.

5.3.1 Product Labeling

A study carried out by the Influence of Lifespan Labeling on Consumers, jointly with the SIRCOME agency, the University of South Brittany, and the University of South Bohemia on behalf of the European Economic and Social Committee showed that lifespan labeling has an influence on purchasing decisions in favor of products with longer lifespans. On an average, sales of products with a label showing a longer lifespan than competing products increased by 13.8%[29].

A comprehensive study by the European Economic and Social Committee (2016) indicates that consumers respond positively to product lifetime labeling.

BOX 5.6 SMARTPHONE REPAIR AND REFURBISHING MARKET IN INDIA[23]

While the global market for refurbished smartphones declined 1% (year-on-year) in 2019, reaching just over 137 million units, India's refurbished smartphone market grew by 9%[24]. Due to the access of internet by Indians in tier-2 and tier-3 cities, disruption led by Reliance Jio and subsequent decline in the prices of data plans by telecom companies have led to unprecedented demand for smartphones[25].

There are several online companies that offer repair and refurbishment of smartphones. Prior to these companies' launch, the most popular place to buy a used smartphone was through the unorganized physical retail market. E-commerce companies like Yaantra, Cashify, Togofogo, and InstaCash offer online marketplaces that are exclusively for the buying and selling used and refurbished smartphones. Companies like Smart Services Pack Advantage (SSP) offer doorstep pick-up and drop facilities for maintenance, upkeep, and repair of smart devices like phones, tabs, and laptops. SSP partners with retailers and brand-owners like Samsung, Micromax, HTC, Croma, and PlanetM, among others, to enhance the consumer experience.[26]

Companies such as Flipkart, Amazon, and Quickr too have started promoting and selling refurbished goods. In August 2018, Flipkart launched a new e-commerce portal called 2GUD. It is an e-commerce site selling only certified refurbished products at cheap prices for value buyers. At present, the refurbished store is selling mobile phones, laptops, smart watches, tablets, and streaming devices. It will soon launch speakers, power banks, smart assistants, hair dyer, hair straightener, TV sets and other 400 product categories.[27] Each electronic product is certified and graded by experts at F1 Info Solutions and Services or other Flipkart partners. All products listed on the website goes through a rigorous 47-step quality check. Flipkart acquired mobile repair chain F1 Info Solutions, marking the entry into the refurbishment market that's often been all about informal low-cost workarounds.

Amazon sells refurbished goods through its globally operated Amazon Renewed program. It is a program where one can buy quality certified refurbished products that have been tested and certified to work and look like new by a qualified manufacturer or a specialized third-party refurbisher. It is reported that the platform's sales of used smartphones in India was growing at a whopping 400% year-on-year[26]. The refurbishment process typically includes a full diagnostic test, replacement of any defective parts, a thorough cleaning and inspection process, and repackaging by the seller or vendor. The product ships with all relevant accessories and is backed by a minimum of 6-month limited warranty offered by the brand or product seller.

Likewise, Quikr is seeing 42% QoQ growth in sale of refurbished phones. Refurbished phones of brands like Apple, Samsung, One plus, Xiaomi, and Motorola seem to be the most popular. Cities like Jaipur, Ahmedabad, Vishakhapatnam, Patna, Raipur, Surat, Madurai, Ludhiana, Guwahati, and Indore are the top cities in terms of demand. Quikr certified smartphones go through rigorous quality checks covering functional, physical, and International Mobile Equipment Identity (IMEI) checks[28].

More research and testing need to be done to study the effectiveness of lifetime labels and develop standardized measurement procedures. If product lifetime standards are based on manufactures' data, they have to be willing to participate, thus incentives need to be introduced. Table 5.2 shows typical requirements on repairs, durability, upgradability, ease of disassembly, and spare parts availability for some products.

5.3.2 Business Initiatives That Build on Product Life Extension

Many companies have expanded their businesses to include not just supply of goods but also new opportunities to do business through repairs and refurbishing. Boxes 5.7 and 5.8 illustrate case studies of Tata Motors in India and GameStop in the USA.

TABLE 5.2

Nordic Label Asking for Assurance on Longer Product Life[1]

	Nordic Label			
	White Goods	Personal Computers	Imaging Equipment	TVs & Projectors
Instruction on maintenance	x		x	x
Instruction on repair				
Performance/durability tests				
Upgradability		x		
Ease of disassembly	x	x	x	x
Priority parts			x	
Spare parts supply	x	x		x
Warranty/Guarantee	x			x

BOX 5.7 CASE OF TATA MOTORS: PROLIFE RECONDITIONED AGGREGATES[30]

Tata Motors in India needed to address its customer needs by providing high quality, low-cost replacement parts. If repaired or overhauled in less than a plant environment, mission-critical equipment like engines could pose high operational risk. Therefore, Tata Motors needed a viable, cost-effective alternative to overhaul vehicle aggregates with a nationwide warranty while maintaining the first life cycle component quality. To respond, Tata Motors established Tata Prolife in 2013 as an after-market product support strategy for Tata Motors' customers.

Tata Prolife offers the customer with reconditioned aggregates in exchange of old aggregates subject to simple acceptance norms. Use of Tata Motors Prolife aggregate ensures performance of the vehicle even after the first life cycle[31].

The customer is provided reconditioned aggregates in exchange for old aggregates subject to simple acceptance norms. Products range from reconditioned Engine Long Block, Gear Box, Power Steering Gear Box, Turbo Charger, Air Compressors to electrical components such as starter motors and alternators. The reconditioned long blocks are also being exported to international markets. In 2016, a total of 23,115 components/equivalent engines were reconditioned[30].

BOX 5.8 GIVING VIDEO GAME PRODUCTS & CONSUMER ELECTRONICS A SECOND CHANCE: GAMESTOP'S REFURBISHING BUSINESS[32]

GameStop headquartered in the USA has been one of world's largest company engaged in gaming hardware and software and consumer electronics. One of GameStop's unique extensions' business has been its ability to cater to a refurbishment business on a scale that distinguishes from its competitors. GameStop operates following a buy-sell-trade model supported by Refurbishment Operations Centers (ROCs). GameStop operates ROCs on a global scale in counties such as Australia, Canada, France, Germany, Ireland, Italy, Sweden, Spain, and the United States.

The ability to receive the video games and consoles that a customer brings in for trade, repair them and return them to the consumer market is something no other retailer can do as well as GameStop. Its largest state-of-the-art facility in Grapevine, Texas has proprietary technology to refurbish the many different brands and operating platforms of smartphones and tablets.

GameStop's unique and robust refurbishment and recycling program has resulted in significant amounts of e-waste being kept out of landfills. Through its trade-in program, GameStop takes in software CDs and electronic devices and accessories that are otherwise destined for landfills and either refurbishes or recycles them. In 2016 alone, GameStop's 10 ROCs around the world refurbished more than 8 million pieces of software CDs and more than 3.7 million consumer electronic devices and accessories and recycled almost 3 million pounds of e-waste that could not be refurbished by cleaning, repairing, or reselling the products in different retail channels.

Several such innovative business models are discussed in Chapter 8.

Product life extension is one of the important upstream strategies that is critical in promoting a circular economy. The business and economics of circular economy equally depend on how effectively we manage to put the waste streams back to the resource pool. Only then the economy will be regenerative. Chapter 6 presents the opportunities and challenges we face in closing the loop.

5.4 Key Takeaways

- Each product has an expected life that it must ensure to the consumer.
- Extending product life through repairs and refurbishing is desirable in the interest to minimize consumption of virgin resources and to reduce environmental impacts.
- How much to extend the product life is, however, a complex decision to take. Tools like LCA could be used, however, their application requires significant data that is not always possible.
- For some products, product durability and recyclability could conflict.
- Planned obsolescence is a barrier and still a challenge to promote product life extension.
- Policies and regulations related to repair, information tools like labeling and initiatives like repair cafes, malls that showcase repaired and refurbished goods and e-commerce websites that offer certified refurbished products will make a difference. Amongst all, consumer education and advocacy have a major role to play.

ADDITIONAL READING

1. **Workshop report: Promoting remanufacturing, refurbishment, repair, and direct reuse** – This report is a summary of the workshop held by the International Resource Panel and the European Commission, under the aegis of the G7 Alliance on Resource Efficiency. They presented preliminary findings of the report by International Resource Panel on remanufacture, refurbishment, repair, and direct reuse. They also discussed measures for overcoming market and policy barriers to promote these circular economy processes. The workshop was held back-to-back with a G7 Meeting on Resource Efficiency and the recommendations on how to advance remanufacturing, refurbishment, repair, and direct reuse fed into discussions about the G7 Roadmap on Resource Efficiency.

Source: European Union. 2017. *Workshop Report: Promoting Remanufacturing, Refurbishment, Repair, and Direct Reuse.* <https://ec.europa.eu/environment/international_issues/pdf/7_8_february_2017/workshop_report_Brussels_7_8_02_2017.pdf> [Accessed 2 November 2020].

2. **Products that last** – This book offers readers an innovative and practical methodology to unravel a product's "afterlife" and systematically scrutinize it for new opportunities. It introduces business models that benefit from the opportunities offered by a much longer product life. It aims to change the way entrepreneurs and designers develop and exploit goods, helping to reduce material and energy consumption over time.

Source: Bakker, C. A., den Hollander, M. C., van Hinte, E. and Zijlstra, Y., 2014. *Products That Last: Product Design for Circular Business Models.* Delft: TU Delft Library.

3. **Reparability criteria for energy related products** – The overall aim of this study is to evaluate and, if possible, quantify the ease of repair for energy-related products (ErPs) considering the economic impact from a consumer perspective. In order to meet this objective, reparability criteria for ErPs are proposed. The developed criteria are in accordance with the ongoing initiative of the CEN-CELEC WG3 working on the standardization of repairability.

Source: Bracquené, E., Brusselaers, J., Dams, Y., Peeters, J., De Schepper, K., Duflou, J. and Dewulf, W., 2018. *Repairability Criteria for Energy Related Products.* Benelux. <https://www.benelux.int/files/7915/2896/0920/FINAL_Report_Benelux.pdf> [Accessed 2 November 2020].

4. **A longer lifetime for products: Benefits for consumers and companies** – This report offers a comprehensive review of the issues surrounding a longer lifetime for products in relation to consumer protection and the economic performance of enterprises. It offers an overview and a balanced representation of views from various stakeholders within the EU. It presents an assessment of the potential impact on the economy throughout Europe as well as reporting on social and environmental issues. It also provides pioneering examples of products with longer lifetimes that are already on the market, the policy context for an initiative on longer lifetimes and possible corresponding policy options to optimize the benefits related to longer product lifetimes in the EU.

Source: European Parliament. 2016. *A Longer Lifetime for Products: Benefits for Consumers and Companies* <https://www.europarl.europa.eu/RegData/etudes/STUD/2016/579000/IPOL_STU%282016%29579000_EN.pdf> [Accessed 2 November 2020].

5. **Promoting product longevity** – This document was prepared for Policy Department A at the request of the IMCO Committee (Internal Market and Consumer Protection). Product longevity can play a useful role in achieving the Paris Agreement goals – material efficiency is an important contributor to energy efficiency and is also important. The product safety and compliance instruments available at European level can contribute to these efforts, if wisely applied. There are suggestions in the literature that product lifetimes are becoming shorter. Hard data is available but is limited and suggests that, if anything, the opposite was the case in the USA in the second half of the Twentieth Century: The average lifetime of durable household goods and of automobiles got substantially longer over time, not shorter.

Source: European Parliament. 2020. *Promoting Product Longevity.* <https://www.bruegel.org/wp-content/uploads/2020/04/IPOL_STU2020648767_EN.pdf> [Accessed 2 November 2020].

6. **Electrical and Electronic Product Design: Product Lifetime** – This report presents the findings from the research commissioned by Waste and Resources Action Programme (WRAP) to explore consumers' current views, attitudes, and perceptions of the lifetimes of electrical products, with a particular focus on providing insight into the nature and extent of "consumer pull" for longer lifetimes. Consumer interest in longer product lifetimes was explored with respect to two different types of electrical products:
   1. Household appliances (fridges, washing machines, and vacuum cleaners) – termed "workhorse products" since they are typically purchased for a lifetime of heavy and prolonged use.
   2. Consumer electronics (televisions and laptops) – termed "up-to-date products" since many consumers look to upgrade periodically to the latest technology in a fast-moving market.

Source: WRAP. 2013. *Electrical and Electronic Product Design: Product Lifetime.* <https://www.wrap.org.uk/sites/files/wrap/WRAP%20longer%20product%20lifetimes.pdf> [Accessed 2 November 2020].

7. **The long view – Exploring product lifetime extension** – The aim of this study is to provide recommendations on the opportunities available to consumers, the private sector, and governments, of developed and developing economies, to address product lifetime extension. The primary focus of the report is on policy-making. This report therefore mainly collates existing data and insights. Expert interviews were conducted to uncover opportunities to address product lifetime extension. Finally, the report collects insightful state of the art examples of policies and private sector initiatives regarding product lifetime extension.

Source: UN Environment. 2017. *The Long View – Exploring Product Lifetime Extension.* <https://www.oneplanetnetwork.org/sites/default/files/the_long_view_2017.pdf> [Accessed 2 November 2020].

Notes

1. Bracquené, E., Brusselaers, J., Dams, Y., Peeters, J., De Schepper, K., Duflou, J. and Dewulf, W., 2018. *Repairability Criteria for Energy Related Products.* Benelux. <https://www.benelux.int/files/7915/2896/0920/FINAL_Report_Benelux.pdf> [Accessed 2 November 2020].

2. Šajn, N., 2019. *Consumers and Repair of Products*. European Parliment. <https://www.europarl.europa.eu/RegData/etudes/BRIE/2019/640158/EPRS_BRI(2019)640158_EN.pdf> [Accessed 2 November 2020].
3. Montalvo, C., Peck, D. and Rietveld, E., 2016. *A Longer Lifetime for Products: Benefits for Consumers and Companies*. European Parliament. <https://www.europarl.europa.eu/RegData/etudes/STUD/2016/579000/IPOL_STU(2016)579000_EN.pdf> [Accessed 1 November 2020].
4. EurekAlert. 2019. *Durability Vs. Recyclability: Dueling Goals In Making Electronics More Sustainable*. <https://www.eurekalert.org/pub_releases/2019-04/giot-dvr040419.php> [Accessed 2 November 2020].
5. Huang, X., Atasu, A. and Toktay, L., 2019. Design Implications of Extended Producer Responsibility for Durable Products. *Management Science*, 65(6), pp. 2573–2590.
6. Marcus, J., 2020. *Promoting Product Longevity. Bruegel*. <https://www.bruegel.org/2020/04/promoting-product-longevity/> [Accessed 2 November 2020].
7. Kerdlap, P., Gheewala, S. and Ramakrishna, S., 2020. To rent or not to rent: A question of circular prams from a life cycle perspective. *Sustainable Production and Consumption*, 26, pp. 331–342.
8. Wrap. 2010. *Environmental Life Cycle Assessment (LCA) Study of Replacement And Refurbishment Options for Domestic Washing Machines*. <http://www.wrap.org.uk/sites/files/wrap/Technical%20report%20Washing%20machine%20LCA_2011.pdf> [Accessed 2 November 2020].
9. UN Environment. 2017. *The Long View – Exploring Product Lifetime Extension*. <https://www.oneplanetnetwork.org/sites/default/files/the_long_view_2017.pdf> [Accessed 2 November 2020].
10. Umweltbundesamt. 2016. *Lifetime of Electrical Appliances Becoming Shorter and Shorter*. <https://www.umweltbundesamt.de/en/press/pressinformation/lifetime-of-electrical-appliances-becoming-shorter> [Accessed 2 November 2020].
11. Bakker, C. A., den Hollander, M. C., van Hinte, E. and Zijlstra, Y., 2014. Products That Last: Product Design for Circular Business Models. Delft: TU Delft Library.
12. European Union. 2019. *Premature Obsolescence Multi-Stakeholder Product Testing Program*. <https://prompt-project.eu/wp-content/uploads/2020/07/PROMPT_20191220_State-of-the-art-existing-testing-rating-systems-and-standards.pdf> [Accessed 2 November 2020].
13. BEUC. 2015. *Durable Goods: More Sustainable Products, Better Consumer Rights*. <https://www.beuc.eu/publications/beuc-x-2015-069_sma_upa_beuc_position_paper_durable_goods_and_better_legal_guarantees.pdf> [Accessed 2 November 2020].
14. Sampere, N., 2015. *Making More Durable and Reparable Products*. European Environmental Bureau. <http://makeresourcescount.eu/wp-content/uploads/2015/07/Durability_and_reparability-report_FINAL.pdf> [Accessed 2 November 2020].
15. Swedish Environmental Protection Agency. n.d. *The Stage Goals*. <https://www.naturvardsverket.se/Miljoarbete-i-samhallet/Sveriges-miljomal/Etappmal/> [Accessed 2 November 2020].
16. European Commission. 2020. *A New Circular Economy Action Plan For A Cleaner And More Competitive Europe*. https://ec.europa.eu/commission/presscorner/detail/en/qanda_20_419> [Accessed 2 November 2020].

17. SGS. 2015. *Built To Last? A Law In France To Combat Planned Obsolescence For Appliances.* <https://www.sgs.com/en/news/2015/07/built-to-last-a-law-in-france-to-combat-planned-obsolescence-for-appliances> [Accessed 2 November 2020].
18. Vice. 2020. *A Right To Repair Law Is Closer Than Ever.* <https://www.vice.com/en/article/k7e7xm/a-right-to-repair-law-is-closer-than-ever> [Accessed 2 November 2020].
19. Restart. n.d. *The Restart Project.* <https://therestartproject.org/> [Accessed 2 November 2020].
20. RREUSE. n.d. *About Us.* <https://www.rreuse.org/about-us/> [Accessed 2 November 2020].
21. Retuna. n.d. About us <https://www.retuna.se/hem/> [Accessed 2 November 2020].
22. Walmart. 2018. *Global Responsibility Report.* <https://corporate.walmart.com/media-library/document/2018-global-responsibility-report/_proxyDocument?id=00000170-ac54-d808-a9f1-ac7e9d160000> [Accessed 2 November 2020].
23. Researched by Aneesa Patel, Internship at Environmental Management Centre, 2018
24. Livemint. 2020. *Refurbished Phone Market Sees 9% Growth In India.* <https://www.livemint.com/technology/tech-news/refurbished-phone-market-sees-9-growth-in-india-11591942078390.html> [Accessed 2 November 2020].
25. Business Today. 2018. *Second-Hand Smartphones Market In India To Get Much Bigger.* <https://www.businesstoday.in/latest/trends/second-hand-refurbished-smartphones-market-in-india-to-get-much-bigger/story/277545.html> [Accessed 2 November 2020].
26. Thomas, A., 2017. *Startup India: Phone Needs Repair? This Startup Will Collect, Fix And Deliver It Back At Your Doorstep.* The Economic Times. <https://economictimes.indiatimes.com/articleshow/60318844.cms?utm_source=contentofinterest&utm_medium=text&utm_campaign=cppst> [Accessed 2 November 2020].
27. Agarwal, N., 2018. *Flipkart Launches 2GUD, New E-Commerce Portal For Refurbished Goods.* mint. <https://www.livemint.com/Companies/vzRSAgDD9LT11k1IWamh1J/Flipkart-launches-2GUD-ecommerce-portal-refurbished-goods.html> [Accessed 2 November 2020].
28. Quikr News. 2018. *Refurbished Smartphone Market Has A Big Scope To Grow In India - VARINDIA.* <http://news.quikr.com/blog/2018/02/06/refurbished-smartphone-market-has-a-big-scope-to-grow-in-india-varindia/> [Accessed 2 November 2020].
29. European Economic and Social Committee. 2016. *The Influence Of Lifespan Labelling On Consumers.* <https://www.eesc.europa.eu/resources/docs/16_123_duree-dutilisation-des-produits_complet_en.pdf> [Accessed 2 November 2020].
30. Circular Economy Guide. n.d. *Refurbishing.* <https://www.ceguide.org/Strategies-and-examples/Make/Refurbishing> [Accessed 2 November 2020].
31. Tata Motors. n.d. *Prolife Business.* <https://www.customercare-cv.tatamotors.com/services/prolife-business.aspx> [Accessed 2 November 2020].
32. Gamestop Corp. n.d. *Innovation.* <http://news.gamestop.com/innovation> [Accessed 2 November 2020].

6

Closing the Loop

Studies suggest that there will be 12 billion metric tons of plastic in landfills by 2050[1]. Every year, almost 90 million tons of extremely useful waste[2] are sent to landfills, which could otherwise be recycled back into the system. Along with the waste, billions of dollars are lost to the landfills due to poor segregation and recycling techniques. Closing the loop of materials is therefore necessary.

Closing the loop requires the application of recycling, recovery, and remanufacturing. A well-established system of reverse logistics and the presence of robust recycling markets are also required. In developing economies, informal sectors, especially women, play a critical role in waste collection and sorting. Their health and safety cannot be compromised.

COVID-19 pandemic has highlighted certain key issues related to recycling, like worker safety, bale purity, labor shortage, and recycling cost. It has also exposed the concerns associated with the risk of recycling and handling hazardous waste. This will be further discussed in Chapter 11, under the section describing the impact of COVID-19 pandemic on the circular economy.

The economics of closing the loop is a complex subject. Pricing of virgin resources, components, and products play a major role in deciding whether it is worth recycling. Often the costs of collection and reverse logistics dominate. If the discarded products contain harmful substances, then technologies are needed for their separation, recovery, or destruction. The application of these technologies requires investments and operations on a scale.

When legislated, the Extended Producer Responsibility (EPR) greatly helps in closing the loop. It is also important that consumers also show their commitment to segregate and return wastes and show preference to buy recycled/remanufactured products through "Extended Consumer Responsibility".

Although market collaborations are vital to advancing modern recycling markets, the governments, private industry, non-profits, and communities play a role in closing the loop. Here, governments have an important role to play in pushing products with recycled content, providing guidance, and enforcing legislation. Recycling associations have to work with the recyclers to create a robust ecosystem of material flows, fair competition, and attract investments to boost the recycling infrastructure. Integrating recycling markets to economic and social development should be the objective. Reducing the consumption of virgin resources, creating, and protecting green jobs, and building capacities of the informal sector should be the agenda.

In this chapter, a few key topics that deserve attention to understand the complex subject of closing the loop will be discussed. To begin with, Box 6.1 describes the three terms – upcycling, downcycling, and closed loop recycling.

BOX 6.1 UPCYCLING, DOWNCYCLING, AND CLOSED-LOOP RECYCLING

UPCYCLING[3]

Upcycling is the process of transforming materials destined to be destroyed into new products of higher value and environmental purpose. However, there is a need that the upcycled products meet the environmental, health and safety requirements and do not pose risks to the ecosystems. It is also necessary that a market exists for the upcycled products.

DOWNCYCLING[4]

Downcycling or cascading is recycling waste where the recycled material is of lower quality and functionality than the original material. Downcycling can help keep materials in use or in re-circulation. Most recycling is actually downcycling, meaning it may reduce the quality of material and value added over time.

CLOSED-LOOP RECYCLING[5]

Closed-loop recycling means that recycling of a material can be done indefinitely without degradation of properties. Conversion of the used product back to raw material allows the repeated making of the same product over and over again. This resonates with the cradle-to-cradle concept described in Chapter 2.

In all of the above, it is to be ensured that:

- The recycled/remanufactured materials should provide the same quality of the product (no deterioration). For example, almost all recycled aluminum from soda cans is suitable to produce the same cans.
- There should be no accumulation of hazardous substances in closing the loop, which can make the recycled/remanufactured product less safe.
- The recycled material can feed the manufacturing process for a different product or industry sector.

The first essential step in closing the loop is to ensure efficient waste segregation. While waste segregation can be an industry practice in manufacturing operations, the situation on waste segregation at the household level can be challenging. Segregation of domestic waste is important as it takes a major share of the total volume of waste generated in the world.

6.1 Segregation of Waste

Recycling, recovery, and remanufacturing have been the principal strategies for closing the loop of waste flow by transforming them into secondary material resources. Unfortunately, many of the waste streams are contaminated and hazardous. These mixed waste streams pose health risks to the workers, especially from the informal sector. Further, mixed waste streams are difficult to sort, clean, and process, thus making the economics of recovery difficult and unviable. Waste segregation at source is, therefore, the key.

When you segregate waste into two basic streams like organic (degradable) and inorganic (non-biodegradable), the waste generated is better understood and consequently recycled and reused with higher potential for recovery. Waste pickers typically use inorganic waste and segregate it further into paper, metal, plastic and then sell them to earn a livelihood. These waste streams get collated through the informal "eco-system" of waste bankers and waste traders who become "material suppliers" to the formal manufacturing sector. As a result, you see products being made from recycled plastic, metals getting recycled to make products, and wastepaper getting mashed into pulp to make recycled paper.

The organic waste component is often converted into compost and/or methane gas using Mechanical Biological Treatment. Compost can replace the demand for chemical fertilizers, and biogas can be used as a source of energy. As a result, much of the waste gets utilized as a resource – benefitting waste pickers, waste traders, small and medium industries, citizens, and the local municipal authority.

If not segregated, waste can pose risks and constraints on the choice of operation of waste processing technologies. Plastic in waste, if incinerated, could lead to the release of toxic dioxins. Household hazardous waste, if not segregated (e.g., spent batteries), can result in compost that is contaminated.

Thus, proper segregation of waste leads to a circular economy, creating green jobs, reducing virgin resources consumption, and promoting investments and innovations. Furthermore, as waste transportation reduces, emissions reduce and life of the landfill increases. Segregated waste also reduces health and safety related risks to waste pickers and to the ecosystems around the waste treatment and disposal sites.

ACTIVITY 6.1 PREPARATION OF CASE STUDY ON WASTE SEGREGATION IN CITIES

Data on waste segregation in cities is not widely available, dated, or reliable. Given this limitation, it is difficult to assess the impact of regulations, economic instruments (such as levying fines), and conducting awareness programs. Prepare case studies of cities where such assessments have been carried out and draw lessons learnt.

Some argue that waste segregation at the household level is very difficult to achieve, and in practice, around 60–70% segregation may be possible. Curbside collection of the segregated waste streams can be expensive. With the advent of advanced sorting technologies, it may be possible to separate waste streams with more than 90% accuracy. Should we, therefore, not push so much on segregation of waste at source, especially at the curbside, and follow the collection of mixed waste as a single stream? What may be the pros and cons of this approach?

6.2 Material Recovery Facility

Waste sorting, storage, and processing is carried out at the Material Recovery Facilities (MRF). MRF is a solid-waste management plant or a facility that processes recyclable materials, so as to sell raw materials for new products to the manufacturers.

It would be good that you refer to the Guidance for MRF prepared by Waste & Resources Action Program (WRAP)[6] in the United Kingdom. MRF's main function is to maximize the quantity of recyclables processed while producing materials that will generate the market's highest possible revenues. MRFs can also function to process organic or biodegradable wastes into a feedstock for biological conversion or into a fuel source for the production of energy[7].

MRFs operate on various scales. Depending on the scale of operations and mechanization level in the facility, MRFs may be classified as manual or mechanized. Usually, small-scale units are manual MRFs and are typically owned, managed, and operated by the communities and informal sector. Here, the material is segregated based on the types of waste (paper, plastic, metal, glass, etc.) and gradation of material within each waste type (e.g., paper segregated into news print, office paper, packaging paper, printed books, etc.). Segregated material is then sold to intermediaries, who supply material in bulk to the recycling industry.

Mechanized MRFs are large facilities with sophisticated equipment to do sorting, shredding, and processing. For example, optical near infrared light and sensors that recognize different types of plastics are being utilized in modern MRFs. These systems accurately separate plastics by resin type. These technologies dramatically increase the potential overall recovery of plastics, for both recycling and energy recovery. Box 6.2 shows illustrations of companies who offer technologies based on Artificial Intelligence (AI).

BOX 6.2 ILLUSTRATIONS OF COMPANIES WHO OFFER SOLUTIONS FOR SORTING BASED ON ARTIFICIAL INTELLIGENCE TECHNOLOGY

Zen Robotics[8]: It offers AI powered solutions for sorting diverse kinds of wastes, ranging from commercial and industrial waste, Municipal Solid Waste, Construction and Demolition waste, plastic, packaging, and scrap metal. Their AI software makes recycling more profitable, efficient, and accurate. ZenBrain is a trainable software programmed into its products – Heavy Picker and Fast Picker, which can sort 6000 and 4000 pieces per hour, respectively.

Heavy Picker is used for sorting bulky materials using various sensors, heavy-duty robot arms and AI. High speed of Fast Picker robot is used for maximizing material recovery and its compact design allows easy installation into existing conveyers and processes.

AMP Robotics[9]: AMP robotics claims that their AI technology applies computer vision for recycling with 99% accuracy rates. They offer solutions for minimizing manual processes and maximizing automation by offering AI based recycling solutions. AMP Cortex is one the products the company offers, which collectively uses AI based learning (brain), vision systems (eyes), and smart sorting robots (hands) to automate and improve recycling. AMP Neuron is an AI based supportive platform, which powers AI based learning and vision system of the Cortex. As the target material is sorted, data is simultaneously being uploaded on the cloud that helps analyze operations (AMP Insights) and take relevant steps to optimize them. This set-up can be easily installed over conveyer belts. The technology is suitable for handling multiple types of waste ranging from municipal solid waste (plastic, aluminum, paper, cardboard, etc.), electronics, and construction and demolition waste.

Mechanized MRFs may be publicly owned and operated, publicly owned, and privately operated, or privately owned and operated. These MRFs serve as an intermediate processing step between the collection of recyclable materials from waste generators and the sale of recyclable materials to markets for use in making new products. Connect with markets is extremely important for the financial viability of the MRFs[7].

6.3 Waste Banks

The concept of a waste bank is close to zero waste shops described in Chapter 2 and resembles the MRFs discussed earlier. A waste bank is a community rooted concept. It creates opportunities for waste recycling, provides green jobs, and builds funds for meeting other social infrastructure needs, e.g., schools and healthcare facilities. The idea of waste banks is now adopted by communities, especially in Asia, Africa, and Latin American Caribbean regions.

The operations at the waste bank are similar to a commercial bank that provides a service of depositing and lending money and conducting, recording, and tracking financial transactions. Here members of the waste bank deposit waste equivalent to depositing cash. Waste is thus given an economic value to incentivize waste deposits. Waste bank working system is based on household collection, by giving rewards to people who manage to sort and deposit the waste.

Managers of the waste bank need to be creative and innovative. They must have an entrepreneurial spirit in order to increase the income of the waste depositors and importantly, raise revenue from waste recycling. The waste bank in a society not only helps to clean the environment, but also generates extra cash for the society. The development of waste bank also assists local governments in empowering communities to manage waste wisely and reduce waste transported to the final disposal (landfill).

The waste bank model encourages people to sort the waste, raising public awareness to process waste wisely in order to reduce waste going into landfills. The deposition of waste with waste banks has become an innovative

ACTIVITY 6.2 PREPARATION OF A BUSINESS MODEL FOR OPERATING WASTE BANKS

Waste bank is a promising concept to achieve waste recycling objectives, providing livelihoods, and reducing consumption of virgin resources. The life of landfills also gets extended when multiple waste banks operate. For the waste clubs' financial viability, payments to the waste depositors, selling price of the segregated materials, and initial investment sharing and profit-sharing formula are very important. Prepare a business model for setting up and operating waste banks and apply on a case study to assess financial viability. The model should make attempt to estimate both tangible and non-tangible benefits. Prepare a statement of risks for your business model and state strategies to address and allocate the risks.

concept at the grassroots level to increase the urban poor's income. A waste bank is a good example of a decentralized waste collection and recycling system in communities.

6.4 Deposit and Refund Schemes[10]

Deposit and Refund Schemes (DRS) provide economic incentives where an amount of money is levied when the product is sold, and then refunded when the product or its container or packaging is returned after use. DRS provide a clear incentive for consumers to return end-of-life products. Most deposit and refund schemes are used effectively to ensure high rates of recovery of drink containers or other packaging, e.g., transport packaging such as boxes or pallets.

DRS needs policy interventions to push separate recovery systems that would not otherwise happen on a commercial basis. DRS complements or supports EPR and helps meet the stipulated targets.

To make the en-masse return of bottles and cans more efficient and convenient, many DRS use automated Reverse Vending Machine (RVM) and sorting machines to analyze and sort containers when they are collected for recycling and reuse. These machines instantly count the number of containers returned, sort away ineligible containers, and pay out the correct deposit refund to recyclers – much faster than is possible through manual, human handling and also further at reduced risks.

RVM can usually be found at convenient places such as supermarkets, allowing users to receive their refund in cash or as vouchers they can use in store. It is also possible to donate the deposit money to charity in some countries, direct from the reverse vending machine.

TOMRA, one of the leading companies in sorting and RVM, operates in more than 40 markets globally, with DRS. The company received more than 40 billion used beverage containers across 84,000 installations worldwide in 2019 alone.

DRS play an important role in changing consumer behavior for recycling reusable material. This holds true even when the deposit value is low in price. TOMRA researched on return rates and deposit values for several countries they operate in. It is seen that the return rates for containers are high regardless of the deposit value. For instance, Germany and Lithuania's return rate is 95% and 92%, respectively, despite Germany's deposit value being EUR 0.25 and for Lithuania EUR 0.10. However, in case of no deposit fee like in the United Kingdom, the return rates are low ranging between 51% and 65%.[11]

ACTIVITY 6.3 DISCUSSION ON DEPOSIT REFUND SCHEMES

Deciding the deposit amount in DRS is a complex exercise. Research on what considerations should be made in deciding this amount. How do DRS intersect with the obligations under EPR?

6.5 Role of Informal Sector

The informal sector consists of small businesses and self-employed persons with little or no legal recognition and poor financial resources. Vast majority of people working informally barely earn enough money needed to survive and are at risk in the collecting, sorting, and recycling operations.

The informal recycling sector, however, as a whole contributes massively to a more circular economy. Furthermore, this sector is one of the most dynamic and adaptive, catering to ever- changing demands in recycled products[12].

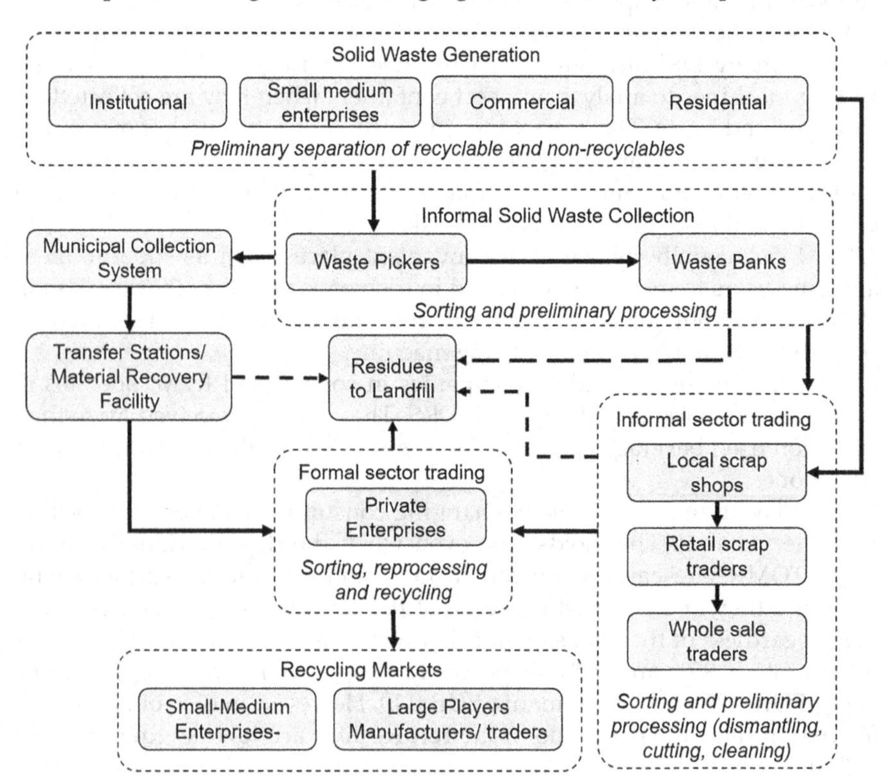

FIGURE 6.1
Waste flow in a municipality and the role of the informal and formal sectors.

There have been notable initiatives to recognize and formalize the informal sector with lead taken by Community Based Organizations (CBOs) and local municipal bodies. Lately, in the interest of EPR, many corporate bodies have partnered, and development institutions have played a facilitatory role. Integration of informal sector with formal sector in the form of partnership is necessary to maximize waste recycling. Figure 6.1 shows flow of waste materials in a municipality and the role of the informal and formal sectors.

Recognizing and incorporating informal workers into circular economies is critically important to bring the required systemic shifts that build long-term resilience, generate business and economic opportunities, and provide environmental and social benefits.

A good 58% of workers in the waste sector worldwide are women and as much as 92% of women work in the informal sector as per the International Labor Organisation (ILO) 2018 report. These women support their families financially while also raising children, looking after other family members, and tending their meagre dwellings. Any solution that is designed through the lens of circularity and inclusivity must be designed with consciousness of gender at the core of its configuration. Non-comprehensive solutions that do not take gender into account are designed to fail. Finally, up-skilling and capacity building is not to be contained only within the Urban Local Body (ULB). These skills are also needed within the circular economy stakeholders themselves[13]. In this perspective, two inspiring case studies are presented in Boxes 6.3 and 6.4 from India.

BOX 6.3 A PARTNERSHIP BETWEEN PUNE MUNICIPAL CORPORATION AND INFORMAL WASTE PICKERS[14]

In 2005, a pilot program implemented by KagazKach Patra Kashtakari Panchayat (KKKP) in collaboration with the Department of Adult Education, SNDT Women's University enabled 1500 waste pickers to become service providers for the door-to-door collection of waste from households in Pune city. This program considerably improved their conditions of work and upgraded their livelihoods, thus, effectively bridging the gap between households and the municipal waste collection service. The pilot was operational since 2006 and Solid Waste Collection and Handling (SWaCH) was formed in 2007.

In 2008, the Pune Municipal Corporation (PMC) signed a 5 year agreement with SWaCH and renewed it in 2016 to decentralize door-to-door collection services for households, shops, offices, and small commercial establishments. The members of the cooperative work in pairs and each pair is in charge of door-to-door waste collection for 200–250 households. Waste pickers receive segregated waste (separated wet or organic waste and dry waste such as plastics, glass, paper, etc.) from households/commercial establishments. They further segregate the recyclables to be sold in the recycling market; and non-recyclable waste is dropped off at feeder points.

SWaCH collects more than 600 tons of MSW per day, about 130 tons are sent for composting and 150 tons are recycled. SWaCH has been authorized by the PMC to collect and channel e-waste according to the rules laid down by the government. Currently, 2600 Waste pickers save the PMC INR 150 million per annum in waste handling costs alone. The model is also energy efficient and environmentally beneficial, as SWaCH waste pickers recycle waste and reduce the quantity of waste sent to landfills. This effort leads to a reduction in Greenhouse Gas emissions (GHG).

BOX 6.4 A PARTNERSHIP BASED PLASTIC WASTE MANAGEMENT PROGRAMME FOR DEVELOPING A SUSTAINABLE MODEL FOR CIRCULAR PLASTIC ECONOMY IN INDIA[15]

United Nations Development Programme (UNDP) India, in partnership with Hindustan Coca-Cola Beverages Private Limited and Hindustan Unilever Limited has built a partnership model to promote closing of loop of plastic waste (especially the litter) in urban local bodies in India. The partnership model promotes collection, segregation, and recycling of all kinds of plastics to move toward a circular economy.

This project aims to:

- Create a socio-technical model for taking plastic waste management from informal to formal economy involving implementing agencies such as CBOs and entrepreneurs interested in waste business.
- Establish MRFs for sustained practices in waste recycling.
- Institutionalize Swachhta Kendras (Cleanup Centers) within governance framework.
- Structure and improve socio-economic conditions of waste pickers.
- Develop technology-supported knowledge management.
- Promote cloud-based traceability, accountability, and digital governance with support of a software firm Mindtree along waste value chain.

In this model, the Safai Sathis (waste pickers) are self-employed having no legally tenable employer-employee relationship either with the municipalities or with the scrap traders. They are not paid by the municipalities. With the waste traders, the relationship is a sale-purchase transaction, at most a patron-client relationship built on years of dealing with the same scrap trader. Waste-pickers suffer from occupation related health issues, face regular harassment and extortion from both the police and the municipal authorities, and no social security benefits are available to workers in this sector.

The project is currently operational in 20 cities, with 22 MRFs. The waste plastic collected and processed so far has already crossed 17,000 metric tons. Through these MRFs, the project has reached out to 1756 Safai Sathis and these waste pickers are connected through Self Help Groups (SHG). Currently, around 700 Sathis are members of the SHG. Regular workshops are organized for them at Swachhta Kendras, where they are given assistance with opening bank accounts, they receive identify cards. Assistance is also given in availing health check-up, and several other personal training sessions.

There are also examples where a corporate has taken the lead. For instance, the Wellbeing Out of Waste (WOW) project undertaken by ITC in India. WOW has achieved significant scale on waste recycling and is operational in cities and large towns connecting with both waste pickers and citizens. The project since its inception has reached 7.7 million citizens in 10 States of India. Between 2007 and 2017, WOW generated livelihoods for 14,500 waste pickers supporting 59 Social Entrepreneurs and diverted 50,196 MT dry waste including 5000+ MT of multi-layered laminates & thin films collected[16].

ACTIVITY 6.4 DISCUSSION ON FORMALIZATION OF INFORMAL SECTOR

Despite the advantages, it has been found that informal sector engaged in waste recycling is averse to formalization. What could be the reasons for a resistance? How could these barriers be overcome?

6.6 Green Dot System

In a sound waste collection and recycling system scale of operation is important if operated at regional as well as national scale. This system needs to be consistent, efficient, and transparent with a charging mechanism that is agreeable to the manufacturers, consumers as well as responsible regional authorities and national governments. One such system was evolved over several years in Europe with lead taken by Germany. Box 6.5 describes in brief the well-known Green Dot System.

BOX 6.5 GREEN DOT SYSTEM[17]

The European Parliament and Council Directive 94/62/EC1 on packaging and packaging waste laid down the foundation for packaging recycling within the European Community. In order to comply with the objectives of this Directive, the Members States had to take the necessary measures. As a result, the Green Dot System was introduced as the main Packaging Waste Management System implemented in almost all European countries[18]. Today, the Green Dot System has evolved into a proven concept in many countries as a model for implementing producer responsibility.

Duales System Deutschland AG was the pioneer company who created the Green Dot System in Germany in 1991. The main objective of this company was to develop schemes to co-ordinate the collection, sorting and recycling of used packaging within Germany. As German manufacturers could not achieve those objectives by themselves, the "Duales System" was established to take care of the whole recycling process.

In 1995, Duales System Deutschland AG decided to expand the Green Dot System in the form of a general license to a European organization. Consequently, it founded the "Packaging Recovery Organization Europe s.p.r.l." (PRO EUROPE), which was domiciled in Brussels. In this way, the European countries could transfer their obligations to this organization, which developed an integrated Packaging Management System to implement the recycling legislations at both national and European levels.

Each country has a private or municipal waste management company that was in charge of running the Green Dot System. PRO EUROPE gave license to the European manufacturers for the use of Green Dot Logotype. Manufacturers from each country established contracts with the specific management enterprise which subsequently became in charge of collecting, sorting, and recycling the packaging material. In short, Green Dot systems contribute to the successful implementation of producer responsibility by obliged companies, e.g., producers and retailers. When consumers see the Green Dot on packaging it means that for such packaging a financial contribution has been paid to a national packaging recovery company that has been set up in accordance with the principles defined in the European Directive 94/62/EC and its national law. The license fee is governed by the number of packaging units put on to the market by each manufacturer and the weight of the applied materials.

The main task of PRO Europe's members is to organize the efficient implementation of adequate national collection and recovery systems, mostly for sales and household packaging. In this way, industrial companies and commercial enterprises are relieved of their individual obligation to take back used sales packaging. The aim is to ensure the recovery and recycling of packaging waste in the most economically efficient and ecologically sound manner.

Moreover, PRO Europe has concluded co-operation agreements with similar systems in the UK ("VALPAK") and Canada ("Green Dot North America"). VALPAK and Green Dot North America are managing use of the Green Dot in the UK and the NAFTA region to ensure that all licensees of the Green Dot may use labeled packaging without problems throughout the world.

More than 95,000 licensees use the Green Dot trade mark. More than 460 billion pieces of packaging marked with the Green Dot are on the market all over the world. This packaging is subsequently disposed of by more than 200 million people via the collection systems set up by the corresponding Green Dot organizations.

6.7 China's Green Fencing

Global annual imports and exports of plastic waste began to rapidly increase in 1993, having grown 723% and 817% in 2016, respectively. In 2016 alone, about half of all plastic waste intended for recycling (14.1 million MT) was exported by 123 countries, with China taking most of it (7.35 million MT) from 43 different countries[19].

There have been some landmark policy decisions that have significantly influenced global recycling business. One such example is China's policy related to import of waste streams called green fencing.

ACTIVITY 6.5 DISCUSSION ON THE GREEN DOT SYSTEM

The Green Dot system has evolved in the EU over two decades. The system has been criticized as monopolistic. Would you agree to this criticism?

ACTIVITY 6.6 DISCUSSION ON IMPACT OF GREEN FENCING

What was the reason for introducing Green fencing? Green fencing resulted into a significant impact on the business of recycling across the world. Research on the impact of green fencing in EU, USA, and Australia and response from these countries to combat the impact. Did Green fencing help boost the circular economies in these countries?

China launched the first Operation Green Fence (OGF) on 1 February 2013 to fend off inflows of illegal waste. The main objective of OGF was to enforce waste trade policies already adopted by China and thereby restrict illegal waste imports. The OGF laid down severe restrictions on the contamination present in the waste streams[20].

Sudden changes in the policies of the Chinese government significantly impacted the global recycling trade. It puts severe pressure on western countries to reconsider their reliance on the cost-effective practice of exporting waste, instead of building domestic recycling infrastructure. There was also a need to create demand for secondary raw materials. The recycling associations had a major role to play with the support of the national governments. China's green fence pushed the agenda on innovation and investment in recycling technology to process domestic waste streams internally and create/expand recycling markets.

6.8 Recycling Waste From Ship Dismantling

Ship dismantling also popularly referred to as "ship recycling" is a process of breaking down a ship when it completes its life tenure. As one of the ship disposal techniques, ship dismantling involves the retired ships to be stripped off their machinery, barring the value-adding materials which are reused for new ships or for other applications. The steel scraps obtained after dismantling an old ship is used again. Moreover, all other parts such as wooden furniture, glass, etc. are also reused for a variety of applications. Ship dismantling is preferred when the maintenance expenses of a vessel keep soaring with time and so ship owners' hand over the old vessel for disposal. Due to the large availability of cheap labor, ship dismantling business is thriving in nations such as Bangladesh, China, Pakistan, and India, around 85% of the world's shipbreaking activities occur in these countries[21].

Alang in India is the biggest ship breaking yard in the world. Several ship breaking contractors have their offices and yards along the coast of Alang, in the State of Gujarat. Traders dealing in different products visit the vessel once the ship breaker is through with clearances from customs, the pollution control board, and the maritime board. Traders negotiate with ship breakers for the entire cache of goods in their category. The products range from furniture to crockery, carpets, consumer goods such as television sets and refrigerators. Because of the bargains that are offered, the place attracts hoteliers, factory owners, art collectors, home makers and others who come looking for the remains of a vessel.

Ship dismantling is regarded to be one of the most dangerous professions in the world as it involves a great amount of risks and hazards. According to The Indian Supreme Court, the incidence of fatal accidents in shipbreaking (two in every 1000 workers) is higher than that in mining (0.34 per 1000 workers)[22].

The focal business of ship breaking in Alang is about steel. The ship breakers prefer the heavy tankers and bulk carriers as more steel can be extracted from them and they are easier to break. Passenger ships with their fancy fittings and hundreds of little rooms take much longer to dismantle. Between 1983 and 2011, the Alang yard produced 36,870,973 tons of steel by dismantling 5508 ships. In 2011–2012, Alang produced 3,856,071 tons of steel—that is about 2% of the annual demand for steel[23].

6.9 Regulating Hazardous Substances in Products

Minimizing hazardous substances in the products is important, not only to protect the consumers but also to make the recycling operations cost-effective and safer. Three gatekeeping regulations first introduced in the European Union (EU) called REACH, WEEE, and RoHS have been illustrated.

REACH[24]

REACH stands for Registration, Evaluation, Authorization, and Restriction of Chemicals. It entered into force in the EU on June 1, 2007.

REACH is a regulation of the EU adopted to improve the protection of human health and the environment from the risks that can be posed by chemicals, while enhancing the competitiveness of the EU chemicals industry. In principle, REACH applies to all chemical substances; not only those used in industrial processes but also in our day-to-day lives, for example in cleaning products, paints as well as in articles such as clothes, furniture, and electrical appliances. Therefore, the regulation has an impact on most companies across the EU.

REACH places the burden of proof on companies. To comply with the regulation, companies must identify and manage the risks linked to the substances they manufacture and market in the EU.

WEEE[25]

WEEE stands for Waste Electrical and Electronic Equipment. To address challenge of managing WEEE in the EU, two pieces of legislation have been put in place: The Directive on waste electrical and electronic equipment (WEEE Directive) and the Directive on the restriction of the use of certain hazardous substances in electrical and electronic equipment (RoHS Directive).

The first WEEE Directive (Directive 2002/96/EC) entered into force in February 2003. The Directive provided for the creation of collection schemes where consumers return their WEEE free of charge. These schemes aim to increase the recycling of WEEE and/or re-use.

In December 2008, the European Commission proposed to revise the Directive in order to tackle the fast-increasing waste stream. The new WEEE Directive 2012/19/EU entered into force on August 13, 2012 and became effective on February 14, 2014.

RoHS[26]

The Restriction of Hazardous Substances legislation also known as RoHS Directive 2002/95/EC, entered into force in February 2003. It restricts the use of hazardous substances in electrical and electronic equipment. The legislation requires heavy metals such as lead, mercury, cadmium, and hexavalent chromium and flame retardants such as polybrominated biphenyls or polybrominated diphenyl ethers to be substituted by safer alternatives. In December 2008, the European Commission proposed to revise the Directive. The RoHS recast Directive 2011/65/EU became effective on January 3, 2013.

The legislation provides for the creation of collection schemes where consumers return their used waste EEE free of charge. The objective of these schemes is to increase the recycling and/or re-use of such products. The RoHS and WEEE directives on Electrical and Electronic Equipment (EEE) were recast in 2011 and 2012 to tackle the fast-increasing waste stream of such products. The aim is to increase the amount of waste EEE that is appropriately treated and to reduce the volume that goes to disposal.

In January 2017, the Commission adopted a legislative proposal to introduce adjustments in the scope of the Directive, supported by the impact assessment. The respective legislative act amending the RoHS 2 Directive,

ACTIVITY 6.7 DISCUSSION ON IMPACT OF REACH, ROHS, AND WEEE

Study the Impact of REACH, RoHS and WEEE in EU's Circular Economy. Which other countries have legislated such legislations and how do they compare with those introduced in the EU?

adopted by the European Parliament and the Council, has been published in the Official Journal on November 21, 2017.

6.10 Recycled Content in the Product

To promote circularity, higher recycled content in products is preferred. Recycled products or products with recycled content however may continue to carry hazardous substances that were previously present in the product or original material. Their presence in excess amounts could cause health risks to the consumer and may even affect the functionality of the product, e.g., strength or durability. In order to address these risks, regulations and standards have been developed on both permissible and mandated recycled content. Box 6.6 shows examples of such regulations.

BOX 6.6 EXAMPLES OF REGULATIONS AND GUIDELINES ON RECYCLED CONTENT IN PLASTIC

In an article published on April 23, 2020, *Sustainable Plastics* cited publication of a new set of guidelines for recycled content in plastic packaging developed jointly by the *British Plastics Federation,* the *Cosmetic, Toiletry and Perfumery Association,* and the *Food and Drink Federation.* The document provides an overview of current European regulations related to the recycled content in plastic packaging as well as technical considerations that can be used by companies planning to include recycled content in their packaging. It also aims to help policymakers better understand current issues within individual sectors. A set of frequently asked questions further introduces and explains related concepts such as functional barriers and testing[27].

The United State Environmental Protection Agency's (US EPA) Comprehensive Procurement Guideline Program defines recycled content preference items and lists items manufactured with recycled material that US EPA deems equivalent to virgin material for standard applications. Under Section 6002 of the Resource Conservation and Recovery Act (RCRA), US EPA is required to designate items that are, or can be, produced with recovered materials and to recommend practices for buying these items. RCRA Section 6002 also requires purchasing agencies to establish Affirmative Procurement Programs for US EPA designated items[28].

There is an emphasis throughout the United States on increasing the uses of Post-Consumer Recycled (PCR) materials, including plastic.

Food and Drug Administration (FDA) is involved when industry collects used polymeric materials (usually food containers) and proposes to recycle these materials to make new food containers. FDA's main safety concerns with the use of PCR plastic materials in food-contact articles are:

- that contaminants from the PCR material may appear in the final food-contact product made from the recycled material
- that PCR material which may not be regulated for food-contact use may be incorporated into food-contact article, and
- that adjuvants in the PCR plastic may not comply with the regulations for food-contact use.

To address these concerns, FDA considers each proposed use of recycled plastic on a case-by-case basis and issues informal advice as to whether the recycling process is expected to produce PCR plastic of suitable purity for food-contact applications. FDA has prepared a document entitled "Guidance for Industry – Use of Recycled Plastics in Food Packaging: Chemistry Considerations", that will assist manufacturers of food packaging in evaluating processes for PCR plastic into food packaging[29].

ISO 14021 defines recycled content as "the proportion, by mass, of recycled material in a product or packaging. Only pre-consumer and post-consumer materials shall be considered as recycled content, consistent with the following usage of the terms:

- Pre-consumer material: Material diverted from the waste stream during a manufacturing process. Excluded is reutilization of materials such as rework, regrind or scrap generated in a process and capable of being reclaimed within the same process that generated it.
- Post-consumer material: Material generated by households or by commercial, industrial, and institutional facilities in their role as end-users of the product, which can no longer be used for its intended purpose. This includes returns of material from the distribution chain.

Apart from regulations, another strategy to address recycled content in products is to adopt a labeling approach. Box 6.7 describes such an initiative called "Remade in Italy". Box 6.7 also shows examples of Green standards in electronics in the USA and Dell's OptiPlex desktops.

BOX 6.7 EXAMPLES OF PROGRAMS FOR RECYCLED PRODUCTS

ReMade IN ITALY[30]

ReMade in Italy® is a non-profit, non-governmental organization aimed to promoting recycled products through independent third-party certification. ReMade in Italy® was found in 2009 and in 2013 Accredia (Italian National Accreditation Body) recognized it as the first certification scheme in Italy and Europe to verify the recycled content in a product. ReMade in Italy represents a useful tool for Public administrations and companies, to identify the recycled products for "Green public procurement", made mandatory in Italy by the law (Legislative Decree 50/2016).

The certification scheme is governed by a non-profit organization, legally recognized, gathering companies that realize recycled products in various category (building, street furniture, office furniture, schools, textiles, and many more). The Association is independent and reinvests revenues in institutional activities supporting Associated Companies for self-promoting on the public and private market and in strengthening the relationship with the competent Institutions.

The "Remade in Italy" label certifies the use of recycled material/reuse in products. The release of the Remade in Italy® certification is subject to a verification process by a third-party body (and therefore independent) for the certification of both management and product systems. The Remade in Italy® label highlights the environmental values of the material/product and is characterized by the assignment of a class, based on the percentage of recycled/reused material present.

GREEN STANDARDS FOR ELECTRONICS IN THE US[31]

Green standards for electronics in the United States establish a consistent set of environmental leadership criteria for the design, use, and end-of-life phases of electronics. Since their initial development, green USA electronics standards have successfully pushed manufacturers to incorporate key performance criteria, including requirements for recycled plastics, the reduction of hazardous materials, end-of-life management, and energy efficiency.

DELL TAKE-BACK PROGRAM[32]

Dell uses the recycled plastics derived from water bottles and old computers that they get through the "take-back" system in their monitors and OptiPlex desktops. Labels are put into the recycled materials for identification of such products and re-using them in the best way possible. The company one of the first ones to get a certification from Underwriters Laboratories Environment, practicing closed loop recycling. Dell's OptiPlex 3030 All-In-One computers are verified to contain a minimum of 10% post-consumer closed loop recycled content. By reusing plastics already in circulation, Dell is cutting down on e-waste, saving resources, and reducing carbon emissions by 11% compared with virgin plastics.

6.11 Recycled Content in Construction Materials

Globally buildings are responsible for a huge share of energy, water, electricity, and materials consumption. The building sector has the greatest potential to deliver significant cuts in GHG emissions at little or no cost. Buildings account for 9 billion tons of GHG emissions annually on CO_2e basis. Infrastructure sector is expected to expand at a steep rate. If recycled building materials and technologies in construction are not adopted during this time of rapid growth, emissions could double by 2050. During construction or at the end of useful life, construction materials and components are often discarded with construction debris accounting for nearly 35% of landfill waste. This waste can be effectively used to increase recycled content in the virgin construction materials[33].

In order to assess contributions to the recycled content by value of a building or infrastructure, design teams need to know the recycled content by mass of products – particularly when looking at options to substitute one brand with another brand of the same type of product within the chosen design specification. Data are therefore needed on a brand by brand basis and must be calculated and recorded on product data sheets in a clear and consistent fashion. To assist, a guide has been developed for WRAP by Building Research Establishment in consultation with product manufacturers and their trade associations. The aim of this Guide is to provide a set of "Rules of Thumb" for the calculation and declaration of recycled content in construction products[34].

6.12 Wastewater Recycling

Wastewater recycling has been practiced extensively across the world right from household or housing society or apartment complex level, city level, and even on a regional scale. There are numerous case studies that demonstrate the environmental, economic, and social benefits of wastewater recycling. Generally, scarcity or non-availability of water and economics of recycling govern the business of wastewater recycling. Advanced technologies such as membrane filtration, adsorption and oxidation have played an important role in delivering quality of treated wastewater fit for various application of reuse. In this section two interesting examples of wastewater recycling are presented.

Purple Pipes in USA[35]

Purple colored pipe is used to identify and distribute reclaimed or recycled water. It was first adopted by the American Water Works Association California-Nevada Section in 1997. It is a non-potable water reuse program geared to reduce potable water consumption guided by the Department of Defense Strategic Sustainability Performance Plan and the Army Net Zero Directive. Water used once for washing and passing through from kitchen sinks, dishwashers, bathroom sinks, tubs and showers is considered gray water. It's gray water that is carried in purple pipe. With minimal treatment, this type of water can be re-used for lower end uses such as flushing toilets or to irrigate landscape. Gray water is household wastewater, except for wastewater from toilets, typically 50–80%. Water from toilets, known as black water, is not used in purple pipe.

In 2003, purple pipe was adopted in the Uniform Color Standard of the American Public Works Association. Since then, plumbing codes have been updated to include more guidance specific to water reuse activities. For example, the International Code Council's 2015 International Plumbing Code (IPC) dedicates an entire chapter to Non-Potable Water Systems. It provides some general guidance on treatment requirements but defers to local jurisdictions with respect to specific levels of purity required for reuse activities. More specific guidance is provided in the IPC on the labeling of pipes and fixtures, backflow and cross-contamination prevention, storage system design, and materials. IPC provides guidance for the following systems: collection, on-site gray water treatment and reuse, rainwater collection and use, irrigation, and reclaimed water. Sources of gray water and reuse activities allowed can vary significantly from state to state. DoD installations are generally required to follow state and local codes in regard to water-reuse practices, as summarized in the Unified Facilities Criteria (UFC) for plumbing systems. UFC 3-420-01 Plumbing Systems states: "Section 602.2.1 Non-potable water exception. A non-potable water supply, when used in an entirely separate system and when approved by the local health department, may be used for flushing water closets and urinals, and for other approved purposes where potable water is not required. Piping containing nonpotable water, that is water not meeting accepted potable water standards, is labeled 'NONPOTABLE WATER, DO NOT DRINK'".

Zero Liquid Discharge

Zero Liquid Discharge (ZLD) is a strategic wastewater management system that ensures that there will be no discharge of industrial wastewater into the environment. It is achieved by treating wastewater through recycling and then recovery and reuse for industrial purpose. Hence ZLD is a cycle of closed loop with no discharge. Although ZLD is a costly process, it paves the way for economic benefits by recovering salts and other chemical compounds[36].

In recent years, greater recognition of the dual challenges of water scarcity and pollution of aquatic environments has revived global interest in ZLD. More stringent regulations, rising expenses for wastewater disposal, and increasing value of freshwater are driving ZLD to become a beneficial or even a necessary option for wastewater management. The global market for ZLD is estimated to reach an annual investment of at least USD 100–200 million, spreading rapidly from developed countries in North America and Europe to emerging economies such as China and India[37].

Although the history of tighter regulations on wastewater discharge can be traced back to the USA Government's Clean Water Act of 1972, India and China have been leading the drive for zero liquid discharge regulations in the last decade.

BOX 6.8 ZERO LIQUID DISCHARGE IN TEXTILE INDUSTRIAL ESTATE, TIRUPUR, INDIA

India was the world's third largest exporter of textiles in 2015, and the sector generates direct employment to 45 million people. In 2015, the federal government took several steps to regulate the sector in an effort to make it more resource efficient. New guidelines, directions, and standards for textile effluents were all influenced by the ZLD approach with the idea to make this mandatory throughout the country.

Tirupur, known as the largest textile exporting cluster in the country implemented world's largest ZLD plant for managing textile effluents[39]. The ZLD system in Tirupur connected 18 Common Effluent Treatment Plants covering 355 dyeing and bleaching industries and 95 individual effluent treatment plants. About 95% of permeate water is recovered from the Reverse Osmosis plant. In addition to recycling of treated wastewater to the textile units, glauber salt is recovered and sold. The salt recovery has helped to ensure financial viability during operations[40].

ZLD system in Tirupur helped in a sustainable growth of the industry while meeting most stringent regulatory norms. Reduction in fresh water demand from the Industry freed up water for meeting the agriculture and domestic demands.

ACTIVITY 6.8 DISCUSSION ON WASTEWATER RECYCLING AND ZERO LIQUID DISCHARGE

Due to difficulties in managing household hazardous waste (arising from cosmetics, medicines, cleaning agents, etc.), wastewater in urban areas has been found to contain objectionable hazardous substances. Stormwaters in urban areas carry contaminants due to indiscriminate waste disposal on the land. This has given a challenge in wastewater recycling as now advanced treatment systems have to be used to ensure safe wastewater reuse. What could be a solution to address this challenge?

While ZLD appears to be a solution to enforce wastewater recycling and material recovery, the ZLD systems have been rather energy intensive and lead to high GHG emissions. There are also issues regarding the disposal of the residues produced in the treatment process. Is ZLD therefore a sustainable solution?

In Europe and North America, the drive toward zero liquid discharge has been pushed by high costs of wastewater disposal at inland facilities. In India, ZLD directives have been issued by the judiciary, e.g., the National Green Tribunal, responding to the complains received from the affected communities[38].

India is taking aggressive actions to curb severe water pollution by issuing ZLD directives. According to a recent technical report, the ZLD market in India was valued at USD 39 million in 2012 and is expected to grow continuously at a rate of 7% from 2012 to 2017. In this market, the textile, brewing and distilling, power, and petrochemical industries are the major sectors of application areas[25].

ZLD is also considered when there is a potential for recovering resources that are present in wastewater. Some organizations target ZLD for their waste because they can sell the solids that are produced or reuse them as a part of their industrial process. For example, lithium has been found in USA oil field brines at almost the same level as South American salars. In another example, gypsum can be recovered from mine water and flue gas desalinization wastewater, which can then be sold to use in drywall manufacturing[34]. Box 6.8 describes a case study of ZLD plants at Tirupur in India.

6.13 R-PET Initiative

PET (polyethylene terephthalate) is the most common type of plastic resin. To create virgin PET, producers extract crude oil and natural gas from the Earth, then process and heat it to form a molten liquid. They spin this liquid into fibers to create polyester fabric, or they mold and solidify it into PET plastic containers[41].

rPET stands for *recycled* polyethylene terephthalate, or recycled PET. When PET packaging is discarded by the consumer, it becomes PET waste. The PET waste then makes it way to a MRF where it is sorted from other materials, baled, and sent to specific PET recycling facilities. At these new facilities, each PET product is washed, and contaminants are removed. They then get sorted according to color and are ground into flakes or made into pellets. These flakes and pellets are then sold as raw material that can be used for a range of polyester products (clothing, carpets, insulation, etc.) or made back into PET products.

The process of converting rPET to a virgin equivalent requires much less energy than glass, aluminum, or other materials. Although PET's feedstocks are derived from crude oil and natural gas, approximately 40% of that energy is trapped within the PET polymer for recapture and reuse every time PET is recycled. This means rPET leads to a greater conservation of raw materials

BOX 6.9 CASE OF GANESHA ECOSPHERE IN INDIA[43]

Ganesha Ecosphere (GESL) is a Kanpur based Post-consumer PET bottle scrap (Poly-Ethylene Terephthalate) recycling company in India. The company has been in the PET recycling business since 1995 and currently is the largest player in the PET recycling industry with a market share of 25%. The company follows strategies such as:

- Technological advancement: The company continuously focuses on deploying advanced technology (imported from Korea, China, and Germany) to consistently narrow the gap in quality between virgin and recycled polyester fiber and increase the variety of value-added products.
- Continuous supply of raw materials through symbiotic relationship with value chain actors: GESL has edge over its peers in sourcing raw materials (majorly post-consumer PET bottles) because of pan-India network of more than 20 collection centers, which provides about 40% of company's raw materials. The company has institutional tie-ups with hotels, malls, restaurants, and exhibition centers. The company has also joined hands with beverage giants like Bisleri and Coca-Cola India for sourcing waste PET bottles.

Finally, the company also engages with informal sector, i.e., mainly waste pickers, to ensure uninterrupted supply of raw materials, i.e., waste.

BOX 6.10 CASE OF PATAGONIA – R-PET

Patagonia began making recycled polyester from plastic soda bottles in 1993. It was the first outdoor clothing manufacturer to transform trash into fleece. Patagonia recycles used plastic bottles, unusable manufacturing waste and worn-out garments (including their own) into polyester fibers to produce clothing. Recycled polyester is used in many of products, including hard shells, boardshorts, fleece, and Capilene® baselayers. For the Fall 2020 season, 84% of Patagonia's polyester fabrics were made with recycled polyester. As a result, company could reduce the CO_2e emissions by 8% compared to virgin polyester fabrics. This amounted to more than 11 million pounds of CO_2e[44].

Through its Common Threads Partnership, Patagonia aims to reduce, repair, reuse and finally recycle. If you have Patogonia clothes that are at the end of their useful life, bring them to a store or mail them in to be recycled. In 2016, it was reported that Patagonia has recycled more than 82 tons of clothing since 2005, sometimes incorporating them into their own new clothes. The company even sells secondhand Patagonia clothing at its Portland, Oregon, store through a trade-in program[45].

and a reduction in GHG emissions by 65% PET virgin bottle resin pellets between 83 and 85 cents a pound, compared to only 58–66 cents a pound for rPET pellets[42].

In 2017, Textile Exchange's Recycled Polyester Round Table announced an rPET Commitment to encourage brands and retailers to publicly commit to accelerating their use of recycled polyester by 25% by 2020. 59 renowned textile, apparel, and retail companies—including major brands such as Adidas, Dibella, Eileen Fisher, Gap Inc., H&M, IKEA, Lindex, MetaWear, Target, and Timberland—committed to or are supporting an increase in their use of rPET by at least 25% by 2020.

The PET recycling industry in India has grown to INR 35 billion (as of 2017), processing ~0.85 million MT of waste, with the organized sector cornering approximately 65% share of the total pie. Box 6.9 describes case of Ganesha Ecosphere in India. Box 6.10 highlights an international example of Patagonia.

6.14 Recycling Organizations

A number of recyclers associations have played an important role in supporting the recycling business, establish an ecosystem of recyclers, collate data, develop knowledge products and do policy advocacy with the national governments. Such organizations are presented in this section. Refer to Box 6.11.

BOX 6.11 SOME IMPORTANT RECYCLING ASSOCIATIONS

SUSTAINABLE RECYCLING INDUSTRIES

The Sustainable Recycling Industries (SRI) program builds capacity for sustainable recycling in developing countries by supporting national initiatives and implementing pilot projects in our partner countries. The program is funded by the Swiss State Secretariat of Economic Affairs (SECO) and is implemented by the Institute for Materials Science & Technology (Empa), the World Resources Forum (WRF) and Ecoinvent. It builds on the success of implementing e-waste recycling systems together with various developing countries since more than ten years[46].

SRI improves local capacity for sustainable recycling together with private and public institutions, as well as the informal sector in Colombia, Egypt, Ghana, India, Peru, and South Africa. SRI facilitates a stakeholder consultation for the development of sustainability criteria for secondary raw materials. It also develops basic data for the assessment of environmental and social life cycle performance for industrial activities through the improvement of local and regional expertise in Brazil, India, and South Africa.

Website: https://www.sustainable-recycling.org/

BUREAU OF INTERNATIONAL RECYCLING

The Bureau of International Recycling is the only global recycling industry association representing around 800 companies and 35 affiliated national recycling federations from 70 different countries. Its members are world leaders in the supply of raw materials and a key pillar for sustainable economic development.

Website: http://www.bir.org

THE INSTITUTE OF SCRAP RECYCLING INDUSTRIES, INC.

The Institute of Scrap Recycling Industries is the voice of the recycling industry promoting safe, economically sustainable and environmentally responsible recycling through networking, advocacy, and education.

Website: https://www.isri.org/about-isri

GLOBAL RECYCLING FOUNDATION

The Global Recycling Foundation supports the promotion of recycling, and the recycling industry, across the world to showcase its vital role in preserving the future of the planet.

GRF's mission is to fund educational and awareness programmes, which focus on the sustainable and inclusive development of recycling, across the world.

Website: https://www.globalrecyclingfoundation.org/

BOX 6.12 REMANUFACTURING INDUSTRIES COUNCIL IN THE USA[47]

The Remanufacturing Industries Council (RIC) established in the United States is a strategic alliance of businesses and academic institutions that works across industry sectors to support the entire remanufacturing industry through a combination of collaboration, education, advocacy, and research. RIC is committed to:

- Promoting the growth of all sectors of the remanufacturing industry.
- Providing a forum for members to identify and address issues of common interest, benchmark and share best practices.
- Building a strong network to stay informed of developments in remanufacturing technology.
- Providing education and training to the industry.
- Working to increase awareness of the benefits of remanufacturing in governments and for the general public.
- Continuing to advocate with policy-makers and government regulators to minimize barriers to US markets and foreign trade.
- Increasing RIC effectiveness by expanding RIC membership to include new businesses and business sectors.

Remanufacturing is a product recovery strategy resulting in end-of-life products being returned to in a new condition or better and receiving a warranty at least equivalent to the original. The responsibility of classifying a product as remanufactured is left to individuals and organizations and so potential exists for products to be incorrectly labeled as remanufactured. Here remanufacturing councils have an important role to play. Boxes 6.12 and 6.13 describe profiles of two important remanufacturing councils in the USA and EU, respectively.

6.15 End of Life of Vehicles

End of Life of Vehicles (ELVs) are classified as hazardous waste and represent a category of waste whose processing is especially difficult because of their complex structure and varied composition.

BOX 6.13 THE CONSEIL EUROPÉEN DE REMANUFACTURE[48]

The Conseil Européen de Remanufacture represents small and large businesses from all remanufactured product sectors in EU. It set the vision to triple the value of Europe's remanufacturing sector to EUR 100 billion by 2030. Its mission is to become the focal point for remanufacturing policy dialogue in Europe.

The Council publishes its annual recommendations on research priorities for national and EU-level innovation funding that will most benefit remanufacturing in Europe. The technical detail of these research priorities has been developed by the European Remanufacturing Network (ERN) of universities and research organizations. ERN was established under the Horizon 2020 Framework by the European Commission. Horizon 2020 is the financial instrument implementing the Innovation Union, a Europe 2020 flagship initiative aimed at securing Europe's global competitiveness.

In 2000, a directive was passed by the EU regarding the treatment of end-of-life vehicles. All companies, domestic and foreign, manufacturing and or selling automobiles in the EU were made to follow and comply with the guidelines stated in the directive. Directive 2000/53/EC (Directive 2000/53/EC – the "ELV Directive") on end-of life vehicles aims at making dismantling and recycling of ELVs more environmentally friendly. It sets clear quantified targets for reuse, recycling and recovery of the ELVs and their components. It also pushes producers to manufacture new vehicles without hazardous substances (in particular lead, mercury, cadmium, and hexavalent chromium), thus promoting the reuse, recyclability and recovery of waste vehicles (see also Directive 2005/64/EC on the type-approval of motor-vehicles with regards to their reusability, recyclability, and recoverability) [49].

In 2017, scrapped passenger cars and light goods vehicles in the EU weighed a total of 5.7 million tons; 94% of parts and materials were reused and recovered, while 88% of parts and materials were reused and recycled[50].

Under the Green New Deal and Circular Economy Action Plan, EU wants to revise the ELV rules. They have committed to creating a roadmap. Currently the feedback period is going on.

- The European Green Deal identifies vehicles as one product where "the Commission will consider legal requirements to boost the market of secondary raw materials with mandatory recycled content".
- The Circular Economy Action Plan also indicates that "the Commission will also propose to revise the rules on end-of-life vehicles with a view to promoting more circular business models by

linking design issues to end-of-life treatment, considering rules on mandatory recycled content for certain materials of components, and improving recycling efficiency".

ELV Law in Japan[51]

The creation of the Japanese Automobile Recycling Law was influenced heavily by European system. There are however a couple of key differences between the two systems. The first difference is in Japan every vehicle owner is obligated to pay the recycling fee in advance and second, the Japanese system relies on "manufacturer responsibility". "Extended Product Liability" is simply the idea that the manufacturer of the vehicle is responsible for disposal and dismantling of the vehicle in a way that does not harm the environment.

ELV in China[52]

Automobile ownership showed a compound annual growth rate (CAGR) of 14.7% in China between 2010 and 2018, and it will continue to rise in the upcoming years, up to 379.2 million units as expected in 2025, despite a decline in both production and sales over the past 2 years[53].

In such a huge automotive market, Chinese ELV and dismantling market springs up. As estimated, there were a total of 8 million end-of-life vehicles in China in 2018, with a scrap rate of 3.5%; but a mere 30% of them were recycled, far lower than virtually 80% in the developed countries. In the first eleven months of 2019, 1.739 million ELVs were recycled in China, a like-for-like spurt of 18.4%, and with the full-year recycling rate expectedly hitting 21%.

Among 731 Chinese car dismantling firms announced by the Ministry of Commerce of China on June 21, 2019, most are small sized with low annual recycling rate of ELVs and scattered resources, though a national ELV dismantling network has already taken shape.

On April 22, 2019, the State Council announced the Measures for the Management of End-of-Life Vehicle Recycling, a policy allowing the recycling and remanufacturing of "five assemblies" (engine assembly, steering assembly, transmission assembly, front and rear axles, and frame) which can be sold to enterprises with remanufacturing capabilities in the light of relevant national regulations.

ACTIVITY 6.9 DISCUSSION ON ELVS

How is the end of life for a vehicle defined? Who decides the end of life? Should ELV be addressed as EPR?

6.16 Circular Business Initiatives in Closing the Loop

Several businesses have emerged in the sector of recycling and close the loop. Presented in this section are three interesting examples. See Boxes 6.14, 6.15, and 6.16.

Banyan Nation is an example that shows how smart cleaning technology, IT based platform and close cooperation with the informal sector can lead to a business that is profitable, providing significant environmental benefits and inclusive. Such business models should be promoted and replicated in closing the loop.

BOX 6.14 BANYAN NATION IN INDIA

Banyan Nation was incepted in 2013 as a start up in Hyderabad and had close to 50 employees. Today, it has recycled more than 7 million pounds of plastics and integrated more than 2000 informal sector waste workers in their value chain.[54]

Banyan Nation has developed thermal cleaning technology that removes contaminants like metals, labels, auto paints, inks, dirt, oils, adhesives, etc. using environment friendly detergents and solvents to supply near virgin quality recycled granules. This segregation is the most critical part as plastic contains different types of resins hence it needs to be separated systematically to avoid the contamination.

Banyan Nation is one of India's first vertically integrated plastic recycling companies that helps global brands use more recycled plastic instead of virgin plastic. Their trademarked plastic cleaning technology converts collected post-consumer and post-industrial plastic waste into high quality recycled plastic granules that they call Better Plastic™. It is comparable in quality and performance to virgin plastic. The brands can use this to make new products and packaging. Its award-winning data intelligence platform integrates thousands of informal recyclers into their supply chain and also helps cities manage their waste more effectively[55].

They do this with the help of the informal sector and the internet of things technology. Banyan Nation built an app that maps more than 1500 stationary recyclers. This enabled them to collect data such as the amount of waste coming out of houses, local efficiencies of collection, and transportation of waste. This enabled them to integrate their supply chain and involve the informal sector in a formal system.

BOX 6.15 BLACK BEAR CARBON[56]

The Black Bear process uses end-of-life tires to produce carbon black. Each year more than one billion tires are removed from vehicles, constituting around 13.5 million tons of solid waste as tires contain around 25% carbon black. This means that this waste-stream contains approximately 3.4 million tons of carbon black. Black Bear partnered with established tire collectors who can use the Black Bear technology to turn this vast waste stream into a valuable raw material. Before the tires enter the Black Bear process, the steel component is removed and recycled, and the rubber is converted into granulate.

Traditional furnace or "virgin" carbon black is produced from fossil fuels, by partially burning crude oil under controlled conditions. Depending on the grade it takes between 1.8 and 2.5 liters of fossil fuels to produce 1 kg of carbon black. The standard way of producing carbon black is thus very polluting. Furnace carbon black manufacturers emit around 38 million tons of CO_2 per year.

Approximately 7.2 million tons of carbon black consumed by the tire industry every year creates an additional environmental issue because after use the tires are mostly burned or landfilled, polluting the environment and wasting the carbon black "resource".

In contrast, the Black Bear process uses end-of-life tires as feedstock to produce consistent, high-quality carbon blacks. This circular economy approach not only massively reduces CO_2 emissions but also helps solve an important waste management problem. Black Bear claims that the recycling or conversion process produces more energy than it consumes.

Box 6.15 describes an interesting business operated on a scale by Black Bear. In this model, the business focuses on tires, establishes a network of used tire suppliers and uses cutting edge technology for reprocessing to produce carbon black. The business model shares commonalities with Banyan Nation.

There are waste streams that are difficult to recycle. Box 6.16 showcases the example of TerraCycle that has demonstrated that recycling of such waste streams is possible, even if not operated on a scale. This example also shows a platform approach that involves multiple stakeholders and mobilizes finance for the benefit of society.

Chapter 5 on extending product life and Chapter 6 on closing the loop, provided an overview of the opportunities in circular economy. These opportunities are specific to each sector. Chapter 7 describes such opportunities in sectors such as textile, steel, agriculture, and food.

BOX 6.16 TERRACYCLE – ADDRESSING WASTE STREAMS THAT ARE HARD TO RECYCLE[57]

TerraCycle was founded in 2001 in Trenton, USA as an organic fertilizer producer, using waste as the resource for production. Over the years, TerraCycle evolved into a world leader in the collection and recycling of waste streams that are traditionally considered not-recyclable.

TerraCycle typically sets up national collection platforms for hard-to-recycle waste streams. This platform is usually funded by consumer product companies, retailers, cities, manufacturing facilities, distribution centres, small businesses that are looking to enhance their environmental performance. Where possible, TerraCycle and their partners focus on how to integrate reclaimed materials into specific products.

To date, more than 80 million people in 21 countries have helped TerraCycle to divert approximately 5 billion pieces of waste from landfills and incineration and in result raise more than 21 million dollars for charities and schools around the world.

6.17 Key Takeaways

- Waste segregation is extremely important to close the loop.
- MRF help in organized waste collection, sorting, processing, and recycling. Creating a demand for recycled materials is however crucial.
- Informal sector, especially in developing and high population countries, is the backbone of circular economy. Business models are needed, that partner with the informal sector and are inclusive and gender sensitive. Skilling and provision of finance, especially the working capital, will play an important role.
- In order to push recirculation of materials, stipulations on recycled content in products are required. At the same time, a precautionary approach needs to be taken by restricting hazardous substances in products. This strategy helps not only to reduce risks to the consumers and the ecosystem but helps in cost-effective recycling.
- Creation of a market for secondary materials and recycled products is the key for economic and social development. Here, the recycling associations should take the lead and play an important role.
- The material flows in closing the loop, often cut across sectors and enter new markets. These cross sectoral strategies (e.g., like rPET) help in longer recirculation of materials.

- Due to globalization, the world of recycling is highly connected. Therefore, policy initiatives taken on waste streams in one country can affect the region or even the world. Green fencing in China is one such example.
- A number of innovative and successful business models exist that can inspire the recycling business. It will be important to analyse these models and provide guidance to the industry.

ADDITIONAL READING

1. **The evolution of mixed waste processing facilities: 1970-today** – This report talks about the advances in waste processing facilities. The key findings of this Report are:
   - Sortation technology continues to evolve and improve. This has enabled significantly higher diversion rates and more recoverable streams.
   - Recovery of high value materials, such as plastics and metals, has the potential to increase significantly via modern mixed waste processing facilities (MWPFs). Recovery rates for lower value materials, such as fiber/paper and glass, are likely to be reduced.

Source: GBB Solid Waste Management Consultants. 2015. *The Evolution Of Mixed Waste Processing Facilities: 1970-Today.* <https://plastics.americanchemistry.com/Education-Resources/Publications/The-Evolution-of-Mixed-Waste-Processing-Facilities.pdf> [Accessed 4 November 2020].

2. **Waste bank: Waste management model in improving local economy** – The aim of this study is to explain waste management model through waste bank and explain how households increase their welfare through this model. This research is a descriptive qualitative research with the purpose to get an explanation through in-depth interview from stakeholders, households and waste bank managers. The results reveal that waste bank management model is not only beneficial in making a clean environment, but also has an impact on local economy by increasing the income of housewives around the waste bank. The community expected more support from the government to improve the mechanism of waste bank and a better pricing model for the waste.

Source: Wulandari, D., Hadi Utomo, S. and Shandy Narmaditya, B., 2017. Waste bank: Waste management model in improving local economy. *International Journal of Energy Economics and Policy*, 7(3), pp. 36–41.

3. **Waste bank as community-based environmental governance: A lesson learned from Surabaya** – This paper discusses an implementation of waste bank as community-based environmental governance. Waste bank as a business is owned by people who consider waste as a valuable economic commodity & savings and have instruments that involve community in waste management. In Surabaya, waste bank grows rapidly and has supported community's livelihood and has encouraged people's self-reliance in environmental management. The objectives of this study are (1) analyze the role of waste bank in supporting community-based environmental governance; and (2) analyze how public engagement (community, government, and private sector) by waste banks implementation creates effective and collaborative environmental management.

Source: Wijayanti, D. and Suryani, S., 2015. Waste bank as community-based environmental governance: A lesson learned from Surabaya. *Procedia – Social and Behavioral Sciences*, 184, pp.171–179.

4. **Deposit systems for one-way beverage containers: Global overview (2016)** – This document provides a comprehensive summary of 38 deposit systems in different jurisdictions around the world. Each deposit market profiled in this review is structured along the following key parameters (where information is available):
   - Mandate (legal basis for the system)
   - Program Scope (beverages and containers included in the system)
   - Deposit and Fees (deposit amounts and different fees)
   - System Operator (composition and tasks of the System Operator)
   - Redemption System (beverage container take-back system)
   - System Results (return rates)
   - Money Material Flow (visual flow chart)

Source: Reloop Platform. 2017. *Deposit Systems For One-Way Beverage Containers: Global Overview.* <https://www.reloopplatform.org/wp-content/uploads/2017/05/BOOK-Deposit-Global-24May2017-for-Website.pdf> [Accessed 4 November 2020].

5. **Container deposits –** South Australia introduced its container deposit legislation in 1977. This official website handled by the South Australia EPA provides the legislation's history and progress up to date.

Source: EPA South Australia. n.d. *Container Deposits.* <https://www.epa.sa.gov.au/environmental_info/waste_recycling/container_deposit> [Accessed 4 November 2020].

6. **Using recycled content in plastic packaging: The benefits –** Consumers are increasingly concerned about the environmental impact of packaging, and many retailers are looking at recycled content in packaging as part of the solution. With this in mind, this document highlights a WRAP project that has demonstrated that recycled PET (rPET) can be successfully used in the production of new retail packaging.

Source: WRAP. n.d. *Using Recycled Content In Plastic Packaging: The Benefits.* <https://www.wrap.org.uk/sites/files/wrap/Using%20recycled%20content%20in%20plastic%20packaging%20the%20benefits.pdf> [Accessed 4 November 2020].

7. **Circular economy model framework in the European water and wastewater sector –** The paper presents a proposition for a new CE model framework in the water and wastewater sector, which includes the six following actions: reduction – prevent wastewater generation in the first place by the reduction of water usage and pollution reduction at source; reclamation (removal) – an application of effective technologies for the removal of pollutants from water and wastewater; reuse – reuse of wastewater as an alternative source of water supply (non-potable usage), recycling – recovery of water from wastewater for potable usage; recovery – recovery of resources such as nutrients and energy from water-based waste, and rethink – rethinking how to use resources to create a sustainable economy, which is "free" of waste and emissions.

Source: Smol, M., Adam, C. and Preisner, M., 2020. Circular economy model framework in the European water and wastewater sector. *Journal of Material Cycles and Waste Management*, 22(3), pp. 682–697.

8. **Toward the implementation of circular economy in the wastewater sector: Challenges and opportunities** – This review manuscript focuses on demonstrating the challenges and opportunities in applying a circular economy in the water sector. For example, reclamation and reuse of wastewater to increase water resources, by paying particular attention to the risks for human health, recovery of nutrients, or highly added-value products (e.g., metals and biomolecules among others), valorization of sewage sludge, and/or recovery of energy.

Source: Guerra-Rodríguez, S., Oulego, P., Rodríguez, E., Singh, D. and Rodríguez-Chueca, J., 2020. Towards the implementation of circular economy in the wastewater sector: Challenges and opportunities. *Water*, 12(5), p. 1431.

9. **Lithium-ion batteries toward circular economy: A literature review of opportunities and issues of recycling treatments** – This paper analyses the current alternatives for the recycling of Lithium-ion batteries, specifically focusing on available procedures for batteries securing and discharging, mechanical pre-treatments and materials recovery processes (i.e., pyro- and hydrometallurgical), and it highlights the pros and cons of treatments in terms of energy consumption, recovery efficiency, and safety issues. Target metals (e.g., Cobalt, Nickel, and Lithium) are listed and prioritized, and the economic advantage derived by the material recovery is outlined. An in-depth literature review was conducted, analysing the existing industrial processes, to show the on-going technological solutions proposed by research projects and industrial developments, comparing best results and open issues and criticalities.

Source: Mossali, E., Picone, N., Gentilini, L., Rodrìguez, O., Pérez, J. and Colledani, M., 2020. Lithium-ion batteries towards circular economy: A literature review of opportunities and issues of recycling treatments. *Journal of Environmental Management*, 264, p. 110500.

10. **Zero-liquid discharge (ZLD) technology for resource recovery from wastewater** – This review examines why a greater focus on environmental protection and water security is

leading to more widespread adoption of ZLD technology in various industries. It highlights existing ZLD processing schemes, including thermal and membrane-based processes, and discuss their limitations and potential solutions. It also investigates global application of ZLD systems for resource recovery from wastewater. Finally, it discusses the potential environmental impacts of ZLD technologies and provide some focus on future research needs.

Source: Yaqub, M. and Lee, W., 2019. Zero-liquid discharge (ZLD) technology for resource recovery from wastewater: A review. *Science of The Total Environment*, 681, pp.551–563.

11. **Ship Dismantling – A Status Report on South Asia** – This report aims to present the prevailing condition in ship breaking yards across the Indian subcontinent, particularly India and Bangladesh. The Pakistani ship breaking industry seems to be in irreversible decline and is therefore hardly mentioned in this report. The report sheds light on the ship breaking business, environmental standards, and working conditions in the yards. The current situation is described through the researcher's personal experience and recent observation of conditions in the yards and through examples. The largest ship breaking yard in India and in the world, Alang, is studied in the greatest detail.

Source: Kumar, R., n.d. *Ship Dismantling: A Status Report on South Asia.* Mott Macdonald, WWF. <https://www.shipbreakingplatform.org/wp-content/uploads/2018/11/ship_dismantling_en.pdf> [Accessed 4 November 2020].

Notes

1. Geyer, R., Jambeck, J. and Law, K., 2017. Production, use, and fate of all plastics ever made. *Science Advances*, 3(7), p. e1700782.
2. Pahl, C., 2020. *How AI and Robotics are Solving The Plastic Sorting Crisis.* Plug and Play Tech Center. <https://www.plugandplaytechcenter.com/resources/how-ai-and-robotics-are-solving-plastic-sorting-crisis/#:~:text=Robotic%20recycle%20sorting%20uses%20artificial,come%20along%20with%20human%20labor> [Accessed 4 November 2020].

3. Looptworks. n.d. *What Is Upcycling?* <https://www.looptworks.com/pages/what-is-upcycling> [Accessed 4 November 2020].
4. MAAR. 2018. *Differences Between Recycling, Upcycling and Downcycling.* <https://www.mariaarrieta.net/post/differences-between-recycling-upcycling-and-downcycling> [Accessed 4 November 2020].
5. Penn State. n.d. *Recycling: Open-Loop Versus Closed-Loop Thinking.* <https://www.e-education.psu.edu/eme807/node/624> [Accessed 4 November 2020].
6. WRAP. n.d. *Guidance for Material Recovery Facilities.* <https://www.wrap.org.uk/collections-and-reprocessing/collections-and-sorting/guidance/guidance-material-recovery-facilities-mrfs> [Accessed 4 November 2020].
7. Cavite Socio-Economic and Physical Profile. 2015. *Environment Sector.* <https://cavite.gov.ph/home/wp-content/uploads/2017/06/22_SEPP2015_Chapter8_Environment.pdf> [Accessed 4 November 2020].
8. ZenRobotics. n.d. *Intelligence With ZENBRAIN.* <https://zenrobotics.com/solutions/intelligence-with-zenbrain/> [Accessed 4 November 2020].
9. AMP Robotics. n.d. *Single-Stream Recycling* <https://www.amprobotics.com/single-stream> [Accessed 4 November 2020].
10. Tomra. 2018. *How Do Container Deposit Schemes Work?* <https://newsroom.tomra.com/how-do-container-deposit-schemes-work/> [Accessed 4 November 2020].
11. Department for Environment, Food and Rural Affairs. 2019. *Consultation on Introducing a Deposit Return Scheme in England, Wales and Northern Ireland.* <https://consult.defra.gov.uk/environment/introducing-a-deposit-return-scheme/supporting_documents/depositreturnconsultdoc.pdf> [Accessed 4 November 2020].
12. WBCSD. 2016. *Informal Approaches Towards A Circular Economy.* <https://docs.wbcsd.org/2016/11/wbcsd_informalapproaches.pdf> [Accessed 4 November 2020].
13. Aich, S., 2019. *Informal Workers: The Front Lines Of Enabling Circular Economies.* Medium. <https://medium.com/s3idf/informal-workers-the-front-lines-of-enabling-circular-economies-29a9e11e992f> [Accessed 4 November 2020].
14. Pune Municipal Corporation. n.d. *Swach.* <https://www.pmc.gov.in/en/swach> [Accessed 4 November 2020].
15. UNDP. n.d. *Plastic Waste Management Programme (2018–2024).* <https://www.in.undp.org/content/india/en/home/projects/plastic-waste-management.html> [Accessed 4 November 2020].
16. ITC Limited. 2018. *Initiatives In Solid Waste Management.* <https://www.itcportal.com/world-environment-day/pdf/WOW_Brochure_Text%20PDF_%20June%202018.pdf> [Accessed 4 November 2020].
17. Guiaen Vase. n.d. *The Green Dot System.* <http://www.guiaenvase.com/bases/guiaenvase.nsf/0/950B6ED17881D76EC1256F250063FAD0/$FILE/Article+Green+Dot+TTZ_+English.pdf?OpenElement> [Accessed 4 November 2020].
18. PRO Europe. n.d. *Member Countries.* <https://www.pro-e.org/member-countries> [Accessed 14 December 2020].
19. Brooks, A., Wang, S. and Jambeck, J., 2018. The Chinese import ban and its impact on global plastic waste trade. *Science Advances*, 4(6), p. eaat0131.
20. Resource Recycling News. 2018. *From Green Fence To Red Alert: A China Timeline.* <https://resource-recycling.com/recycling/2018/02/13/green-fence-red-alert-china-timeline/> [Accessed 4 November 2020].

21. Euroshore. n.d. *Environmental Ship Dismantling.* <https://euroshore.com/policy-statements/environmental-ship-dismantling> [Accessed 4 November 2020].
22. NGO Shipbreaking Platform. n.d. *Where Ships Go To Die.* <https://shipbreakingplatform.org/spotlight-swiss-focus/> [Accessed 4 November 2020].
23. Thakkar, M., 2012. *Alang Ship Recycling Yard Turns Into An Unlikely Shopping Destination For Bargain Hunters And Collectors.* The Economic Times. <https://economictimes.indiatimes.com/alang-ship-recycling-yard-turns-into-an-unlikely-shopping-destination-for-bargain-hunters-and-collectors/articleshow/17434674.cms?utm_source=contentofinterest&utm_medium=text&utm_campaign=cppst> [Accessed 4 November 2020].
24. European Chemicals Agency. n.d. *Understanding REACH.* <https://echa.europa.eu/regulations/reach/understanding-reach> [Accessed 4 November 2020].
25. European Commission. n.d. *Waste Electronic Equipment.* <https://ec.europa.eu/environment/waste/weee/index_en.htm> [Accessed 4 November 2020].
26. European Commission. n.d. *RoHS Directive.* [online] Available at: <https://ec.europa.eu/environment/waste/rohs_eee/index_en.htm> [Accessed 4 November 2020].
27. Boucher, J., 2020. *Guidelines On Recycled Content And Recyclability.* Food Packaging Forum. <https://www.foodpackagingforum.org/news/guidelines-on-recycled-content-and-recyclability> [Accessed 4 November 2020].
28. US General Services Administration. n.d. *Recycled Content Products.* <https://www.gsa.gov/governmentwide-initiatives/sustainability/buy-green-products-services-and-vehicles/buy-green-products/recycled-content-products> [Accessed 4 November 2020].
29. U.S. Food and Drug Administration. n.d. *Recycled Plastics In Food Packaging.* <https://www.fda.gov/food/packaging-food-contact-substances-fcs/recycled-plastics-food-packaging> [Accessed 4 November 2020].
30. ReMade in Italy. 2017. *Remade In Italy.* <https://www.remadeinitaly.it/wp-content/uploads/2017/01/Remade-in-Italy-EN.pdf> [Accessed 4 November 2020].
31. The Repair Association. n.d. *Standards Summary.* <https://repair.org/standards> [Accessed 5 November 2020].
32. Dell Technologies. n.d. *Recycled Materials.* [online] Available at: <https://corporate.delltechnologies.com/en-us/social-impact/advancing-sustainability/sustainable-products-and-services/materials-use/recycled-materials.htm> [Accessed 5 November 2020].
33. Reddy, S., 2016. Recycled and Recyclable Content Green Materials for Buildings – For Climate Protection. *International Journal of Engineering Sciences & Research Technology.* <http://www.ijesrt.com/issues%20pdf%20file/Archive-2016/October-2016/77.pdf> [Accessed 4 November 2020].
34. WRAP. n.d. *Calculating and Declaring Recycled Content In Construction Products.* <https://www.wrap.org.uk/sites/files/wrap/Rules_of_Thumb1.pdf> [Accessed 4 November 2020].
35. Whole Building Design Guide. n.d. *Purple Pipe.* <https://www.wbdg.org/FFC/ARMYCOE/TECHNOTE/technote26.pdf> [Accessed 4 November 2020].
36. Amutha, K., 2017. Sustainable chemical management and zero discharges. *Sustainable Fibres and Textiles,* pp. 347–366.
37. Tong, T. and Elimelech, M., 2016. The Global Rise of Zero Liquid Discharge for Wastewater Management: Drivers, Technologies, and Future Directions. *Environmental Science & Technology,* 50(13), pp. 6846–6855.

38. Saltworks Techologies. 2018. *What Is Zero Liquid Discharge & Why Is It Important?.* <https://www.saltworkstech.com/articles/what-is-zero-liquid-discharge-why-is-it-important/> [Accessed 4 November 2020].
39. Grönwall, J. and Jonsson, A.C. 2017. The Impact of 'zero' coming into fashion: Zero Liquid Discharge uptake and socio-technical transitions in Tirupur. *Water Alternatives* 10(2), pp. 602–624
40. Tamil Nadu Pollution Control Board. 2019. *Best Practice – Zero Liquid Discharge (ZLD) System In Textile Processing Units.* <https://tnpcb.gov.in/pdf_2019/TextileCETP_ZLD15519.pdf> [Accessed 4 November 2020].
41. EarthHero. 2017. *What's The Deal With Rpet?* <https://earthhero.com/whats-the-deal-with-rpet/> [Accessed 4 November 2020].
42. Waiakea Springs. n.d. *What Is Rpet Plastic?* <https://waiakeasprings.com/blog/what-is-rpet-plastic/> [Accessed 4 November 2020].
43. TERI. 2018. Circular Economy: A Business Imperative for India. < http://wsds.teriin.org/2018/files/teri-yesbank-circular-economy-report.pdf> [Accessed 4 November 2020]
44. Patagonia. n.d. *Recycled Polyester.* <https://www.patagonia.com/our-footprint/recycled-polyester.html> [Accessed 4 November 2020].
45. Lozanova, S., 2016. *How Patagonia Is Recycling Bottles Into Jackets.* Earth 911. <https://earth911.com/business-policy/how-patagonia-is-recycling-bottles-into-jackets/> [Accessed 4 November 2020].
46. World Resources Forum. n.d. *Sustainable Recycling Industries.* <https://www.wrforum.org/projects/sustainable-recycling-industries-sri/> [Accessed 4 November 2020].
47. Remanufacturing Industries Council. n.d. *About RIC.* <http://www.remancouncil.org/about-ric> [Accessed 4 November 2020].
48. Remancouncil. n.d. *Conseil Européen De Remanufacture.* <https://www.remancouncil.eu/index.php> [Accessed 4 November 2020].
49. European Commission. n.d. *End of Life Vehicles.* <https://ec.europa.eu/environment/waste/elv/index.htm> [Accessed 4 November 2020].
50. EuroStat. 2020. *End-Of-Life Vehicle Statistics – Statistics Explained.* [online] Available at: <https://ec.europa.eu/eurostat/statistics-explained/index.php/End-of-life_vehicle_statistics> [Accessed 4 November 2020].
51. 3R Corporation. n.d. *Japanese Automobile Recycling Law.* <https://www.3-r.co.jp/contents/en/en_recycle.html#:~:text=The%20Automobile%20Recycling%20Law%2C%20promulgated,in%20an%20eco%2Dfriendly%20manner.&text=In%202000%20a%20directive%20was,end%2Dof%2Dlife%20vehicles> [Accessed 4 November 2020].
52. PRN Newswire. 2020. *China End-Of-Life Vehicle (ELV) And Dismantling Industry Report, 2019-2025.* <https://www.prnewswire.com/news-releases/china-end-of-life-vehicle-elv-and-dismantling-industry-report-2019-2025-300985556.html#:~:text=In%20such%20a%20huge%20automotive,80%25%20in%20the%20developed%20countries.> [Accessed 4 November 2020].
53. Report Linker. 2019. *China End-Of-Life Vehicle (ELV) And Dismantling Industry Report, 2019-2025.* <https://www.reportlinker.com/p02590391/China-End-of-Life-Vehicle-ELV-and-Dismantling-Industry-Report.html?utm_source=PRN> [Accessed 14 December 2020].

54. Accenture. 2018. *Accenture Announces Winners of the Fourth Annual Circulars In Davos.* <https://newsroom.accenture.com/news/accenture-announces-winners-of-the-fourth-annual-circulars-in-davos.htm> [Accessed 4 November 2020].
55. Banyan Nation. n.d. *Banyan Nation.* <http://banyannation.com/#ourwork> [Accessed 4 November 2020].
56. Black Bear. n.d. *About Us.* <https://blackbearcarbon.com/about-us/> [Accessed 4 November 2020].
57. R to Pi Project. 2019. *TERRACYCLE: A Circular Economy Business Model Case.* <http://www.r2piproject.eu/wp-content/uploads/2019/05/TerraCycle-Case-Study.pdf> [Accessed 4 November 2020].

7

Circular Economy in Select Sectors

7.1 Textile

7.1.1 Overview

Global fiber production doubled over the last 20 years to 107 million meter in 2018 and is expected to reach 145 million meter in 2030[1]. The global textile market size was valued at USD 961.5 billion in 2019[2].

Clothing/apparel represents more than 60% of the total textiles used. The clothing industry employs more than 300 million people along the value chain; cotton production alone accounts for almost 7% of all employment in some low-income countries. In the last 15 years, clothing production has approximately doubled, driven by a growing middle-class population globally, and increased per capita sales in mature economies. The latter rise is mainly due to the "fast fashion" phenomenon – with a quicker turnaround of new styles, increased number of collections offered per year, and often lower prices[3].

Based on raw material, the clothing market has been segmented into cotton, chemical, wool, silk, polyester, hemp, and others. Natural fibers such as cotton and linen as well as synthetic fibers including polyester and acrylic polyamides, are primarily used for the manufacturing of household textiles. Synthetic fibers have dominated the fiber market since the mid-1990s when they overtook cotton volumes. Polyester had a market share of around 51.5% (55.1 million meter) of total global fiber production in 2018. Market share of recycled polyester increased from 8% in 2008 to around 13% in 2018[1].

The offshoring and outsourcing model that emerged with trade liberalization and expanded with the end of the textile quota system in 2005, resulted in rapid growth in the industry in Asia and other parts of the developing world. At the same time, Europe and North America experienced significant employment losses, and concerns about working conditions in global supply chains have persisted.

The apparel supply chain is global, comprised of millions of small, medium, and large manufacturers in every region of the world, all operating under pressure to hold down costs, innovate products and deliver on tight deadlines[4]. However, this type of business strategy not only brings the benefits of lower production costs, but also a series of risks, in the form of obligations. Consumers and the public ultimately view these retail companies as being responsible for the products, irrespective of who made them, and where[5].

Incorporating sustainability into the supply chain is becoming a key priority for many textile and apparel companies. For example, H&M, Patagonia, and The North Face have incorporated various approaches to enhance their levels of sustainable supply chain management. Circular supply chains have a great role to play to achieve competitiveness, consumer expectations especially on transparency and traceability in the textile business.

Environmental issues related to textile and garment sector include consumption of chemicals that are used in the cultivation of natural fibers, emissions in the production of synthetic fibers and hazardous chemicals released in the wastewater. When washed, some garments release plastic microfibers, which account for 0.5 million tons every year contributing to marine ocean pollution[6]. High water consumption has been another major concern.

To ensure sustainable practices, textile industries need to shift toward circular business models.

A circular textile demands[7]:

- Use of mono materials where possible and ensure that products made from multiple materials can be easily disassembled to aid product recyclability.
- Assess what substances and materials of concern are used in production that cause pollution and/or prevent recycling then work with suppliers to remove them.
- Consider how other waste in the supply chain from textile off cuts to packaging can be captured then reused or recycled through internal processes or working with partner organizations.
- Keep garments in use and reuse as long as possible through developing or participating in collection schemes and supporting the development of technologies to recycle used textiles back to 'good as new' raw materials. Provide complete and transparent information about where and by whom materials are sourced, transformed and assembled and improve traceability.

7.1.2 Circular Economy Opportunities

This section describes CE related opportunities and a few examples to explore at each stage of the life cycle.

7.1.2.1 Raw Material

New materials are being developed to replace or complement existing resource intensive raw materials and restrict chemicals that are hazardous posing difficulties in safe recycling of used clothing.

A Greenpeace Germany report released in 2018, called "Destination Zero: seven years of Detoxing the clothing industry", found that of 80 international brands that committed to reduce the amount of hazardous chemicals in their clothing production, all achieved significant progress. The companies represented 15% of global clothing production and included fashion, sportswear and luxury retailers, outdoor brands, and suppliers. Puma, Adidas, and Nike were among the first brands to react to Greenpeace's call to action, signing up to a "detox commitment"[8].

New materials and product innovation in the industries are mostly driven by large multinational enterprises in partnerships with small start-ups and supported by governments in some countries. LAUNCH NORDIC is an example of a partnership that has brought together government entities and 12 large brands (such as IKEA Group, Nike, and Novozymes) with the goal of accelerating innovation in sustainable materials[9].

Re:newcell, a start-up company supported by LAUNCH NORDIC, has found new ways of turning used cotton and viscose into biodegradable fibers, yarn and fabrics, and Qmilk has developed a technology to convert unused milk into a silk-like, biodegradable fabric. Swiss brand Qwstion created Bananatex using the fibers of the stem of the abaca tree – a tree of the banana family. They cultivate plants of the banana tree family known locally as "Banana Hemp" or "Abacá" in the Philippines within a natural ecosystem of sustainable forestry. They then process them into a material that is a viable alternative to synthetic fabric. The fabric is made from 100% natural banana fibers and is topped with a natural beeswax coating for a water-resistant finish[10]. Markets for the alternate natural fibers are however small and are expected to gradually grow.

Similarly, the fiber manufacturer Lenzing has developed a new fiber, based on cotton waste. The waste comes from the textile giant Inditex, which includes brands such as Zara, Pull&Bear, Massimo Dutti, and Bershka. The garments are sold in Inditex shops[11].

Clothing manufacturer NAECO is committed to keep oceans clean. In addition to transforming ocean plastic waste into luxury garments, they organize beach cleans, avoid single-use plastics, and in a stand against fast fashion, offer a 5-year guarantee on each of their clothing products. Each garment is also made from their unique and fully recyclable material which encourages consumers to carry out conscious end of product life disposal. NAECO operate on a small and transparent supply chain that is largely located within the United Kingdom. The company was a winner of the DHL Fashion Potential Award 2019[12]. DHL Award for International Fashion Potential in partnership with British Fashion Council, to support the global ambition of British fashion designers.

7.1.2.2 Manufacturing

Big players in the denim business, where water usage is especially high, have introduced new technologies and made fresh commitments to cut down on water usage[10]. For example, ozone is used to remove redeposited indigo that reduces subsequent washing and rinsing steps. Mechanical techniques, including hand-sanding have replaced stone washing processes with pumice stones and enzymes, and processing steps have been combined saving water and reducing time.

In June 2019, denim manufacturer Wrangler launched the first denim collection using its new Indigood foam-dyeing technology that is claimed to eliminate almost 100% of the water typically used in the indigo-dyeing process. Similarly, another US denim producer, Levi's, announced in August 2019, a new water action strategy that will focus on areas where water is needed the most and reduce consumption. The company has set a goal to reduce its cumulative water use for manufacturing by 50% in water-stressed areas by 2025[13].

Weseta Textil AG has developed a terry towel that weighs only 380 g/m^2 and at the same time has the absorbency of a 600 g/m^2 towel. This means that 45% less energy is required for washing[11].

Membrane technologies have made a headway for separation toward chemical recovery. The applications include dye bath recovery and reuse, spin finish and size recovery, etc.

Ecofoot has developed hybrid pigments composed of a dye chemically linked to a polymer particle that reacts with cellulose fibers at temperatures as low as 25°C. Together with hybrid pigments and auxiliaries, more than 50% of water in the intermediate and final rinses can be saved in the total process of preparation and dyeing[14].

Officina+39, an Italy based company, developed the sustainable dye range Recycrom using recycled clothing, fiber material, and textile scraps. It developed a sophisticated eight-step system (patent pending) in which all the fabric fibers are crystalized into an extremely fine powder that can be used as a pigment dye for fabrics and garments made of cotton, wool, nylon, or any natural fiber. Recycrom can be applied to the fabrics using various methods such as exhaustion dyeing, dipping, spraying, screen printing, and coating. Recycrom is applied as a suspension and hence can be easily filtered from water, thus reducing the environmental impact[14].

An economic and sustainable innovation comes from the Swedish company WeAre SpinDye[11]. In contrast to the previous practice of spinning polyester fibers and dyeing them afterwards with many environmental risks, the fiber is already dyed during the spinning process. The dyeing of yarns or fabrics becomes obsolete, which saves resources and costs.

The concept of the Pepwing company, on the other hand, is to merge polyester with the masterbatch without water. The result of this dye free masterbatch dyeing process is a dyed chip, which is then pressed out into yarn. The advantage of this innovation is that the in-process water consumption can be reduced by 50% and the fabrics produced have a higher color fastness. In addition, the end product is recyclable[11].

"Right first time" is a decisive term in textile dyeing process. It refers to getting the right shade of fabric the first time without the uncalled for need to re-shade or washing off to reduce the depth of shade. It has been found that by applying optimized dyeing recipes and practices, less dye, auxiliaries, and energy are used.

High right first time has many advantages[15]:

- Considerable savings in terms of energy, labor, and reprocessing costs.
- Reduction in pollution because of less dye in the effluent.
- Increased productivity as there is no question of re-processing, so adequate time to dye more production.

With the advent of new developments in dyes and modern dyeing machines with control, the right first time in dyeing process has considerably improved setting benchmarks of more than 90%.

Textile industry is one of the highest energy consuming sectors. Use of renewable energy in the textile sector is now increasing. A textile company in Atlanta in USA has prepared seven of its manufacturing plants to use 100% renewable electricity; and 89% of its overall electric power is renewably sourced, including power received directly from the grid, purchased as renewable energy certificates, or generated on-site[16]. Similarly, Sunshot Technologies is an on-site solar power solutions company in India, which provides comprehensive rooftop solar power plant solutions for commercial and industrial customers. One of their key installations in the textile sector include rooftop solar power plant set-up for Rieter, a Swiss textile machinery manufacturing brand located near Pune, India. The 2 MWp capacity rooftop solar power plant is expected to generate around 2.9 million units of electricity annually and will contribute to the environment by reducing 2500 tons of CO_2 every year[17].

7.1.2.3 Packaging

Approximately 14% of the plastic packaging used globally makes its way to recycling plants, and only 9% is recycled – while an estimated 33% is left in fragile ecosystems, and 40% ends up in landfill[18]. Unlike rigid plastics, flexible plastic packaging lacks a circular solution because it is often made by blending several materials, and too light weight for separation and recycling[19].

Compostable and recyclable alternatives to conventional plastic packaging help designers and fashion houses reinforce sustainability ethos throughout their supply chain.

Apparel company PVH has ambitious sustainability goals that include using 100% sustainably and ethically sourced packaging by 2025. In 2018, the company, reported that 74% of its packaging is now recyclable. Moreover, the PVH Dress Furnishings Group saved nearly 200 tons of plastic by reducing the thickness of its packaging polybags[20].

7.1.2.4 Transportation

The impact of items travelling repeatedly to and from delivery centers to customers' homes can add significantly to the carbon footprint of a garment. The emission footprint of garments has considerably expanded due to long transportations and hence requires consideration for how to deliver garments in the most efficient and sustainable manner possible.

Under the sharing economy concept, reverse logistics can be arranged on internet platforms to solve the lack of scale of returns trip and increase the likelihood of transportation operating at full capacity and by using the same trucks as for deliveries in the return trip. See Box 7.1 that describes the example of Mr. R.

Considering the ecological and social concerns, eco-labels were introduced in the market. Ecolabels are a specific type of product labeling that certifies the environmental and social performance of products and services. To be certified with an ecolabel, the product or service must demonstrate that it can reduce the overall environmental impact of its production or use through by fulfilling specific, predefined criteria. Lately, ecolabels that declare carbon footprint have emerged. There is plethora of eco-labels in the textile sector. Some focus on organic textile such as Global Organic Textile Standard (GOTS)[22] and consortiums like Oeko-TEX who have standards

BOX 7.1 REVERSE LOGISTICS BUSINESS OF MR R IN CHINA[21]

Mr. R (the registered trademark) is a company dealing with reused clothing collecting, sorting, reselling, and recovery in the Chengdu area. Chengdu is a large city in China with more than 16,045 thousand population in 2017.

Mr. R's Reverse Logistics System is based on mobile application (APP) (Abbreviation of the SAAS [Software as a Service] software) which it has developed. The APP aggregates strategic alliance partners on the platform by sharing their deliveries capacity to lower transportation costs and also greenhouse gas (GHG) emissions.

In Mr. R's Reverse Logistics System, there are more than 240 convenience stores, 80 laundries, 260 express stations, and 45 post offices, this amounts to a total of about 625 alliance members in Chengdu and covering more than 200 communities. This system provides convenience for locals who want to dispose of end-of-life clothing whenever and wherever. The annual collection of recycled clothing is more than 3000 tons.

and certification schemes that aim to give confidence to the customer. Requirement of eco-labels by brand owners has influenced the supply chains especially on the choice of chemicals, manufacturing processes, and codes of conduct, raising thereby the sustainability quotient.

7.1.2.5 Consumption

The textiles value chain operates in an almost completely linear way. Large amounts of non-renewable resources are extracted to produce clothes that are often used for only a short time, after which the materials are mostly sent to landfill or incinerated. It is estimated that more than USD 500 billion of value is lost every year due to clothing underutilization and the lack of recycling. Less than 1% of material used to produce clothing is recycled into new clothing[3]. Most of this recycling consists of cascading to other industries and use in lower-value applications, for example, insulation material, wiping cloths, and mattress stuffing – all of which are currently difficult to recapture. In addition, there are significant costs for landfilling. Even though some countries have high collection rates for reuse and recycling (such as Germany, which collects 75% of textiles), much of the collected clothing in such countries is exported to low- and middle-income countries with no collection infrastructure of their own and hence leading to pressure on the landfill systems.

The resale market in the garment industry is on the rise. According to a report by resale platform ThredUp, it has grown 21 times faster than first-hand fashion retail over the past three years and is expected to grow from 24 billion US dollars to 51 billion US dollars in the next five years. According to the company statistics, 100 million garments have been processed so far resulting in approximately 403 million kgs of CO_2e displaced and USD 2.8 billion off estimated retail value[23].

Exclusive Sample Sales has a similar mission. Launched by Lauren Rogers and Rachel Amis in 2016, the company offers a service for fashion retailers to sell their excess inventory via a sample sale[10].

7.1.2.6 Post-Consumer Use

Recycled fibers can be created either from pre-consumer waste or post-consumer waste. One of the biggest hurdles that arises from the latter is that the process can be complicated, energy intensive and expensive to separate blends - a mix of two or more types of textile material.

The H&M Foundation teamed up with research institute Hong Kong Research Institute of Textiles and Apparel, to explore methods to separate and recycle fiber blends made from the three main types of textile material: synthetics, cellulose, and animal fibers. In 2017, the partnership discovered a hydrothermal recycling system that could fully separate and recycle cotton and polyester blends into new fibers and cellulose powder[10].

Italian company Aquafil is also on the frontline of textile upcycling. It uses a regeneration and purification process to turn nylon from waste such as fishing nets, fabric scraps, carpet flooring, and industrial plastic from landfills and oceans into Econyl regenerated nylon, an upcycle material that has been used to create catwalk collections for high fashion brands such as Gucci. It is part of an initiative called the Healthy Seas that has helped collect 375 tons of fishing nets to be recycled between 2013 and 2018[10].

7.1.3 International Initiatives

In 2017, the Ellen MacArthur Foundation launched the Circular Fiber initiative together with H&M, Nike and Lenzing and other key stakeholders to bring a circular economy approach to scale in the textile industry. In a report published in 2017, the initiative specifically calls for[3]:

- Phasing out of microfiber plastics and other substances of concern.
- Increased clothing utilization.
- Improved recycling by transforming clothing design, collection, and reprocessing.
- More effective use of resources and a move to renewable energy.

In 2017, Textile Exchange's Recycled Polyester (rPET) Round Table created an rPET Commitment to encourage brands and retailers to publicly commit to accelerating their use of recycled polyester by 25percent by 2020. 59 renowned textile, apparel and retail companies – including major brands such as adidas, Dibella, Eileen Fisher, Gap Inc., H&M, IKEA, Lindex, MetaWear, Target and Timberland—committed to or are supporting an increase in their use of rPET by at least 25% by 2020[24]. In India, Reliance Industries as part of its commitment to circular economy has launched R. Elan – Fashion for Earth where the fibers are made out of 100% PET bottles[25]. The brand was unveiled at Lakme Fashion Week in 2018.

In 2018, fashion stakeholders, under the auspices of UN Climate Change, worked to identify ways in which the broader textile, clothing, and fashion industry can move toward a holistic commitment to climate action. They created the Fashion Industry Charter for Climate Action, containing the vision to achieve net-zero emissions by 2050[26].

The United Nations Alliance for Sustainable Fashion is an initiative of United Nations agencies and allied organizations designed to contribute to the Sustainable Development Goals through coordinated action in the fashion sector. Specifically, the Alliance works to support coordination between UN bodies working in fashion and promoting projects and policies that ensure that the fashion value chain contributes to the achievement of the Sustainable Development Goals' targets. The scope of the Alliance's work extends from the production of raw materials and the manufacturing of garments, accessories, and footwear, to their distribution, consumption, and disposal[27].

7.2 Steel

7.2.1 Overview

Steel is an alloy of iron and carbon containing less than 2% carbon and 1% manganese and small amounts of silicon, phosphorus, sulfur, and oxygen[28]. Steel's durability is one of the key properties that makes it the world's most important engineering and construction material.

Study by the World Steel Association (worldsteel) shows that the steel industry contributes an estimated 3.8% of global GDP. Of the 3.8% of global GDP, less than 20% is generated in the steel industry itself. The remaining 80% is generated in industries that supply to the steel industry, or industries that depend on steel as a key input to its production of goods and services. Globally, it is estimated the steel industry enables 96 million jobs. This is about 3% of all globally employed persons according to the International Labour Organisation[29].

Steel is the most recycled material in the world, more than aluminum, paper, glass, gas, and plastic combined. Steel scrap from lower value can be converted into high value steels by using appropriate processing and metallurgy[30]. Recovered co-products can be reused during the steelmaking process or sold for use by other industries. This prevents landfill waste, reduces CO_2 emissions, and helps preserve natural resources. The recovery and use of steel industry co-products have contributed to a material efficiency rate of 97.6% worldwide[31]. On average new steel products contain 37% recycled steel. Although all available steel scrap is recycled, there is not enough scrap available to meet demand for new steel products. This is due to the long life of steel products, given steel's strength and durability. Around 75% of steel products ever made are still in use today[32].

Over the years, due to technological improvement in steelmaking and strict environmental regulations, emphasis on raw material quality and new markets coupled with innovative ideas on waste reduction and rescue have resulted in drastic reduction in the quantity of waste generated in steel works from 1200 kg to less than 200 kg per ton of crude steel and recycling rates have reached to 95–97% in some parts of the world. Few steel industries have approached to cent percent waste utilization without discharge of any waste to environment[33].

In steel industry all three types of waste materials are generated, i.e., solid, liquid, & gaseous wastes. The generation quantity of various types of waste materials differ from one steel plant to other depending upon the steelmaking processes adopted and pollution control equipment installed. The most common type of wastes generated in steel plant are as follows[33]:

- Solid wastes like, hot metal pre-treatment slag, dust, sludge, mill scale, refractories, scrap, muck & debris, etc.
- Liquid wastes like industrial effluent, oil, grease, etc.
- Gaseous wastes like flue gases, fume extraction, etc. However, waste disposal and dump are very big issue for environment today and therefore, these wastes are being tried for reuse through recycling & utilization.

Until the last decade, the slag, dust and sludge generated by integrated steel plants was called waste, but now this term has been replaced with "by-product" due to intensive re-utilization of these wastes[34]. Slag is a waste which is generated during manufacturing of pig iron and steel and is classified as a potential by-product[35].

7.2.2 Circular Economy Opportunities

The lifecycle of steel broadly involves raw material extraction, steel production, consumption, reuse & remanufacture, and steel scrap recycling. Each lifecycle stage involves various stakeholders. For example, the stakeholders of the raw material extraction stage include the local communities around mines, Local, State, & National governments and steel producers who source iron ore from the mines[36].

7.2.2.1 Raw Material Extraction

According to the World Steel Association, it typically takes 1.6 MT of iron ore and around 450 kg of coke to produce a ton of pig iron, the raw iron that comes out of a blast furnace[37]. Steel scrap is also one of the industry's most important raw materials. It comes from demolished structures, End-of-Life Vehicles (ELVs), machineries as well as from the yield losses in the steelmaking process.

ArcelorMittal BioFlorestas (a company in the ArcelorMittal group) has been cultivating renewable eucalyptus forests on areas once degraded by cattle farming activities. The effect has been to rebuild soils, increase biodiversity, and improve the overall health of the ecosystem. The fast-growing trees, ready for harvesting in 5–10 years, are then converted to charcoal to create "carbon neutral steel", so called as the carbon sequestration during the growth of the forest (not only trees, but other plants) matches or even exceeds the carbon released during combustion for the steel production process.

Brazil, through ArcelorMittal's activities, is the first and only country in the world using renewable biomass fuel from certified FSC forests. The FSC certification also guarantees that decent working conditions are in place for the laborers involved in the timber harvesting and charcoal production[38].

7.2.2.2 Steel Production

Based on the manufacturing process, there are two key steelmaking routes globally:

a. Blast Furnace-Basic Oxygen Furnace (BF-BOF): The primary steelmaking route where steel is produced from raw materials like iron ore, coal, and limestone.
b. Electric Arc Furnace/Induction Furnace (EAF/IF): The secondary steelmaking route where steel is produced from steel scrap, sponge iron or Direct Reduced Iron.

The EAF route is more environmentally efficient due to its lower carbon footprint, material efficiency and energy efficiency. The EAF-route enables steel scrap recycling, and thereby drives participation in a resource efficient circular economy. Improved efficiency in steel recycling is a critical opportunity[39].

Promoting EAF technology through economic instruments such as tax benefits and increasing scrap recycling through a dedicated policy[40] addressing the stakeholders such as the informal sector are important elements in increasing resource efficiency and circularity in steel.

7.2.2.3 Steel Recycling

Steel recycling uses 74% less energy, 90% fewer virgin materials and 40% less water. It also produces 76% fewer water pollutants, 86% fewer air pollutants and 97% less mining waste.

Co-products from the steel industry have many uses within the industry itself, in other industries and in wider society. In some cases, it is the physical properties that determine the use, such as steelmaking slag used as aggregates in road construction; and sometimes it is the chemical composition e,g. process gases used as fuel to produce heat and/or electricity. Valuable nonferrous metals can also be recovered from slags, dust and sludge when the concentrations are sufficient[41].

Some examples of common uses of steel industry co-products:

- Blast furnace slag – substitute for clinker in cement-making.
- Steelmaking slag – aggregates in road construction, soil improvement.

- Process gases – heat and electricity production.
- Dust and sludge – internal and external use of iron oxides and alloying elements.
- Petrochemicals from coke making – tar, ammonia, phenol, sulfuric acid, and naphthalene for the chemical industry.
- Emulsions from mills and used oil – reducing agent in blast furnaces or used in coke plants.
- In all cases, using a steel industry co-product as a substitute for an equivalent product will improve resource efficiency and contribute to the circular economy.

JSW Steel is a noteworthy example. JSW Steel is India's largest steel exporter, shipping to more than 100 countries across 5 continents. The Steel Sustainability Champions programme has recognized JSW Steel among six companies that have set the example in making tangible impacts in supporting sustainable development and a circular economy.[42]

Box 7.2 provides an example of recycling efforts taken by Tata Steel.

BOX 7.2 RECYCLING AT TATA STEEL[43]

Tata Steel is amongst the largest steel re-processors in the world. In South Wales, Tata Steel has a department dedicated to recycling packaging steel. It offers an end-market for used packaging steel, ensuring that the company plays a part in closing the steel recycling loop. Nearly, 650 million tons of steel are recycled annually.

Tata Steel in India has significantly increased its LD slag utilization in Sinter making and other various industrial applications during 2020. Further, the company has achieved 100% utilization of fly ash through cement plants, brick, & paver block making and using for highway road construction.

Tata Steel is commissioning a scrap recycling plant of 0.5 MnTPA capacity, in Rohtak, Haryana, in India. The initiative aims to provide the much-needed raw material fillip to the steel industry by making available quality processed ferrous scrap, streamlining the currently unorganized scrap supply chain, lowering the dependency on imports, and enhancing the transparency and efficiency in the entire value chain. The scrap would be procured from various market segments such as ELVs scrap, obsolete household Scrap, construction & demolition scrap, industrial scrap, etc. This scrap would be processed and supplied to Electric Arc Furnace, Induction Furnace, and Foundries for downstream steel making., satiating their longstanding demand[44].

The automobile sector, especially ELVs are a key source of recycled steel. From light vehicles, steel accounts for 65% of the weight of the vehicle. Therefore, their efficient recovery is of importance[45]. Chapter 6 describes various ELV related regulations in China, Germany, and India.

In many countries, the ELV sector is largely informal and is concentrated in hubs. A major challenge faced by ELV operators is that their capacity is hampered by their limited access to professional development and financial opportunities for business expansion and development[46]. The key concern in attaining the circular economy for steel is the existing value chain, which is unorganized and fragmented:

a. Scrap is collected from various sources at a common site where sorting, segregation and basic processing (all manual operations) are performed. These operations currently do not conform to any safety and environmental regulations.
b. The lack of processing infrastructure results in low quality scrap from domestic sources. This scrap is unclean, has low density and is high in contaminants. Scrap processing includes steps like shredding, baling, shearing, etc.
c. Hence, high quality steel scrap is imported by large players. Having a consistent domestic supply will allow for import substitution.

7.2.3 International Initiatives

Worldsteel has a number of initiatives related to recognizing best sustainable practices in the steel industry.

- In November 2020, worldsteel published Sustainable steel – Indicators 2020. The publication features the steel industry's sustainability performance via its 8 sustainability indicators and focuses on 3 key steel applications from a life cycle perspective: automotive, construction and packaging. A total of 104 steel companies representing 1.1 billion tons of crude steel production contributed data – this covers nearly 60% of global crude steel production[47].
- Additionally, since 2018, worldsteel is updating its life cycle inventory data for 17 steel products on an annual basis. This ensures that customers and stakeholders have the most up-to-date data available when making their material choices[48].
- Every year, during its annual conference in October, worldsteel hosts the Steelie Awards to recognize the contributions and achievements of companies and individuals in seven categories including technology, education and environment – all areas that are vital for the industry to remain sustainable. The Excellence in LCA award recognizes companies that have played a key role in establishing

and guiding the work of worldsteel in LCA demonstrating their commitment to the ethical and pro-active use of LCA and shaping the debate in the public and policy-making arenas. Member nominations are called for by worldsteel and judged by an external expert panel[49].

- worldsteel has also launched an industry-wide sustainability recognition program called Champions Program, to encourage steel companies to increase their efforts, set higher standards and make further progress. The program is an effective vehicle to demonstrate stronger commitment and proactive engagement of the industry to sustainable development. Program objectives[50]:
    - To have a holistic picture of the sustainability effort of the industry and steel companies around the world.
    - To demonstrate stronger commitment and proactive engagement of the steel industry.
    - To go beyond participation-only recognition – looking at commitment, measurement, and actions all together.
    - To encourage steel companies to aim higher and do more in their sustainability activities.

The Association of European Producers of Steel for Packaging (APEAL) is committed to working with all relevant stakeholders to ensure understanding and support for steel as a sustainable and resource efficient packaging solution. Founded in 1986, APEAL members – ArcelorMittal, Liberty Liège-Dudelange, Tata Steel, thyssenkrupp Rasselstein and U.S. Steel Košice – employ over 200,000 workers in Europe, 15,000 of whom are employed directly the production of steel for packaging across ten dedicated manufacturing sites. Their objective is to[51]:

- Contribute positively to the development of EU policy related to steel for packaging, particularly in the areas of packaging, waste, recycling, and recovery.
- Monitor technical developments to ensure industry compliance.
- Document, support, and communicate the environmental, social and economic benefits of steel for packaging.

The European Steel Association (EUROFER) represents almost 100% of EU steel production. Founded in 1976, EUROFER's headquarters is in Brussels. It is the voice of the European steel industry to policy makers, civil society, and relevant stakeholders. EUROFER's members are steel companies and national steel federations based throughout the EU. The national steel federations and major steel companies of Switzerland and Turkey are also associate members. Amongst other actions, the association releases white papers and publications to spread awareness regarding circular economy and sustainability in the steel industry[52].

7.3 Agriculture and Food

7.3.1 Overview

Agriculture is crucial to economic growth. In 2018, it accounted for 4% of GDP and in some developing countries, it can account for more than 25% of GDP[53]. In 2017, an estimated 866 million people were officially employed in the agricultural sector[54].

7.3.2 Challenges in the Agriculture and Food Sector

The current food production and consumption habits are unsustainable. Food production generates various environmental impacts, such as eutrophication and increased GHG emissions[55]. Agriculture, forestry, and land use change are responsible for about 25% of global GHG emissions. Agriculture accounts for 70% of fresh-water use. One third of food produced globally is either lost or wasted at different stages of the food system[53]. Current inefficiency in the food economy leads to lower productivity, higher consumption of energy, and natural resources, and higher costs of food waste disposal. In low-income countries, most losses occur during primary stages of production, but high-income countries frequently waste more than 40% at the consumption stage[55]. A 2020 report found that nearly 690 million people – or 8.9% of the global population – are hungry, up by nearly 60 million in five years[56]. According to the UN's Food and Agriculture Organization (FAO), the inefficiencies of the food economy cost, globally, as much as USD 1 trillion a year, or even USD 2 trillion when social and environmental costs are included[55].

In 2017, world total agricultural use of chemical or mineral fertilizers was 109 Mt nitrogen (N), 45 Mt phosphate (P_2O_5), and 38 Mt potash (K). With respect to 2002, this represented increases of 34%, 40%, and 45%, respectively. Due to excessive use however, this has led serious environmental issues such as global warming, air pollution, water quality degradation, and soil acidification. Moreover, increasing residues of pesticides in food have been recent concerns contributing to human exposure.

7.3.3 Circular Economy in the Agriculture and Food Sector

There is great potential to implement circular agriculture as part of strategy to foster sustainability in the food system. In circular agriculture, waste is seen as a raw material to produce new valuable products, including crops, food, feed, and energy. The potential solutions include nutrient cycling which includes recovering nutrients from manure, recovering, and reusing nutrients in sewage sludge, cascading use of materials, as well as supporting local farms and de-specialized agricultural holds.

Food brands, retailers, chefs, and other food providers have a major influence on what we eat. In circular models, they have an important role in designing food products, recipes and menus that are healthy for people and the environment. This extends to food packaging that preserves food and is also compostable, so it be recycled as nutrients into the soil. Marketing and branding are also important to tell the story of the food to consumers so that benefits such as increased margins and new markets can be realized.

According to the European Community, Europe's circular economy market currently exceeds 2 trillion euros, providing 22 million jobs in different sectors such as agriculture, forestry, food, chemicals and bioenergy, accounting for about 9% of the EU workforce[57].

Farming, food production, food distribution, and food consumption and food waste are important aspects of the agricultural system that have opportunities in circular economy.

7.3.3.1 Farming[58]

Precision agriculture: it is based on the optimized management of inputs in a field according to actual crop needs. It involves gathering, processing, and analyzing temporal, spatial (including satellite positioning systems like Global Positioning System [GPS] and remote sensing) and individual data combining them with other information to support management decisions to manage crops and the use of fertilizers, pesticides, and water at "the right amount, at the right time, in the right place". By doing so, optimum performance is achieved with less quantities of inputs and resources needed leading to minimal environmental impact.

Water reuse: Cities generate wastewater that has potential to reuse in agricultural production. Apart from its value as water, wastewater may also contain nutrients that benefit agricultural production. Further, several other agricultural activities can also benefit from local recycling schemes. For example, animal production operations generate a substantial amount of wastewater, which is rich in organic matter containing macro and micronutrients important to agriculture.

Biofertilizer: organic wastes including food waste, crop stalks and stubble (stems), leaves, seed pods, and animal waste are usually produced throughout farming activities. These wastes are the cheapest resource that can be used by farmers when it is managed safely and converted into biofertilizer products. The use of biofertilizer can add nutrients to promote plant growth, maintain soil fertility and sustainability, ensuring the production of safe and healthy food, providing an economically viable support to farmers for realizing the goal of increasing productivity and sustainability. It is also a cost-effective and renewable source that can potentially supplement or even substitute inorganic fertilizers.

Local sourcing: Local sourcing can play a significant role in supporting the development of a distributed and regenerative agricultural system. It allows cities to increase the resilience of their food supply by relying on a more diverse range of suppliers (local and global). By understanding their existing peri-urban production, cities can demand food that is not only grown regeneratively, but also locally – when it makes sense – and support diversification of crops by selecting varieties best fitting local conditions, thereby building resilience[59].

Lufa Farms are a Montreal food company who are pioneers in urban farming. In 2011, Lufa planted the first seeds in the world's first commercial rooftop hydroponic greenhouse. One year later, the vegetables harvested from this 0.75-acre area was sufficient to feed 2000 local inhabitants[60].

7.3.3.2 Food Production and Distribution

Bioenergy is energy produced from biofuels. It comprises electricity, heat and a wide range of transportation fuel. Biofuels is energy produced directly or indirectly from biomass. Biofuels can include for example, liquid biofuels, i.e., fuel derived from biomass for transportation uses, gaseous biofuels such as methane gas, and solid biofuels like fuelwood, charcoal, etc. Sources of biomass include energy crops, agricultural and forestry wastes and by-products, manure or microbial biomass. Some examples are leaves, residues, cutover residues, sawdust, bark, chip, and corn husks among others. Bioenergy developments offer the opportunity for enhanced energy security and access by reducing the dependence on fossil fuels and providing a localized solution. Increased energy security in turn can have positive effects on food security[58].

Regenerative food production employs techniques that replenish and improve the overall health of the local ecosystem. Examples of regenerative practices include shifting from synthetic to organic fertilizers, employing crop rotation, and using greater crop variation to promote biodiversity. Farming types such as agroecology, rotational grazing, agroforestry, conservation agriculture, and permaculture, all fall under this definition. Regenerative practices support the development of healthy soils, which can result in foods with improved taste and micronutrient content[61].

7.3.3.3 Food Consumption and Food Waste

The most effective ways for affluent societies to reduce the environmental impact of their diets are to reduce consumption of meat and dairy products, to favor organic fruits and vegetables, and to avoid goods that have been transported by air on both individual and institutional levels (e.g., public procurement, public catering). Additionally, various policy interventions such as information-based instruments, market-based initiatives, direct regulations, and "nudges" can be used to spread awareness and change food

BOX 7.3 FOOD WASTE AT INDIAN WEDDINGS

In India, weddings are a pompous affair for the rich and the poor. There are approximately 10 million weddings a year and each of them waste 10–20% of total food served during the wedding. Food waste worth about USD 14 billion in losses annually[63]. There are a number of ways that this waste can be reduced:

- Choose caterers who take up responsibility of food waste management.
- Send food waste to composting units or biogas plants.
- Limit the number of guests.
- Tie up with NGOs who manage post-event waste such as - Feeding India, Robin Hood Army, No Food Waste (described in Chapter 8).
- Ask guests to confirm their attendance.
- Avoid buffet dinner to eliminate food wastage.
- Choose menu wisely to include locally sourced abundant ingredients.

consumption behaviors[62]. Food loss and waste can be designed out along the entire food supply chain. Designers can develop products and recipes that use food by-products and food waste as ingredients, and those which, by avoiding certain additives, can be safely returned to the soil, or used in other ways. Box 7.3 highlights the food waste problem at Indian weddings and a few ways to curb the same.

Implementing a food waste reduction program requires people to change their behavior. Companies like IKEA believe that the "human factor" is key to the success of implementation. IKEA aims to halve the food waste at all IKEA stores by end of the 2020 fiscal year. Box 7.4 describes the Food is Precious Initiative of IKEA.

One of the best approaches is to prevent or reduce the generation of waste in the first place. Prevention can be achieved by improving the efficiency of the work chain, which means less food is lost during the production process. Reduction is tackled in several ways. For example, companies are donating excess of food to those needy and are leading campaigns to educate people about household food loss due to excess. Another option to reduce waste is to send the leftovers of the raw material to animal feed. To prevent it from ending up in a landfill people are getting creative by processing food waste for production of beer, cement, biogas, and ethanol are just some of the examples that excess food can be turned into[65]. Box 7.5 shows example of Woolworths Supermarkets in Australia.

BOX 7.4 "FOOD IS PRECIOUS" INITIATIVE OF IKEA[64]

IKEA launched a global program on reducing food waste in December 2016 focusing on reducing food waste. In the first phase of the program, "smart scales" were used to measure the food thrown away in IKEA's direct operations at restaurants (in the kitchens), bistros, and Swedish Food Markets. Co-workers used the scale to weigh the food waste, categorize it, and identify the reason behind the waste. The system then calculated the cost of the waste. Using this method, co-workers could identify common factors behind food waste and develop solutions to reduce waste.

The data collected when weighing and registering food waste helped IKEA identify ways to prevent food waste. Analyzing the data enables IKEA to be more efficient in forecasting and in their use of raw materials. It also provided a holistic and informed view of the company's food operations, which allows IKEA to analyze how different parts of the operation can affect food waste. Measuring food waste at the company level also raised awareness among co-workers.

Between December 2016 until end of January 2019, the program led to reduction in 1,786,605 kg of food waste, saved 4,003,896 meals and avoided 7,682,229 kg of GHG emissions.

BOX 7.5 WOOLWORTHS SUPERMARKET INITIATIVE ON SUSTAINABLE FOOD WASTE MANAGEMENT[66]

Woolworths Supermarkets is Australia's largest chain of grocery stores. It has 995 stores operating around the nation with around 115,000 staff in stores, distribution centers, and head office.

The chain has enlisted not-for-profit partners to redirect food still fit for human consumption into the mouths of those who need it the most. Any food that is no longer fit for human consumption is diverted to farms to be used as animal feed or commercial organic composting.

The company's Stock feed for Farmers program has been running for more than 10 years. Woolworths have also partnered with Tribe Breweries to create a circular economy beer – Loafer. This limited-edition pale ale has been created with more than 350kg of leftover bread and each sale represents a meal being delivered to vulnerable Australian's through its partnership with Feed Appeal. With these programs in place, Woolworths has recorded more than 55,000 tons of food being diverted from landfill, and 10 million meals have been delivered to Australians in need across the country.

Similarly, Winnow, is a software as a service company founded by former management consultant Marc Zornes in 2014, based on the insight that a potential USD 252 billion of resource saving was possible by 2030 through the reduction of food waste. Winnow provides very basic hardware allowing the simple collection of data (weight and food type) in large commercial kitchens. A subscription model provides daily, weekly, and bespoke reporting on food waste patterns and trends. Winnow's hardware and reporting is currently provided to 1000 kitchens in 30 countries saving customers GBP 9 million/year as well as significant associated GHG emissions[60].

7.3.4 Role of Cities[67]

In a circular economy, food is designed to cycle, so the by-products from one enterprise provides input for the next. Cities can make the most of food by redistributing surplus edible food, while turning the remaining inedible by-products into new products, ranging from organic fertilizers for regenerative peri-urban farming, to biomaterials, medicine, and bioenergy. Rather than a final destination for food, cities can become centers where food by-products are transformed, through emerging technologies and innovations, into a broad array of valuable materials. These could range from organic fertilizers and biomaterials, to medicine and bioenergy, thereby driving new revenue streams in a thriving bioeconomy. Besides ensuring that edible food is distributed to citizens, the choice of the "best" option depends on the local context, including the type of available feedstock, and the demands for particular products in that specific region. In Chapter 8, wasted food delivery

BOX 7.6 THE FOOD INITIATIVE[67]

The Food Initiative will engage 20+ cities on a journey to a circular economy for food with London, New York, and São Paulo as Strategic Partners. Members – Almere, Barcelona, Bogotá, Kyoto, Lisbon, Milan, Phoenix, Porto, Rio de Janeiro, Salvador, Sevilla, Toronto, and Torres Vedras (with more to be announced) will further accelerate implementation efforts. Municipalities, local and global businesses, and resource managers will work together in new ways to drive real systemic change.

The Food initiative will activate unprecedented collaboration to mobilize the vision laid out in the Cities and Circular Economy for Food report. Food brands, producers, retailers, governments, innovators, waste managers, and other food players are all working toward three main ambitions based on circular economy thinking. Achieving these three ambitions in cities could generate annual benefits worth USD 2.7 trillion by 2050.

program of Annakshetra Foundation in India is described that follows the principles of collaborative consumption.

Following the publication of the *Cities and Circular Economy for Food* report at the World Economic Forum in Davos (January 2019), the Ellen MacArthur Foundation launched the Food Initiative. Over the next three years, it will bring together key actors to stimulate a global shift toward a regenerative food system based on the principles of a circular economy. Box 7.6 further describes the Food Initiative.

ACTIVITY 7.1 DISCUSSION ON INTERSECTIONS BETWEEN MULTIPLE SECTORS

In this chapter, circular economy opportunities were described for three sectors such as textile, steel, and food and agriculture. Do you see any intersections between the sectors on material flow and cascading of waste as resources?

Can material flow analysis (MFA) be used to map such intersections between multiple sectors? How should the policies and regulations be designed that such material flows ensure sustainability?

ADDITIONAL READING

1. **Textile:**
   a. **State of fashion 2020** – For the fourth year in a row, the Business of Fashion and McKinsey & Company have teamed up to provide an annual picture of The State of Fashion. This is now a knowledge base that they build on every year, identifying the key themes and business imperatives shaping the industry while tracking the ways in which fluctuations in the world economy feed through into fashion. The report also includes the fourth readout of the industry benchmark, the McKinsey Global Fashion Index (MGFI): its extensive database of companies allows to analyse and compare the performance of individual companies against their peers, by category, segment, or region.

Source: McKinsey and Company. 2020. *The State of Fashion 2020.* <https://www.mckinsey.com/~/media/McKinsey/Industries/Retail/Ourpercent20Insights/Thepercent20statepercent20ofpercent20fashionpercent202020percent20Navigatingpercent20uncertainty/The-State-of-Fashion-2020-final.ashx> [Accessed 17 December 2020].

b. **Mapping sustainable textile initiatives** – This report aims to chart a plan for a coordinated Nordic effort toward sustainable development in textiles and identify ongoing initiatives in the area. The aim was an ambitious plan with a potential for significant reductions in environmental pressures, but also green growth. To reach these goals, we staked out four regions a Nordic plan should include – replace fast fashion, reduce resource input, redirect global vs local, rethink for whom.

Source: Norden. 2015. *Mapping Sustainable Textile Initiatives: And a Potential Roadmap for a Nordic Actionplan.* <https://norden.diva-portal.org/smash/get/diva2:840812/FULLTEXT01.pdf> [Accessed 17 December 2020].

c. **Fashion & environment** – This paper is created by Julie's Bicycle and Centre for Sustainable Fashion (CSF) at London College of Fashion, University of the Arts London, on behalf of the British Fashion Council, enabled by DHL. It offers an overview of the environmental impacts of the fashion industry, a presentation of good practice in the UK, and how the industry can actively explore new definitions of good design and great business, including in:
   - Design and materials Integrating sustainable design principles into product, service and system creation
   - Green technology Fiber innovation, enzymology and molecular biology
   - Manufacturing and processing Supply chain transparency, blockchain technology, water use and energy efficiency
   - Packaging and delivery Innovation in shipping and logistics, reducing plastic and packaging waste
   - Education and engagement Pioneering new curriculum, changing cultures and driving demand for sustainable action
   - Strategy Setting carbon-based targets in alignment with the Paris Agreement; engaging with all above elements

Source: DHL. 2019. *Fashion & Environment: An Overview of Fashion's Environmental Impact & Opportunities for Action.* [online] Available at: <https://inmotion.dhl/uploads/content/2019/03_Fashion/whitepaper.pdf> [Accessed 17 December 2020].

d. **A new textiles economy: Redesigning fashion's future** – This report presents a positive new vision for a system that works and summons the creative power of the fashion industry to build it. In a new textiles economy clothes would be designed to last longer, be worn more and be easily rented or resold and recycled, and would not release toxins or pollution. Exploring new materials, pioneering business models, harnessing the power of design, and finding ways to scale better technologies and solutions are all needed to create a new textiles economy.

Source: Ellen Macarthur Foundation. 2017. *A New Textiles Economy: Redesigning Fashion's Future.* <https://www.ellenmacarthurfoundation.org/assets/downloads/publications/A-New-Textiles-Economy_Full-Report_Updated_1-12-17.pdf> [Accessed 17 December 2020].

2. **Agriculture and food:**
   a. **Regenerative agriculture** – This article goes through the importance of regenerative/sustainable agriculture and the how various methods have evolved over the years. It also holds up an argument regarding the sustenance of regenerative farming methods.

Source: Jeffries, N., 2019. *Regenerative Agriculture: How it Works on the Ground.* [online] Medium. Available at: <https://medium.com/circulatenews/regenerative-agriculture-how-to-grow-food-for-a-healthy-planet-9a5f637c0f3e> [Accessed 17 December 2020].

   b. **Cities and circular economy for food** – The report aims to highlight the often-underappreciated role urban food actors can play to drive food system transformation, and to spark a global public private effort to build a circular economy for food. Shifting to a circular economy for food presents an attractive model with huge economic, health, and environmental benefits across the food value chain and society more broadly.

Source: Ellen Macarthur Foundation. 2019. *Cities and Circular Economy for Food.* <https://www.ellenmacarthurfoundation.org/assets/downloads/CCEFF_Full-report-pages_May-2019_Web.pdf> [Accessed 17 December 2020].

c. **World food and agriculture statistical pocketbook** – This publication is meant to be an easily accessible, quick reference to selected key indicators on agriculture and food security, which are presented along four main themes.

The Setting describes the main trends in the use of agricultural resources such as land, labor, investment, fertilizers, and pesticides. This section also highlights the pressure on food systems caused by demographic and macroeconomic development.

- The Hunger dimension is monitored through two essential SDG indicators: the prevalence of undernourishment and the prevalence of moderate or severe food insecurity, based on the Food Insecurity Experience Scale (FIES). This section presents the state of food insecurity and malnutrition in the world along four dimensions – availability, access, stability, and utilization.
- Food supply offers vital information on the nature and quantity of world agricultural production and trade, including their utilization, such as food consumption, feed, and other uses.
- The Environment examines the interactions of agriculture with the ecosystem

Source: Food and Agriculture Organization of the United Nations. 2019. *World Food And Agriculture Statistical Pocketbook.* <http://www.fao.org/3/ca6463en/ca6463en.pdf> [Accessed 17 December 2020].

d. **FAOSTAT** – This web resource created by the FAO provides the latest data on all things related to agriculture across the globe.

Source: Food and Agriculture Organization of the United Nations. n.d. *FAOSTAT.* <http://www.fao.org/faostat/en/#data> [Accessed 17 December 2020].

e. **Life cycle analysis in the framework of agricultural strategic development planning in the Balkan region** – Agricultural sector should be considered, as one of the main economic development sectors in the entire world, while at the same time is responsible for important pollution. The life cycle assessment (LCA) procedure was involved in the agricultural strategic development planning for Balkan region, as a useful tool to identify and quantify potential environmental impacts from the production of apple juice, wine, and pepper pesto in three selected sites in Greece, North Macedonia, and Bulgaria. The results indicate that changes in the cultivation and the production must be considered in order to optimize the environmental footprint. Moreover, the whole approach could be useful for agricultural stakeholders, policy makers and producers, in order to improve their products ecological performance, reduce food loss and food waste and increase the productivity of the agricultural sector, while at the same time can improve the three pillars of sustainability through strategy development.

Source: Tsangas, M., Gavriel, I., Doula, M., Xeni, F. and Zorpas, A., 2020. Life cycle analysis in the framework of agricultural strategic development planning in the Balkan region. *Sustainability*, 12(5), p. 1813.

f. **Fighting food waste: Using the circular economy** – This report highlights information on food waste in Australia and globally, principles of the circular economy, as well as leading businesses in Australia that have adopted circular economy principles, some up to 90 years ago, and have developed successful businesses that manage food waste.

Source: KPMG. 2019. *Fighting Food Waste: Using the Circular Economy.* [online] Available at: <https://assets.kpmg/content/dam/kpmg/au/pdf/2019/fighting-food-waste-using-the-circular-economy-report.pdf> [Accessed 17 December 2020].

Notes

1. Textile Exchange. 2019. *Preferred Fiber & Materials Market Report 2019.* <https://textileexchange.org/wp-content/uploads/2019/11/Textile-Exchange_Preferred-Fiber-Material-Market-Report_2019.pdf> [Accessed 14 December 2020].

2. Grand View Research. 2020. *Textile Market Size, Share & Analysis – Industry Report, 2027.* <https://www.grandviewresearch.com/industry-analysis/textile-market#:~:text=Basedpercent20onpercent20rawpercent20materialpercent2Cpercent20the,sharepercent20ofpercent2039.5percent25percent20inpercent20 2019.&text=Woolpercent2Dbasedpercent20textilepercent20accountedpercent20 for,termspercent20ofpercent20volumepercent20inpercent202019.> [Accessed 14 December 2020].
3. Ellen Macarthur Foundation. 2017. *A New Textiles Economy: Redesigning Fashion's Future.* <https://www.ellenmacarthurfoundation.org/assets/downloads/A-New-Textiles-Economy_Summary-of-Findings_Updated_1-12-17.pdf> [Accessed 14 December 2020].
4. International Finance Corporation. n.d. *Global Apparel Supply Chain.* <https://www.ifc.org/wps/wcm/connect/industry_ext_content/ifc_external_corporate_site/manufacturing/blogs+and+articles/manufacturing_textiles> [Accessed 14 December 2020].
5. Bank J. Safra Sarasin Ltd. 2014. *Supply Chains in the Clothing Industry – A House of Cards?!* <https://www.eticanews.it/wp-content/uploads/2014/09/Report-Bank-J-Safra-Sarasin-Supply-Chains-in-the-Clothing-Industry.pdf> [Accessed 14 December 2020].
6. UNEP. 2019. *Fashion'S Tiny Hidden Secret.* <https://www.unenvironment.org/news-and-stories/story/fashions-tiny-hidden-secret> [Accessed 16 December 2020].
7. Lissaman, C., 2019. *What Is Circular Fashion?* Common Objective. <https://www.commonobjective.co/article/what-is-circular-fashion> [Accessed 16 December 2020].
8. Greenpeace International. 2018. *Destination Zero - Seven Years of Detoxing the Clothing Industry.* <https://www.greenpeace.org/international/publication/17612/destination-zero/> [Accessed 16 December 2020].
9. International Labour Organisation. 2019. *The Future of Work In Textiles, Clothing, Leather and Footwear.* <https://www.ilo.org/wcmsp5/groups/public/—ed_dialogue/—sector/documents/publication/wcms_669355.pdf> [Accessed 16 December 2020].
10. Fashion United. 2019. *Future Of Fashion: Production - Sustainable, High-Tech And On-Demand.*: <https://fashionunited.uk/case/future-of-fashion-production-sustainable-high-tech-and-on-demand> [Accessed 14 December 2020]. https://www.ilo.org/wcmsp5/groups/public/—ed_dialogue/—sector/documents/publication/wcms_669355.pdf> [Accessed 16 December 2020].
11. Emprechtinger, F., 2019. *Sustainable Solutions In The Textile Industry.* Lead Innovation. <https://www.lead-innovation.com/english-blog/sustainable-solutions-in-the-textile-industry> [Accessed 14 December 2020].
12. DHL Guide. 2020. *Educating Customers About the True Value of Your Product.* <https://dhlguide.co.uk/educating-customers-about-the-true-value-of-your-product/> [Accessed 16 December 2020].
13. Levi Strauss. 2019. *Levi Strauss & Co. Announces New Global Water Action Strategy, Heightening Focus on Most Water-Stressed Areas in Supply Chain.* <https://www.levistrauss.com/wp-content/uploads/2019/08/Water-Strategy-Press-Release.FINAL_.August-22-2019.pdf> [Accessed 14 December 2020].
14. Mogilireddy, V., 2018. *Sustainable Dyeing Innovations: Greener Ways To Color Textiles.* PreScouter. <https://www.prescouter.com/2018/11/sustainable-dyeing-innovations-greener-ways-color-textiles/> [Accessed 14 December 2020].

15. Pigments and Solvent Dyes. 2019. *What Exactly Is 'Right First Time' Dyeing?.* <https://xhpigments.wordpress.com/2019/03/06/what-exactly-is-right-first-time-dyeing/> [Accessed 14 December 2020].
16. Fiber2Fashion. 2013. *Eco Friendly Renewable Energy Source for Textile Machinery.* <https://www.fiber2fashion.com/industry-article/7069/a-facelift-to-textile-machinery-with-renewable-energy> [Accessed 16 December 2020].
17. The Indian Textile Journal. 2019. *Textile Industry Eyeing Benefits of Renewable Energy Like Solar.* <https://indiantextilejournal.com/interviews/Textile-industry-eyeing-benefits-of-renewable-energy-like-solar> [Accessed 16 December 2020].
18. Unilever. n.d. *Rethinking Plastic Packaging – Towards A Circular Economy.* <https://www.unilever.com/sustainable-living/reducing-environmental-impact/waste-and-packaging/rethinking-plastic-packaging/> [Accessed 16 December 2020].
19. T I P A. n.d. *Overview.* <https://tipa-corp.com/about/overview/> [Accessed 16 December 2020].
20. GOMES, L., 2019. *8 Brands Turning to Responsible Packaging Solutions.* Current Daily <https://thecurrentdaily.com/2019/10/15/8-brands-turning-to-responsible-packaging-solutions/> [Accessed 16 December 2020].
21. Yan, G., 2019. Reverse Logistics in Clothing Recycling: A Case Study in Chengdu. *International Journal of Environmental and Ecological Engineering,* 13(5).
22. Global Organic Textile Standard - https://www.global-standard.org/
23. ThredUP. n.d. *Our Impact.* <https://www.thredup.com/impact> [Accessed 14 December 2020].
24. Textile Exchange. 2020. *Recycled Polyester Commitment.* <https://textileexchange.org/recycled-polyester-commitment/#:~:text=Inpercent202017percent2Cpercent20Textilepercent20Exchange'spercent20Recycled,polyesterpercent20bypercent2025percent25percent20bypercent202020.> [Accessed 16 December 2020].
25. Textile Focus. 2019. *R|Elan™: New Range of Innovative Fabric.* <http://textilefocus.com/relan-new-range-innovative-fabric/> [Accessed 16 December 2020].
26. UNFCC. n.d. *Fashion for Global Climate Action.* <https://unfccc.int/climate-action/sectoral-engagement/fashion-for-global-climate-action> [Accessed 16 December 2020].
27. UNEP. 2019. *UN Alliance For Sustainable Fashion Addresses Damage Of 'Fast Fashion'.* <https://www.unenvironment.org/news-and-stories/press-release/un-alliance-sustainable-fashion-addresses-damage-fast-fashion> [Accessed 16 December 2020].
28. Worldsteel Association. n.d. *About Steel.* <https://www.worldsteel.org/about-steel.html> [Accessed 16 December 2020].
29. Worldsteel Association. 2020. *The Steel Industry in Modern Society.* <https://www.worldsteel.org/media-centre/press-releases/2020/edwin-basson-presentation-gfsec-october-2020.html> [Accessed 16 December 2020].
30. Steel Recycling Institute. n.d. *Steel is the World's Most Recycled Material.* <https://www.steelsustainability.org/recycling> [Accessed 16 December 2020].
31. Worldsteel Association. 2018. *Fact Sheet: Steel Industry Co-Products.* <https://www.worldsteel.org/en/dam/jcr:1b916a6d-06fd-4e84-b35d-c1d911d18df4/Fact_By-products_2018.pdf> [Accessed 16 December 2020].
32. Worldsteel Association. n.d. *Steel Recycling.* <https://www.worldsteel.org/steel-by-topic/sustainability/materiality-assessment/recycling.html5> [Accessed 16 December 2020].

33. Ambasta, D., Pandey, B. and Saha, N., 2016. *Utilization of Solid Waste from Steel Melting Shop*. MECON Limited. <http://www.meconlimited.co.in/writereaddata/MIST_2016/sesn/tech_4/3.pdf> [Accessed 16 December 2020].
34. Steel Technology. n.d. *Waste Disposal and Recycling in Steel Industry*. <https://www.steel-technology.com/articles/wastedisposal> [Accessed 16 December 2020]
35. Government of India Ministry of Mines Indian Bureau of Mines. 2017. *Slag- Iron and Steel*. <http://ibm.nic.in/writereaddata/files/03202018150040Slag_Iron_Steel_AR_2017.pdf> [Accessed 16 December 2020].
36. Section adapted from: NITI Aayog and EU Delegation to India, 2019. *Resource Efficiency and Circular Economy: Current Status and Way Forward* < https://www.eu-rei.com/pdf/publication/NA_EU_Status%20Paper%20&%20Way%20Forward_Jan%202019.pdf> [Accessed 14 December 2020].
37. Worldsteel Association. n.d. *Raw Materials*. <https://www.worldsteel.org/steel-by-topic/raw-materials.html> [Accessed 16 December 2020].
38. Ellen Macarthur Foundation. n.d. *Case Studies: Arcelormittal Brasil*. <https://www.ellenmacarthurfoundation.org/case-studies/new-entry> [Accessed 16 December 2020].
39. Also recommended by 'Strategy Paper on Resource Efficiency in Steel Sector through Recycling of Scrap & Slag', 2018, Circular Economy and Resource Efficiency, NITI Aayog. Website: http://niti.gov.in/content/circular-economy-and-resource-efficiency
40. A policy on scrap recycling has been recently drafted by the Ministry of Steel and is expected to be approved in the near future.
41. Worldsteel Association. n.d. *Steel Industry Co-Products Position Paper*. <https://www.worldsteel.org/publications/position-papers/co-product-position-paper.html> [Accessed 16 December 2020].
42. JSW Group. n.d. *JSW Steel: Champions in Steel-Making, Champions of Sustainability*. <https://www.jsw.in/steel/champions-sustainability> [Accessed 17 December 2020].
43. Tata Steel BSL Ltd. 2020. *Waste Management*. <https://tatasteelbsl.co.in/WasteMangmt.html> [Accessed 16 December 2020].
44. Tata Steel. 2020. *Tata Steel Flags-Off The 1St Raw Material Consignment of Ferrous Scrap At Its Steel Recycling Plant Being Set-Up In Rohtak, Haryana*. <https://www.tatasteel.com/media/newsroom/press-releases/india/2020/tata-steel-flags-off-the-1st-raw-material-consignment-of-ferrous-scrap-at-its-steel-recycling-plant-being-set-up-in-rohtak-haryana/> [Accessed 17 December 2020].
45. GIZ. 2017. *Training Manual for Resource Efficiency in Automobile Component Manufacturing Companies in India*. <http://re.urban-industrial.in/live/hrdpmp/hrdpmaster/igep/content/e64918/e64922/e67075/e67088/GIZAutoTrainingManualFull.pdf> [Accessed 16 December 2020].
46. Central Pollution Control Board. 2015. *Analysis of End-of-Life Vehicles (ELVs) Sector in India*. <https://cpcb.nic.in/openpdffile.php?id=TGF0ZXN0RmlsZS9MYXRlc3RfMTE0X0FuYWx5c2lzT2ZFTFZzSW5JbmRpYTFfLnBkZg> [Accessed 16 December 2020].
47. Worldsteel Association. 2020. *Sustainable Steel - Indicators 2020 And Steel Applications*: <https://www.worldsteel.org/media-centre/press-releases/2020/sustainable-steel–indicators-2020-and-steel-applications.html> [Accessed 16 December 2020].

48. Worldsteel Assocation. n.d. *Life Cycle Inventory Database for Steel Industry Products* <https://www.worldsteel.org/en/dam/jcr:828020e8-14f9-43c5-8b2d-3e8d5c96d1e6/worldsteel_LCA_FAQ_web_Apr%25202020.pdf> [Accessed 16 December 2020].
49. Worldsteel Association. 2019. *Steelie Awards.* <https://www.worldsteel.org/about-us/steelie-awards.html> [Accessed 16 December 2020].
50. Worldsteel Association. 2020. *CHAMPIONS PROGRAMME 2019.* <https://www.worldsteel.org/steel-by-topic/sustainability/steel-sustainability-champions.html> [Accessed 16 December 2020].
51. APEAL. n.d. *About APEAL.* <https://www.apeal.org/about-us/> [Accessed 16 December 2020].
52. EUROFER. n.d. *About the European Steel Association.* <https://www.eurofer.eu/> [Accessed 16 December 2020].
53. The World Bank. 2020. *Agriculture and Food.* <https://www.worldbank.org/en/topic/agriculture/overview> [Accessed 16 December 2020].
54. Global Agriculture. n.d. *Industrial Agriculture and Small-Scale Farming.* <https://www.globalagriculture.org/report-topics/industrial-agriculture-and-small-scale-farming.html> [Accessed 16 December 2020].
55. Jurgilevich, A., Birge, T., Kentala-Lehtonen, J., Korhonen-Kurki, K., Pietikäinen, J., Saikku, L. and Schösler, H., 2016. Transition towards Circular Economy in the Food System. *Sustainability*, 8(1), p.69.
56. Food and Agriculture Organization of the United Nations. 2020. *The state of food security and nutrition in the world 2020.* <http://www.fao.org/3/ca9692en/online/ca9692en.html#chapter-Key_message> [Accessed 16 December 2020].
57. European Commission. 2020. *Commission Staff Working Document.* <https://ec.europa.eu/environment/circular-economy/pdf/leading_way_global_circular_economy.pdf> [Accessed 16 December 2020].
58. Food and Agriculture Organization of the United Nations. 2020. *Circular Economy: Waste-To-Resource & COVID-19.* <http://www.fao.org/land-water/overview/covid19/circular/fr/#:~:text=Apercent20'circularpercent20agriculturepercent20economy'percent20proposes,dischargespercent20(i.e.percent20wastewater)percent20andpercent20surface> [Accessed 16 December 2020].
59. Ellen Macarthur Foundation. n.d. *Food and the Circular Economy.* <https://www.ellenmacarthurfoundation.org/explore/food-cities-the-circular-economy> [Accessed 17 December 2020].
60. Jeffries, N., 2018. *A Circular Economy for Food: 5 Case Studies.* Medium. <https://medium.com/circulatenews/a-circular-economy-for-food-5-case-studies-5722728c9f1e> [Accessed 17 December 2020].
61. Sitra. 2019. *Cities and the Circular Economy For Food.* <https://www.sitra.fi/en/publications/cities-circular-economy-food/> [Accessed 17 December 2020].
62. Reisch, L., Eberle, U. and Lorek, S., 2013. Sustainable food consumption: an overview of contemporary issues and policies. *Sustainability: Science, Practice and Policy*, 9(2), pp. 7–25.
63. Food Tank. 2019. *To have and to throw: tackling indian wedding food waste.* <https://foodtank.com/news/2019/06/to-have-and-to-throw-tackling-indian-wedding-food-waste/#:~:text=India's%20estimated%2010%20million%20weddings,at%20weddings%20goes%20to%20waste.> [Accessed 18 December 2020].
64. Food Loss and Waste Protocol. 2017. *IKEA Food: "Food Is Precious" Food Waste Initiative.* <https://flwprotocol.org/case-studies/ikea-food-food-precious-food-waste-initiative/> [Accessed 17 December 2020].

65. Unilever. n.d. *Going Beyond Zero Waste to Landfill.* <https://www.unilever.com/sustainable-living/reducing-environmental-impact/waste-and-packaging/going-beyond-zero-waste-to-landfill/> [Accessed 16 December 2020].
66. KPMG. 2019. *Fighting Food Waste.* <https://assets.kpmg/content/dam/kpmg/au/pdf/2019/fighting-food-waste-using-the-circular-economy-report.pdf> [Accessed 16 December 2020].
67. Ellen Macarthur Foundation. n.d. *Food and the Circular Economy.* <https://www.ellenmacarthurfoundation.org/explore/food-cities-the-circular-economy> [Accessed 16 December 2020].

8

Business Models in a Circular Economy

In Chapter 2, the concept of the Life Cycle was introduced. In Chapter 4, the "12Rs" were presented with illustrative examples. In Chapters 5 and 6, examples of businesses that operated for product life extension and closing the loop were showcased. Chapter 7 described opportunities in circular economy in sectors such as textile, steel, agriculture, and food. In this chapter, eight different business models that operate in a circular economy mapping with the life cycle are presented.

Figure 8.1 shows the eight business models. Many of these models intersect and, in some cases, operate as a hybrid or in combination depending on the

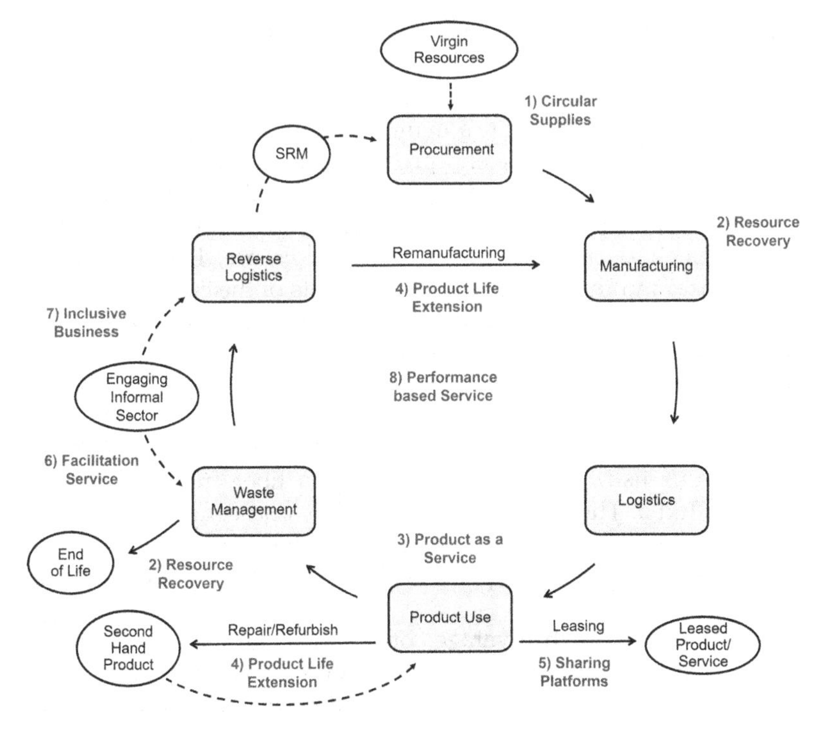

FIGURE 8.1
Business models in circular economy mapping with life cycle.

context and the opportunity. Therefore, these business models should not be considered silo and can be applied at more than one stage of the product life cycle.

8.1 Circular Supplies

Businesses do not always have to use virgin raw materials and depending on techno-commercial viability and supply availability, many businesses have started sourcing Secondary Raw Materials (SRM). In some cases, this strategy has given an economic advantage on a "total cost" basis. It also reduces the footprint of Greenhouse Gas (GHG) emissions of the product, thereby giving a brand advantage. Box 8.1 illustrates two examples of companies in

BOX 8.1 EXAMPLES OF CIRCULAR SUPPLIES IN INDIA

Ecoware – Ecoware, based in Delhi National Capital Region (NCR), India, is engaged in the manufacturing of disposable tableware (plates, bowls, cups, etc.), cutlery, and food takeaway packaging.

All Ecoware products are compostable in 90 days, as the raw materials are drawn from natural sources. Raw material (bagasse fiber) is sourced from paper mills that use bagasse as a substitute for wood to produce paper and board. At Ecoware's manufacturing facility, the bagasse is pulped and molded into the end-product using a thermoforming process, post which the products are trimmed, packed, and dispatched.

Ecoware not only sells its products across India, but also exports to countries such as Sri Lanka, Kuwait, Peru, France, Ecuador, Kenya, South Africa, and the Caribbean. Major clients of the company include retailers, large corporates involved in food and beverage manufacturing, and individual customers.

Visit https://ecoware.in/

Saathi Pads – Saathi, established in 2015, manufactures compostable sanitary pads using locally sourced banana fiber from the State of Gujarat in India. The pads are biodegradable and compostable. These fibers provide leak-proof outer layers of the napkin. When disposed, Saathi pads are claimed to degrade within 6 months, at a rate 1200 times faster than plastic pads.

It is estimated that the average conventional sanitary pad contains 3.4 g of plastic. On an average, women would generate approximately 60 kg of waste plastic from sanitary pads alone over a lifetime.

Visit https://saathipads.com/

India that are built on circular supplies. There are several such examples in the sustainable packaging industry.

ACTIVITY 8.1 DISCUSSION ON CIRCULAR SUPPLY MODEL-BASED BUSINESSES

For circular supply-based business, what certifications may have to be obtained to give confidence to the consumer and secure business? Refer to the various regulations described in Chapter 6 regarding the recycled content. Research on case studies where application of Life Cycle Sustainability Assessment (LCSA) was done to take a decision on circular supplies, especially on the replacement of materials.

8.2 Resource Recovery

In this model, businesses can buy useful waste from manufacturing operations, agricultural activities, restaurants, and landfills and process the wastes for recovery of resources.

Recycling infrastructure such as Material Recovery Facilities (MRF) includes businesses that provide opportunities for waste sorting, recycling, and recovery on a scale. These businesses are set up as an independent facility servicing waste management for an industrial cluster or for urban areas. Sometimes, the infrastructure is developed in various models of Public-Private Partnership (PPP) and may include a service of providing a market to the recycled products, or in some cases, offer reverse logistics for scrap/waste collection.

Waste to Energy business has been growing despite controversies regarding its circularity and mixed experience on the ground. There have been challenges in addressing variations in the waste composition and in managing compliance with emission standards. Resources recovered in waste to energy plants include thermal energy and energy products such as briquettes and pellets.

As per the Basal Convention, various wastes, including hazardous wastes, can be disposed of in an environmentally safe and sound manner through the technology of co-processing in cement kiln[1]. Co-processing is the use of waste as a source of energy or raw material or replacing natural resources and fossil fuels. Waste materials used for co-processing are referred to as Alternate Fuels and Raw materials (AFR). As an energy-intensive industry (especially thermal energy), cement production has an excellent opportunity to consume hazardous waste as AFR. Cement kiln being operated with a high temperature of around 1400 °C, destroys the hazardous waste, and there will be no residue left; still, the inorganic content gets fixed with the clinker. The dual factor of resource conservation and reduced GHG emissions makes a strong case for considering co-processing hazardous waste in cement kiln as a better alternative for hazardous waste disposal. Many cement companies offer services for the co-processing of hazardous wastes to the market as a business.

Waste to energy plants following the biological route generates biogas for heating and cooking purposes and electricity generation. Biogas after cleaning and compressing can be used for running vehicles and buses.

GENeco Bio-Bus in the United Kingdom showcased an inspiring example. After running a successful trial route between Bath and Bristol Airport, the GENeco Bio-Bus is used in regular service within the city of Bristol[2].

BOX 8.2 BIOBUSES IN PUNE, INDIA[3]

Pune-based Noble Exchange (NEX) started supplying bio-CNG created out of city waste to about 100-odd public buses. The project is planned to be run in collaboration with Pune Mahanagar Parivahan Mahamandal (PMPML).

This project was showcased at World Economic Forum 2020 at Davos in January, where participants discussed ideas to tackle the climate emergency arising out of sustained global warming.

Biogas is produced through a process of anaerobic decomposition of municipal waste, generating methane content of about 94–96%. In comparison, conventional CNG offers about 85–92% methane. As a result, it is claimed that with its better air-to-fuel ratio, Compressed Biogas (CBG) offers lower engine heating and better torque, leading to enhanced combustible efficiency and mileage for the vehicles.

Each day, at NEX's Pune plant, about 100 tons of bio-waste are processed, helping generate 3000 kg of CBG. The CBG is filled in special cylinders for supply to clientele, which includes India's largest oil manufacturing company, Indian Oil Corporation in Pune. Mahindra CIE Automotive, a subsidiary of CIE Automotive Group of Spain, picks up the remaining gas. The excess CBG will meet the forging needs of Mahindra's plant near Chakan, in Pune.

Government of India announced the "Sustainable Alternative Towards Affordable Transportation" initiative towards end of 2018. Further, under the new biofuel policy, it was announced in 2018 that CBG has been included among the biofuels for automobiles, offering assured purchase from private entrepreneurs by the government oil marketing companies. These policy reforms brought financial viability to the CBG business.

The use of gas-powered vehicles reduces emissions and helps in improving urban air quality compared to emissions from diesel. Secondly, the solution addresses the need to manage wastes through the bio-methanation and helps in the reduction of GHG emissions. Thirdly, use of biogas showcases a sustainable and renewably sourced alternative to fossil fuels.

The bio-bus business is getting traction in India. Box 8.2 presents a case of company Noble Exchange in India. This case study shows how Government policies play an important role in the financial viability of the business.

It would be good to read reports on business models on waste to energy that illustrate interesting global examples[4].

Urban mining has emerged as a business in the last two decades for waste recovery from dumpsites and landfills. In this model, businesses buy waste from landfills or get appointed by the governments on specific contracts to clear landfills by segregating and collecting all items that can be recycled/reused/remanufactured. In both cases, the end results lead to better landfill management and resource recovery.

Box 8.3 presents the case of urban mining in the city of Kumbakonam in India.

Recovering resources from food wastes have been a growing business. Companies like Entocycle[6] in London upcycle local food waste like – fruit and vegetable reject from supermarkets, coffee grounds, and brewer's grains by feeding them to insects who convert them into proteins. This not only

BOX 8.3 BIOMINING AT KUMBAKONAM IN INDIA[5]

Kumbakonam Municipality in India implemented a biomining project under the Design, Build, Finance, Own, and Operate concept. This project was developed by company Zigma for clearing 131,250 cubic m^3 of municipal waste spread in more than 7.5 acres of land with a capacity to process 350 m^3 a day.

The biomining concept involves the application of composting bio cultures on loose waste heaps, followed by conventional aerobic windrows on the site. The waste is then sterilized, stabilized, and readied for segregation using machinery to separate substances that are sent for recycling, re-using, or composting. Aggregates such as coconut shells, plastics, wood, rubber, glass, inert, and soil-enriching bio earth are collected. While coconut shells and wood are sold as fuel, rubber and glass are sent for recycling industries. Plastic is supplied to recycling plants and cement plants.

The Swatch Bharat program in India supports biomining projects. After the success at Kumbhakonam, several municipalities in India are implementing bio-mining projects with the city of Mumbai implementing one of world's largest biomining projects.

BOX 8.4 EXAMPLES OF RECOVERY FROM FOOD WASTE

Agro Loop BioFiber[7] – The company transforms crop food waste into scalable high-value natural fiber products for manufacturing high quality textiles for the fashion industry. They predominantly use the Banana tree, Oilseed flax, Oilseed hemp, Pineapple leaves, Cane Bagasse, Raw Straw as raw material. Apart from high quality yarn and fabric, the Agraloop biorefinery technology is modular and the set up requires low capital expenditure. The technology can be implemented across the globe for making sustainable and resilient communities.

Bio-bean[8] – Bio-bean works with logistics and waste management companies to collect used and discarded coffee beans and upcycle them into valuable products before they reach landfills. The coffee beans are recycled into biomass fuels for industries by converting them into Coffee logs and pellets, and also used for extracting flavors, dyes, and pigments.

Toast Ale[9] – Toast Ale is preventing bread waste from reaching the landfills by brewing beer from them. By brewing beer from more than 1 million bread slices, the company has avoided 42 tons of CO_2 emissions and raised more than 1 million toasts so far.

helps recycle food waste, but the protein produced acts as a sustainable alternative to traditional protein sources (e.g.: Soy and Fishmeal), which otherwise lead to deforestation and overfishing. Box 8.4 presents more examples of recovery from food waste.

Around the world, 700,000 tons of recycled rare metals are traded annually. The figure is projected to reach 1.1 million tons in fiscal year 2026. Urban mines are becoming more attractive to major nonferrous producers, because they contain larger amounts of rare metals than conventional mines and because conventional mining is growing more costly.

Companies like Mitsubishi Materials, Dowa Holdings and several others are developing projects where mountains of discarded electronics are processed to recover gold, platinum, palladium, and other rare metals. With the new investments, Mitsubishi expects to increase its rare metals recycling capacity from 140,000 tons a year in 2017 to 200,000 tons in fiscal year 2021, making it one of the world's largest collectors of second-hand rare metals. The company will also start collecting rare metals from lithium-ion car batteries. Dowa Holdings is increasing its ability to collect and recycle catalyzers, used to clean auto emissions from gasoline-engine vehicles. Catalyzers contain precious metals such as platinum, palladium, and rhodium. The company plans to set up an electric furnace that can melt catalyzers – a step Dowa expects to help it increase the urban mine's production by 40% by 2020[10].

ACTIVITY 8.2 DISCUSSION ON RESOURCE RECOVERY MODEL-BASED BUSINESSES

Resource recovery has been one of the widely used business model. PPP models are often used by municipal governments, draw on finance and expertise of the private sector against a concession such as provision of land. Research on some of the PPP models-based resource recovery models in water, waste, and energy sectors. Do these models always work? What are the risks that need to be addressed and allocated in the PPP contracts?

8.3 Product as a Service (PaaS)

This business model has been popularized over the past decade. Even though the intention of creating this business model was not based on a circular economy, it fits closely with its principles. Here, instead of purchasing a commodity, buyers are encouraged to lease it on a short-term basis or need basis. This strategy not only saves cost and time, but also reduces resource consumption and improves net utilization of resources. Box 8.5 illustrates two examples of PaaS.

BOX 8.5 EXAMPLES OF PRODUCT AS A SERVICE

MUD Jeans[11] – MUD Jeans provides jeans on a subscription-based model. It was founded by Bert van Son in 2012 and has its headquarters in Almere, Netherlands. In 2017, it had a turnover of EUR 822,000 (75% sales: 25% lease). It also has a partnership with RE:Pack – a returnable packaging company that helps MUD Jeans collect pairs of jeans from customers. The repairs are free, and users can swap their jeans for a new pair. MUD Jeans are made from 23–40% recycled material which is derived from discarded jeans. This model creates more flexibility for customers in terms of choices and at a lower cost and creates a predictable material supply chain for MUD Jeans. MUD Jeans also has a lower environmental impact associated with production and consumption of jeans.

Philips "Pay per Lux"– The "Pay per Lux" concept consists of providing the exact amount of light for workspaces and rooms that employees need when using them for specific tasks – no more, and no less. Also, maintenance is included. Whenever the lighting needs a change, Philips can either adapt the existing system further to the client's wishes, or simply reclaim its materials and recycle them via LightRec, Philips' partner, responsible for the re-use of lighting components[12].

In association with architects Kossmann dejong and Philips Design, lighting fixtures were specially developed for Amsterdam Airport (Schiphol) that will last 75% longer than other conventional fixtures as the design of the fixtures improved the serviceability and therefore improved the lifetime[13].

Equipment leasing is a good example of product as a service. Here equipment suppliers work on a contract basis to lease out heavy machineries required during construction and even in waste management as it is not always feasible to buy equipment, especially for small scale businesses. The customer can also opt for wet leasing where leasing of the equipment comes along with a contract of repairing/refurbishing it. Example of a leasing service by Orix in India for construction equipment and vehicles is illustrated in Box 8.6.

Too often, the lack of cooperation between suppliers and users of chemicals leads to an unnecessary over-consumption of chemicals and the generation

BOX 8.6 BUSINESS OF ORIX IN LEASING OF CONSTRUCTION EQUIPMENT AND VEHICLES[14]

ORIX offers equipment Leasing solution to small and medium enterprises (SME) to large and Multinational Corporations (MNC) with an opportunity to upgrade businesses or undertake large projects without having to bear the burden of initial payment. With their equipment leasing option, the technology obsolescence risk is managed by ORIX and the end-of-life waste management is taken care by ORIX with the return of assets at the end of lease period. Customers may choose schemes between owning the asset at the end, to returning the asset once the lease is over.

Under the full-service lease, leasing solution for vehicles can be used to fund the vehicle finance costs as well as the costs like, registration, maintenance, accidental repair costs, insurance and more. Under this service, customer's monthly lease rental includes periodic and breakdown servicing, accidental repair costs, replacement of car tyre and battery cost, pick up and drop facility, towing, and roadside assistance and 24-h helpline service. At the end of the lease term, the user has an option to extend the Lease or surrender the vehicle and opt for a new car.

of hazardous waste. This style of volume-based business can hinder innovation and may lead to unsustainable management of chemicals.

Chemical Leasing is based on an alternative approach that turns the business model upside down: profit does not depend on more volume, but less. Chemical consumption becomes a cost rather than a revenue factor. Box 8.7 provides further explanation on chemical leasing as a service.

BOX 8.7 CHEMICAL LEASING AS A SERVICE[15]

Since 2004, UNIDO in cooperation with Austria and Switzerland launched a Global Chemical Leasing Program[16], promoting a performance-based business model. UNIDO defines chemical leasing as a service-oriented business model that shifts the focus from increasing sales volume of chemicals towards a value-added approach. Thus, in the chemical leasing model the supplier does not sell chemicals in quantities. The supplier sells the function of the chemical. The functions performed by the chemical and functional units are the main basis for payment. Within chemical leasing business models, the responsibility of the user and the supplier is extended and may include management of the entire life cycle and so this service as a great potential in circular economy.

Chemical leasing strives for a win-win situation. It aims at increasing the efficient use of chemicals while reducing the risks of chemicals and protecting human health. It improves the economic and environmental performance of participating companies and enhances their access to new markets. Key elements of successful chemical leasing business models are proper benefit sharing, high quality standards and mutual trust between participating companies.

The financial industry is resistant to support the business model of chemical leasing as the costs of managing hazardous waste and ecological risks still remain external to the financial business case. Writing legal contracts on risk sharing is also a complex task.

Two examples for chemical leasing are illustrated below:

Example 1: A producer of automotive parts needs solvents to clean and degrease them. The company pays the chemical supplier for the functions performed by the chemical, that is, the cleaned metal parts. The company does not pay according to the amount of solvent used.

Example 2: A car producer needs surface protection for its cars. This includes car body pre-treatment, surface activation and the application of a system of coatings. Under chemical leasing the company pays per car body protected. It does not pay according to the amount of chemicals used.

The chemical leasing model allows users of chemicals to concentrate on their core business while benefiting from the services and know-how of the supplier. Generally, the chemical user is responsible for monitoring the process, documenting records on units to be paid and respecting the process parameters defined by the supplier. In this model, the user reduces procurement costs of the chemicals and improves productivity and resource efficiency. The handling and storage of chemicals becomes safer with reduced risk and improved health and safety conditions to the workers. This is achieved through substitution, process optimization, and new equipment for improved handling of chemicals.

Chemical leasing model depends a lot on the partnership between chemical producer and user. This partnership helps in continuous improvement and compliance with international regulations such as REACH.

Processes that are part of a chemical leasing plan are usually non-core processes in the company, such as: conveyor lubrication in the beverage industry, water treatment, cleaning and disinfection in hotels and hospitals, surface protection and treatment in metal finishing or car production and bonding of packaging in the food sector.

ACTIVITY 8.3 DISCUSSION ON PRODUCT AS A SERVICE MODEL-BASED BUSINESSES

Product as a service is gaining popularity in the business of circular economy. Research on total cost benefit analyses of this business model against the conventional option of ownership. Are there any advantages of the ownership model, especially in operating infrastructure assets that are used over a long term?

8.4 Product Life Extension

This business model is based on earning profits by repurposing products once it is discarded by consumers. In Chapter 6, few examples covering repair and refurbishing related businesses were presented.. In this section, a case study of PanurgyOEM is highlighted (see Box 8.8).

Refurbishing is a growing business. A 2018 catalogue of top refurbishing companies in the Silicon Valley in the USA shows promising examples[18].

BOX 8.8 REPAIR AND REFURBISH BUSINESS BY PANURGYOEM[17]

PanurgyOEM have developed a business on repairs supported by a comprehensive reverse logistics system over 30 years. Its 90,000 square-foot repair facility includes dedicated work areas and technicians trained to repair a variety of products such as consumer electronics, computers, and office equipment, etc.

The solution offered is comprehensive, as it covers fixing of the product at an assembly, sub-assembly, or even a component level, backed by an extensive quality control process. The business therefore functions as a natural extension to clients manufacturing team, saving the time, energy, and cost required to track returns, manage repairs, hire a team of repair technicians, and get back repaired and refurbished products. The company benefits, as there is a faster turnaround on repairs with higher quality, also by saving money by not buying new products and addressing the challenge of stocking impaired products in the warehouse as a liability for disposal.

ACTIVITY 8.4 DISCUSSION ON PRODUCT LIFE EXTENSION MODEL-BASED BUSINESSES

Can the business of product life extension be integrated with PaaS and with inclusive business models (discussed later in this chapter) in circular economy. Look for examples that follow such an integrated strategy.

8.5 Sharing Platforms or Collaborative Consumption

Sharing of platforms is synonymous to concepts of collaborative consumption or shared economy. The sharing model is a service compensation model in which the owner sells access to underutilized assets to subsequent customers. Owners are responsible for maintenance and service quality. This business is generally driven through digital platforms where a product or an asset is used and reused by stakeholders on a need basis, the concept is close to the business of product as a service. One of the main differences with leasing is that the typical period of usage for sharing platforms is much shorter. Also, the number of users of assets in a sharing platform is much greater, justifying the name of this change in behavior[19].

AirBnB is an online platform that allows sharing rooms and properties/ homes with travelers. It acts as an affordable alternative to a hotel. AirBnB has been operational since 2008. The host and guest community have together generated a direct economic impact of USD 100 billion in 2018. This is an example of a scale that is looked for in a shared economy.

Carpooling is a good example to differentiate. Renting or leasing a car is an example of a PaaS but sharing the car ride with other passengers is an example of sharing a platform.

Carpooling is the shared use of a car by the driver and more than one passenger for commuting. Carpooling is a business offered by companies with digital platforms (mostly on mobile apps) and have connected car owners with interested commuters and corporates. There are possibilities of regular carpools (on a daily or weekly basis) for commuters who have a common work destination. These carpools operate on a repeated schedule. Casual carpools (carpools at specific date) are offered for weekends, exceptional trips, vacations on an as-needed basis.

Carpools are attractive business models as they reduce communing costs to the customer or the corporate. More importantly, fuel consumption and emissions released per customer also reduce and so also the traffic congestions on the road. Carpooling also makes a case for expanding professional network while commuting. There are several businesses now that offer carpooling solutions, even in countries across the globe (e.g., Zify[20]) and not limiting to passenger vehicles but also buses and scooters (e.g., Sride[21], Redbus[22]).

Sharing resources is an important strategy towards sustainable consumption. Food waste is one potential sector for opportunities. Box 8.9 shows interesting examples of food waste initiatives based on collaborative consumption.

BOX 8.9 FOOD WASTE RELATED INITIATIVES BASED ON COLLABORATIVE CONSUMPTION IN INDIA

Annakshetra Foundation[23]: Established in November 2010 by the Centre for Development Communication, Jaipur, Annakshetra Foundation Trust is a non-profit which collects unused surplus food left over after weddings, anniversaries, birthday parties and other socio-religious functions (much of which would otherwise be wasted) and makes it available to those in need through a network of volunteers. It is the first organization of its kind in India, which has taken up an initiative towards "Zero Food Wastage". It is a good example of "collaborative consumption".

The foundation has developed a large network with hotels, marriage halls and other associations which provide surplus food to be redistributed in slums, orphanages, and poor areas of the city. The Foundation does not use an electronic platform, but a system based on phone calls is used. Once collected, the food is stored in the deep freezer and tested for nutrient value by experts. After being tested, the food is distributed in slums and orphanages. Since its inception, the organization has served or distributed more than 3.5 million meals to the needy people in the city saving more than INR 30 million.

Feeding India – The initiative was launched in 2014 in India with the mission of "better food for more people" and "zero hunger". Currently it has a network of more than 26,000 volunteers working across more than100 cities in India. To date, the NGO has served more than 123 million meals to the underprivileged[24]. In 2019, the food delivery app, Zomato acquired this NGO to support their mission to achieve zero hunger in India. While Feeding India will remain an NGO, Zomato will fund the entire salaries of the team – and some core initiatives. For example, Zomato will fund the development of the "Feedi.ng" app – this app will connect donors and volunteers at a scale never seen before, and it is hoped that in a few years, it will be the platform to serve at least 100 million underprivileged people every month[25].

Robin Hood Army[26] – The Robin Hood Army is a volunteer based, zero-funds organization that works to get surplus food from restaurants and the community to serve less fortunate people. It was founded in Delhi, India in August 2014 taking inspiration from Portugal's Re-Food Program. Currently the organization has its presence in 11 different countries, including India. As of 2019, it had served more than 28 million meals across 159 cities globally. Local chapters are run by friends and colleagues, who hope to create a difference in their own unique way. The idea is to create self-sustained chapters across the world who will look after their local community.

No Food Waste[27] – No Food Waste is an organization which aims to redistribute excess food from weddings, parties, events to those who are hungry. The idea took root in Coimbatore, where the founder started with 2 Shopper Bags and a volunteer to collect the surplus food and deliver to the homeless through Public Transportation System from October 16, 2014.

In case doners want to donate excess food, they can call the No Food Waste volunteers who will collect the excess food and distribute it among people who need it. The excess food will be checked for quality because lack of proper refrigeration and storage tend to spoil food. Every day the excess food collected is used to feed an average of 1400 people, across 10 cities in South India, who lack food. In terms of figures, nearly Rs 8 crore worth of food loss has been prevented by this organization.

ACTIVITY 8.5 DISCUSSION ON SHARING PLATFORM MODEL-BASED BUSINESSES

Collaborative consumption is aligned to sustainable consumption. Businesses offering shared services must influence their customers by communicating the environmental and social benefits and take a pride to influence other customers. This often requires a behavioral change. Have you come across communication campaigns by the businesses offering shared services? What can be the barriers?

8.6 Facilitation for Circular Economy as a Service

There is now a growing demand to look for facilitation in delivering services in a circular economy. These faciliatory services include providing stakeholder connections through platforms, helping in aggregation (especially of waste), reverse logistics and supporting materials management over the life cycle. Here digital platforms play an increasing role.

Aggregation is a service that helps to "connect the dots" especially when waste is to be collected from waste generators to reach to the waste processor for sorting, cleaning, and recycling operations. Box 8.10 shows an example of mobile app "Too Good To Go" used to ensure that food wasted can be potentially sold to the customers at a low price. Box 8.11 presents case studies on Kabadiwalla Connect and Recykal in India.

BOX 8.10 MOBILE APP BY TOO GOOD TO GO[28]

Too Good To Go addresses the challenge of food waste by providing waste minimization services through an app. This app connects restaurants, cafés, bakeries, and food outlets with locals. The foodservice providers can enlist unsold food items at lower rates on their dashboard for locals to choose from.

The app has connected more than 3 million people who save meals by collecting bags of food that remain unsold. They have managed to connect with more than 4000 food store partners. The company originally started operations in Europe and has now launched its services in the USA.

BOX 8.11 AGGREGATION USING DIGITAL PLATFORM

Kabadiwalla Connect[29]– The informal ecosystem drives the recovery of post-consumer waste in cities in the developing world. Kabadiwalla Connect (KC), a technology-based social enterprise based in Chennai, established a unique business process and award-winning technology, to integrate the informal ecosystem into the reverse logistics supply chain. This system helped municipalities, brands, and waste management companies recover postconsumer waste efficiently and more inclusively in the developing world.

KC uses Information and Communications Technology (ICT) and Internet of Things (IoT) based technology to leverage the already existing informal infrastructure toward a more efficient waste management system. The KC platform makes the informal ecosystem more accessible to other players. Municipalities can get help from the informal sector to bring down their collection costs. Corporates can meet their obligation on extended producer responsibility. The company offers following four service areas:

- Mapping: Spatially enabled, industry compliant data-collection on informal and formal waste infrastructure in cities in the developing world.
- Digitalization: Know Your Customer (KYC) and transaction-based material tracking and traceability across stakeholders in the formal/informal supply-chain.
- Sourcing: SRM guarantees for processors/Producer Responsibility Organisations (PRO) in cities in the developing world through informal sector procurement.
- Municipal Collection: Reverse logistics solutions for post-consumer waste collection and powered by local informal scrap-shops and their waste-pickers.

Recykal[30]– Based in Hyderabad (India), Rapidue Technologies Pvt. Ltd. is a technology company that specializes in digital cloud-based technology for waste management since 2017. Through their brand "Recykal", they create an ecosystem for the stakeholders in the waste management space, which includes waste generators (businesses and consumers), waste processors (aggregators, informal sector, and urban local bodies) and recyclers to connect and transact plastic waste and e-waste. Bulk waste generators can also avail Recykal's EPR services through the platform.

Scanning of available SRM in the region is often the starting step before checking the techno-commercial viability. Availability of portals/directories that list vendors with details on available waste streams as SRM with contact details play an important role. Such B2B systems could be operated by a private entity. Generally, lead is taken by the industry associations, especially by the bodies such as recyclers associations. Box 8.12 presents an example of waste recovery platform in Ghana.

BOX 8.12 WASTE RECOVERY PLATFORM IN GHANA[31]

In Ghana, UNDP facilitated the establishment of a "Waste Recovery Platform" that connects stakeholders and stimulates partnerships, so as to address waste management data, identify and address policy implementation gaps; with the ultimate goal of promoting a transition towards a circular economy.

The Waste Recovery Platform has four dimensions[32]:

- A physical convening mechanism that brings together all the stakeholders on a periodic basis to connect, discuss issues of common interest and forge partnerships for effective waste management.
- A digital platform that includes a number of tools (e.g., waste resource map, compendium of technologies, etc.) to provides real time information/data on waste management and facilitate material exchange.
- Act as a promoter of innovation with catalytic support for innovative R&D and businesses that demonstrate, contribute knowledge, and raise awareness on opportunities for waste recovery or minimization (e.g., Waste Recovery Innovation Challenge).
- A communication dimension that creates awareness and build knowledge for the general behavioral change needed to make waste recovery systems effective.

Since July 2018, more than 300 stakeholders have actively participated in the co-designing process, had their voices heard, and contributed to the definition and development of the various components of the Platform.

The Waste Recovery Platform is not a project, but an initiative. The Platform was established with seed funding from UNDP's Country Investment Facility (USD 500,000). The first edition of the Waste Recovery Innovation Challenge was co-funded by the Embassy of the Netherlands (USD 100,000). Its activities for the year 2020 are supported by a partnership with the Coca-Cola Foundation.

BOX 8.13 CASE STUDY ON REVERSE LOGISTICS BY DHL EXPRESS[34]

JD Wetherspoon in UK operated more than 900 pubs. Approximately 2000 roll cages containing food, drink and other supplies are delivered to Wetherspoon pubs every day. Some 1600 roll cages are then filled with waste, that are collected by DHL and backhauled to the National Distribution Centre in Daventry for sorting and processing.

Initially, twenty-nine waste streams were collected and recycled including cardboard, plastics, cooking oil, Waste Electrical and Electronic Equipment (WEEE), milk bottles, aluminum & steel cans, wood, furniture, clothing, and catering equipment. Tetra Pak was added later and food waste at some sites, while contaminated plastics were also added to the materials to be collected. All non-recyclable waste was sent to produce energy from waste, ensuring that no waste was sent to the landfill.

The operation saw year-on-year growth in recycled waste from 5426 tons in 2008 to 8489 in 2014. And the mileage saved in collecting waste streams at the same time as delivering helped drive a GHG emission reduction of 24% per million cases delivered, equivalent to 44 tons each year.

Reverse logistics is the process of collecting and aggregating products, components, or materials at the end-of-life for reuse, recycling, and recovery. Take-back programs, warranties and product defect returns all require reverse logistics to get the product from the consumer back to the manufacturer or a waste management facility.

Deutsche Post DHL Group, along with Cranfield University and the Ellen Macarthur Foundation developed a new circular economy model for logistics. They released the report, "Waste not, Want not: Capturing the value of the circular economy through reverse logistics" [33].

Box 8.13 provides a case study from DHL where delivery of products was made synchronous to the collection of waste streams, minimizing therefore vehicle mileage, reduce costs and lower GHG emissions.

Companies are now offering solutions for material management, acting as an intermediary between businesses. Apart from working as an aggregator, these businesses also offer extended services such as preparation of contracts during the procurement, manufacturing and waste management stages of a product life cycle including in reverse logistics operations.

A materials manager's role is important throughout the lifecycle in ensuring resource efficiency and minimizing waste to landfills. A materials manager can advise on producer responsibility practices, recycling methods as

BOX 8.14 MATERIAL MANAGEMENT FOR CLOSING LOOP – CASE OF CIRCUL8[35]

Circul8 is a cloud-based software suite for the Circular Economy – it is the industry-proven software solution encompassing multiple regulations and waste streams as well as supporting a wider range of stakeholders in the Circular Economy. Being deployed in 20 countries, across 4 continents, managing 8 million tons of waste, and EUR 2.7 billion of revenue. It is built to manage every step in the process from individual producer responsibility take-back through to the final recycling of resources, that can be fed back into the manufacturing process, closing the loop.

Circul8 provides an online, real-time portal, which is accessible to every user in the process. This portal can perform tasks such as: request a pick-up, declare put-on-market data, confirm execution of a pick-up order, obtain traceability information, validation, or confirmation of processes, and generate customized reports.

well as standards set by governments, and maximize the use of materials procured and products throughout the life cycle. This area of service is considered as an emerging area of business in circular economy. Box 8.14 presents the facilitation service provided by Circul8.

There are several consulting companies that have emerged focusing on circular economy taking a faciliatory role. Thinking Circular[36] offers consulting service for a wide range of topics in circular economy, from technical advice to product re-designing including material sourcing, Sphera[37] offers technical advice on how to improve existing products, modify the supply change, generate new ideas and form new partnerships so that the organization can move towards a circular economy.

ACTIVITY 8.6 DISCUSSION ON FACILITATION MODEL-BASED BUSINESSES

Facilitation using digital platform is playing increasing role in the business of circular economy. This service can be offered as a business or it can support any of the business models described in this chapter. Few examples illustrating this business model giving examples of aggregation, B2B portals, reverse logistics, life cycle material management have been provided. What may the basis of the commercials of these models, e.g., a subscription or fees based on product/waste materials reversed?

8.7 Inclusive Business

This business model is particularly important and relevant in developing economies where the informal sector plays a major role in waste picking and recycling operations. Working with the informal sector on an inclusive basis, waste management companies can collect the waste for upcycling, while ensuring that the livelihoods, health, and safety of waste pickers is looked after. Several such initiatives were illustrated in Chapter 6. In this section, two examples that work in waste recycling as an inclusive business are highlighted. See Box 8.15.

BOX 8.15 NEPRA – AN EXAMPLE OF INCLUSIVE BUSINESS IN INDIA[38]

NEPRA is a waste management company primarily engaged in the sorting of dry solid waste. The company was founded with a goal to be a one stop, single contact for dry solid waste management for stakeholders such as waste generators, waste pickers, transporters, and recyclers. This would streamline the supply chain of solid waste and ensure greater levels of recycling, hence achieving a circular economy. The overall business operation involves the collection of dry waste from waste generators, segregation of recyclables through a combination of mechanical and manual processes, and sale of recyclables to authorized recyclers.

NEPRA's workforce comprises of more than 55% of female staff, many of whom were earlier informal waste pickers. Approximately 90% of the waste sorted by NEPRA is recycled to manufacture various goods and commodities, and the remaining 10% is sent to cement kilns for use as refuse derived fuel. Currently, NEPRA has its facilities in Ahmedabad and Indore which together manage more than 100,000 tons of solid waste annually.

As of 2018, NEPRA procures waste from more than 1700 waste pickers – a ten times increase since 2013. Waste pickers are typically poor, illiterate, and unskilled – with many being migrants and/or ranking lowest in caste hierarchy. NEPRA provides these workers with fair prices, transparent service, and immediate cash payments. On an average, waste pickers have seen a 30–40% increase in their income levels with NEPRA. As NEPRA's operations have grown, the company has been able to offer formal employment, including at its MRF, and, as of September 2018, employs 210 people. All employees undertake regular health and safety training and are offered retirement benefits and health insurance. Furthermore, the company is promoting financial inclusion by ensuring all employees have a bank account to which pay checks are deposited as opposed to being paid in cash as was in the initial stage of operations.

ACTIVITY 8.7 DISCUSSION ON INCLUSIVE BUSINESS

Including the interest of informal sector in operating business on circular economy is very relevant in developing economies where waste pickers take a major share in the material circulation. Private sector players, especially those who have EPR obligations have recognized the importance to be inclusive. Partnership models between the "formal" and "informal" sector are at the core of such inclusive business models. Research more such business models where you see that inclusion also includes skill building and microfinance, to promote innovations and entrepreneurship.

8.8 Performance Based Service

Selling performance is the most profitable and most material-efficient business model of the circular economy[39]. The fees in this service depend on the success or on delivered performance, incentivizing the service provider and averse the risk to the contractor because of failure or poor performance. An example is – fees determined based on the percentage of energy saved through efficiency practices advised by the consultant.

In many ways, the performance-based business model follows "consumer pays principle". It focuses on resource efficiency, optimization while addressing contextual sustainability to provide financial advantages on both sides (i.e., contractor and service provider), leading to higher competitiveness. Energy, waste, and water sectors have considerably matured in performance-based contracts.

Energy Performance Contracting (EPC) is a form of "creative financing" for capital improvement, allowing funding energy upgrades from savings accrued through cost reductions. Under an EPC arrangement, an external organization such as Energy Service Companies (ESCO), implements a project to deliver energy efficiency, or a renewable energy project, and uses the stream of income from the cost savings, or the renewable energy produced, to repay the costs of the project, including the costs of the investment. Essentially the ESCO will not receive its payment unless the project delivers energy savings as expected[40].

The approach is based on the transfer of technical risks from the client to the ESCO based on performance guarantees given by the ESCO. In EPC ESCO remuneration is based on demonstrated performance; a measure of performance is the level of energy savings or energy service. EPC is a means to deliver infrastructure improvements to facilities that lack energy engineering skills, manpower or management time, capital funding, understanding

BOX 8.16 PERFORMANCE CONTRACT FOR WASTE MANAGEMENT IN CHENNAI IN INDIA[41]

The Corporation of Chennai signed a concession agreement with Ramky Enviro Engineers in November 2011 for management of Municipal Solid Waste in the city. The project scope primarily involves door-to-door collection of household waste, segregation of waste, transport, and disposal of non-recyclable waste to the designated dumping grounds and street sweeping. Contractor payment involved a tipping fee to per ton of waste collected and transported; 50% of the fee is paid automatically by weight while the remaining 50% is linked to contractor performance as measured by certain performance parameters captured daily. Minor penalties (0.25–2%) are levied for service deficiencies relating to public complaint frequencies, worker safety, non-compliance with certain aspects of collection and vehicle/personnel deployment. The reference cited for this case study showcases several such examples in India.

ACTIVITY 8.8 DISCUSSION ON PERFORMANCE BASED SERVICE MODELS

Performance based business models play an important role in circular economy. These models demand innovation and lead to continuous improvement in re-circulation of the resources. The key in these models is the clear definition of success or performance criteria. These criteria must be objective and measurable. Describe the performance criteria you would propose for providing a service on waste to energy, wastewater recycling and energy efficiency improvement projects. Is it possible to include "outcome" or "impact" indicators in the performance contracts? Is it feasible?

of risk, or technology information. Cash-poor, yet creditworthy customers are therefore good potential clients for EPC[41].

Box 8.16 illustrates an example of performance-based waste management contract in Chennai, India.

8.9 Hybrid Business Models

The eight models of business in circular economy do have overlaps and hybrid versions are found in many practical applications. In this section two examples are provided, one of PRO and second by Bundles on PaaS.

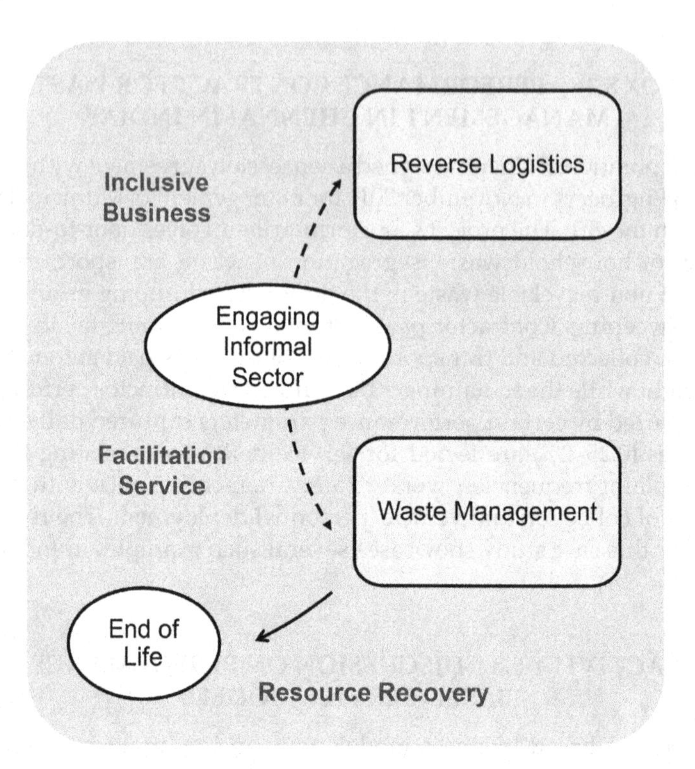

FIGURE 8.2
The PRO model.

PRO is defined as a professional organization authorized or financed collectively or individually by producers, which can take the responsibility for collection and channelization of waste (typically e-waste and plastic waste) generated from the "end-of-life" of their products to ensure environmentally sound waste management. A PRO is a professional organization that helps producers/brand owners meet their EPR targets through various processing technologies for plastic waste including Multi Layered Plastic (MLP) or its end-of-life applications like waste to energy, waste to fuel, waste to road and waste to cement kiln in the country[42]. Figure 8.2 shows how a PRO model make use of business models such as inclusive, resource recovery and providing services of facilitation.

In India, PRO is expected to assist the producers upon engagement in one or more of the following tasks[43].

- Achieving collection targets. *(facilitation service)*
- Establishment of collection mechanisms such as door to door collection. *(inclusive business)*

- Implementation of buy back/take back. *(facilitation service)*
- Establishment of collection centers/points. This may include setting up of collection godowns or operating through warehouses. *(facilitation service)*
- Logistics arrangement. *(facilitation service)*
- Ensuring traceability of the waste collected and channelized & disposal of processed waste at end of life. *(facilitation service)*
- Ensuring environmentally sound and safe dismantling and recycling. *(resource recovery)*
- Conducting awareness program among consumer's/bulk consumers/producers for collection and channelization of waste. *(facilitation service)*
- Helping producers in filing of quarterly/annual returns as per the rules. *(facilitation service)*
- Providing an extended producer's responsibility plan as legally required *(facilitation service).*

For managing plastic waste, scope of services offered by PRO include:

- Collection, sorting and processing of MLP and/or recyclable plastic and/or beverage carton.
- Dispatch of MLP for co-processing in cement kilns in accordance with guidelines.
- Dispatch of recyclable plastic and/or beverage carton or recycling in accordance with guidelines.
- Provide take back credits of MLP and/or recyclables plastic and/or beverage carton equivalent to the amount of waste diverted towards co-processing/recycling.
- Provide a diversion certificate stating quantum of waste diverted from landfilling, tons of CO_2e emissions mitigated and energy saved which can be submitted to the regulator.

Some PROs follow a business model by attracting partners on a franchising mode and provide technical assistance for processing of waste plastic and guarantee the market for recycled products. Resource recovery business is thus included. PRO India for example claims more than 1000 such franchise across India who are involved in plastic recycling[43].

In all the eight business models described above, financing and policy frameworks play a vital role. Chapter 9 discusses the subject of financing in circular economy. Chapter 10 provides an overview of the policy frameworks.

Box 8.17 provides further information on the hybrid business model by Bundles and Miele in the Netherlands.

BOX 8.17 HYBRID BUSINESS MODEL BY BUNDLES AND MIELE IN THE NETHERLANDS[44]

Bundles offers subscriptions on white goods such as washing machines, tumbler dryers, dishwashers, and coffee machines. The machines are made by Miele who follows principles of sustainable product design, especially design for disassembly and design for repairability. Through its collaboration with Bundles, Miele is encouraged to design goods more circularly by developing products that are upgradable, can be easily disassembled and have long lifespans.

The machines are installed at no additional costs at customers premises. The old machines are taken by for recycling if so, wished by the customer. So, take back is practiced as a part of the service.

Instead of ownership, the customers pay Bundles for access and use. The time that a washing machine is out of order, is compensated. Above all, Bundles guarantee a high quality and well-functioning appliances with endless guarantee.

Bundles thus follows a hybrid PaaS model, with an option of take back of old white goods (resource recovery), providing its customers with long-life products (product life extension) and charging them a fixed fee for access to the asset (product lease) and a variable fee per washing cycle on the basis of pay-per-use (as PaaS). In some instances, assets are shared (product sharing).

A device is attached to the washing machines which monitors its usage. Statistics gathered from the machine are displayed on the *Wash-App* and translated into tips and insights to reduce the overall cost of doing laundry, including energy, water, and detergent consumption. The App also gives feedback and advice on the effect of different sorting, dosing, and programming schemes. Further, Miele washing machines are connected to the internet and therefore, it is possible to detect the most effective proportion between the washing load and amount of soap. Bundles thus uses digital technology for facilitation.

Due to the long-life span of the machines Bundles uses, Bundles could save more than 2000 machines from disposal in cooperation with more than 1000 users. Bundles customers save an average of 91 kWh of energy per year, more than 10 l of detergent and more than 3000 l of water through efficient equipment and advice. That is the equivalent of 220 tons of GHG emissions saved.

8.10 Key Takeaways

- There are several business models available in a circular economy. While each business model is unique, there are overlaps and in practice business models are operated in a hybrid basis. The business of PRO is an example.

- The business of facilitation is gaining popularity. Here, digital platforms are playing a major role.
- PPP have a great potential in the business of resource recovery in a circular economy. These models are very relevant to the municipal bodies and industrial clusters.
- The business models on product life extension and product-as-a-service intersect. These models when integrated could be more effective. The product related regulations extending over life cycle and efforts towards behavioral change of the consumers will play an important role in the success of these business models. Similarly, business models based on sharing will work only when there is an understanding and commitment from the customers or the consumers.
- Inclusive business models have recently emerged, especially in the developing economies. These models provide significant social benefits apart from economic and environmental advantages.
- The performance-based business models should be promoted. These models push innovation, set new benchmarks and result in continuous improvement in resource efficiency and resource re-circulation.

ADDITIONAL READING

1. **Performance-based contracting for services: A new mechanism for increasing water utility efficiency:** IWA (International Water Association) launched its Task Group on Performance-based Contracting (PBC) in 2011 to provide thought leadership on this new topic. This article discusses what PBC is about and its relevance to the water industry today, present the few projects that have been implemented so far, and how the new Task Group intends to support the development of PBC in upcoming years.

Source: Marin, P., Janssens, J., Williams, T. and Castro-Woolbridge, V., 2015. Https://Www.Iwa-Network.Org/Wp-Content/Uploads/2015/12/1464301513-PBC-IWA-19Nov12-V4.Pdf. The International Water Association. <https://www.iwa-network.org/wp-content/uploads/2015/12/1464301513-PBC-IWA-19Nov12-v4.pdf> [Accessed 7 November 2020].

2. **OECD guidelines for performance-based contracts between municipalities and water utilities in Eastern Europe, Caucasus, and Central Asia (EECCA):** The requirements for performance-based contracts, as contemplated in these Guidelines, are provided as a starting point for further and improved contractual arrangements in the water sector in EECCA. These Guidelines are relevant for municipalities wishing to establish a contractual relationship with a public or a private utility.

Source: OECD. 2006. OECD Guidelines For Performance-Based Contracts Between Municipalities And Water Utilities In Eastern Europe, Caucasus And Central Asia (EECCA). <https://www.oecd.org/environment/outreach/36397942.pdf> [Accessed 7 November 2020].

3. **Waste not, want not report:** As part of the Circular Economy 100, a cross-industry and multi-disciplinary platform, several global, circular-minded companies, have identified both a knowledge and performance gap regarding one of the key circular economy enablers: reverse logistics. To address this need, Deutsche Post DHL Group and Cranfield University joined forces with selected CE100 member companies to develop a model for assessing and improving reverse logistics' processes across product groups and even related sectors. Based on company interviews, exploratory workshops, applied logistics expertise and scientific method, the Reverse Logistics Maturity Model presented in this report provides a roadmap for meeting the challenges of reverse logistics – from complexity and regulatory requirements to the dilemma of return forecasting – and devising effective return management programs.

Source: Deutsche Post DHL. n.d. *"Waste Not, Want Not" - Reverse Logistics For A Circular Economy.* <https://www.dpdhl.com/en/trends-in-logistics/studies/circular_economy.html> [Accessed 7 November 2020].

Notes

1. UN. 2011. Basel Convention <http://www.basel.int/Portals/4/download.aspx?d=UNEP-CHW.10-6-Add.3-Rev.1.English.doc> [Accessed 7 November 2020].
2. GENeco. n.d. *Case Study: Bio-Bus.* <https://www.geneco.uk.com/Case_study_bio_bus/> [Accessed 7 November 2020].

3. Abidi S. 2020. *City Waste to Fuel 100 Bio-CNG Buses in Pune*. Autocar Professional. <https://www.autocarpro.in/feature/city-waste-to-fuel-100–biocng-buses-in-pune-56521> [Accessed 7 November 2020].
4. Innovation Policy Platform. n.d. *Business Models for Converting Waste to Energy*. <https://www.innovationpolicyplatform.org/www.innovationpolicyplatform.org/system/files/3%20Waste-to-Energy_Apr6/index.pdf> [Accessed 7 November 2020].
5. YouTube. 2018. *Bio-Mining At Kumbakonam*. <https://www.youtube.com/watch?v=X9GANY9mqxo> [Accessed 7 November 2020].
6. Entocycle. n.d. *About*. <https://www.entocycle.com/protein> [Accessed 7 November 2020].
7. Circular Systems. n.d. *Agraloop Biofibre*. <https://circularsystems.com/agraloop/#what-is-agraloop> [Accessed 7 November 2020].
8. bio-bean. n.d. *Elements*. <https://www.bio-bean.com/elements/> [Accessed 7 November 2020].
9. Toast Ale. n.d. *Home*. <https://www.toastale.com/> [Accessed 7 November 2020].
10. Nikkei Asia. 2017. *Japanese Companies Digging For Gold In Urban Waste*. <https://asia.nikkei.com/Business/Japanese-companies-digging-for-gold-in-urban-waste> [Accessed 7 November 2020].
11. Ellen Macarthur Foundation. n.d. *MUD Jeans: Pioneering A Lease Model for Organic Cotton Jeans*. <https://www.ellenmacarthurfoundation.org/case-studies/pioneering-a-lease-model-for-organic-cotton-jeans> [Accessed 7 November 2020].
12. Circulator. n.d. *Philips Pay Per Lux*. <https://www.circulator.eu/browse-the-cases/detail/philips-pay-per-lux#:~:text=The%20'Pay%20per%20Lux'%20concept,Also%20maintenance%20is%20included.> [Accessed 7 November 2020].
13. Philips. 2015. *Philips Provides Light as a Service to Schiphol Airport*. <https://www.philips.com/a-w/about/news/archive/standard/news/press/2015/20150416-Philips-provides-Light-as-a-Service-to-Schiphol-Airport.html> [Accessed 7 November 2020].
14. ORIX. n.d. *Leasing*. <https://www.orixindia.com/leasing.php> [Accessed 7 November 2020].
15. Grineva, M., 2018. *Chemical Leasing - The Solution to Managing Chemicals Sustainably: 12 Things You Need To Know*. LinkedIn. <https://www.linkedin.com/pulse/chemical-leasing-solution-managing-chemicals-12-things-grineva/> [Accessed 17 November 2020].
16. Global Chemical Leasing Program - www.chemicalleasing.org
17. PanurgyOEM. n.d. *Repair and Refurbishment Services*. <https://www.panurgyoem.com/repair-refurbishment-services/> [Accessed 7 November 2020].
18. Silicon India. 2018. *20 Most Promising Refurbishment Companies*. <https://enterprise-services.siliconindia.com/ranking/refurbishment-companies-2018-rid-360.html> [Accessed 7 November 2020].
19. Circular Economy Guide. n.d. *Sharing Platforms*. <https://www.ceguide.org/Strategies-and-examples/Sell/Sharing-platforms> [Accessed 7 November 2020].
20. Zify <http://zify.co/en-EU/>[Accessed 7 November 2020]
21. Sride <https://sride.co/>[Accessed 7 November 2020]
22. Redbus <https://www.redbus.in/rpool>[Accessed 7 November 2020]

23. Annakshetra. n.d. *About Us.* <http://annakshetra.org/pages/about_us> [Accessed 7 November 2020].
24. Feeding India. n.d. *Our Work.* <https://www.feedingindia.org/ourwork> [Accessed 14 December 2020].
25. Goyal, D., 2019. *Welcoming Feeding India To Zomato.* Zomato. <https://www.zomato.com/blog/fighting-hunger-and-wastage> [Accessed 14 December 2020].
26. The Robin Hood Army. n.d. *Home.* <https://robinhoodarmy.com/> [Accessed 14 December 2020].
27. No Food Waste. n.d. *About.* <https://nofoodwaste.org/about/> [Accessed 14 December 2020].
28. Too Good To Go. n.d. *Home.* <https://toogoodtogo.com/en-us> [Accessed 7 November 2020].
29. Kabadiwalla Connect. n.d. *About.* <https://www.kabadiwallaconnect.in/> [Accessed 7 November 2020].
30. Recykal. n.d. *About Us.* <https://www.recykal.com/> [Accessed 7 November 2020].
31. UNDP. 2020. *Transitioning to a Circular Economy With a Multi-Stakeholder Platform in Ghana.* <https://www.africa.undp.org/content/rba/en/home/presscenter/articles/2020/transitioning-to-a-circular-economy-with-a-multi-stakeholder-pla.html> [Accessed 7 November 2020].
32. UNDP. 2020. *Multi-Stakeholder Waste Recovery Initiative.* <https://www.gh.undp.org/content/ghana/en/home/projects/waste_initiative.html> [Accessed 7 November 2020].
33. Deutsche Post DHL. n.d. *"Waste Not, Want Not" – Reverse Logistics For A Circular Economy.* <https://www.dpdhl.com/en/trends-in-logistics/studies/circular_economy.html> [Accessed 7 November 2020].
34. Edie. 2015. *What Makes A Sustainability Leader? Case Study: JD Wetherspoon & DHL.* <https://www.edie.net/library/What-makes-a-Sustainability-Leader–Case-study–Wetherspoons-and-DHL/6625> [Accessed 7 November 2020].
35. Circul8. n.d. *The Circular Economy Software.* <https://circul8.eu/pdf/circul8_brochure_web.pdf> [Accessed 7 November 2020].
36. Thinking Circular. n.d. *Consulting.* <https://thinking-circular.com/consulting/> [Accessed 7 November 2020].
37. Sphera. n.d. *Circular Economy Consulting.* <https://sphera.com/circular-economy-consulting/> [Accessed 7 November 2020].
38. EMPEA Institute. 2018. *Case Study: Nepra Resource Management Pvt. Ltd.* <https://www.empea.org/app/uploads/2018/10/CaseStudy_Nepra_WEB.pdf> [Accessed 7 November 2020].
39. Stahel, W., 2013. Policy for material efficiency – Sustainable taxation as a departure from the throwaway society. *Philosophical Transactions of the Royal Society A: Mathematical, Physical and Engineering Sciences,* 371(1986), p. 20110567.
40. European Commission. n.d. *Energy Performance Contracting.* <https://e3p.jrc.ec.europa.eu/articles/energy-performance-contracting> [Accessed 7 November 2020].
41. Srinivasan, S., 2015. *Performance-Based Contracting In Sanitation Delivery.* Centre for Development Finance. <http://www.susana.org/_resources/documents/default/3-2284-22-1437031496.pdf> [Accessed 7 November 2020].

42. Mondaq. 2019. *Producer Responsibility Organization*. <https://www.mondaq.com/india/waste-management/786196/producer-responsibility-organization#:~:text=PRO%20is%20defined%20as%20a,management%20of%20such%20e%2Dwaste.> [Accessed 7 November 2020].
43. PRO India - https://proindia.net/
44. Sustainable Finance Lab. 2019. *The Circular Service Platform*. <https://sustainablefinancelab.nl/wp-content/uploads/sites/334/2019/04/The-Circular-Service-Platform-White-Paper.pdf> [Accessed 17 November 2020].

9

Innovation and Financing in Circular Economy

9.1 Innovation

Innovation is often considered as a driver to circular economy. The entrepreneurs conceive innovations, and with the support of finance, these ideas get translated to material innovations, sustainable product designs, and circular business models. Finance for innovation plays an important catalytic role. Processes of appraisal and due diligence in financing (especially by angel investors and venture capitalists) serve as a peer mechanism to look at the innovation through a practical lens. Therefore, we must build an ecosystem for innovators, entrepreneurs, and the financing communities to move the wheel of a circular economy.

The ecosystem for innovations presents an opportunity to establish a regenerative economic model that leads not just the economic and environmental benefits but social benefits as well and on a scale it deserves. It is also critical that we make every effort to bring in the behavioral change in society to appreciate and support the innovations and participate and be part of the solution.

Innovation in a circular economy is not to be limited to "material and technological innovations" but also in business models and in devising policy frameworks. While innovations are desirable, we must assess them by examining disruptions that some may cause across the life cycle. Impact assessment of disruptive technologies is seldom done.

The corporations' sustainability commitments, zero waste goals by local and regional authorities, and the national governments' commitments have been some of the key drivers to innovation in a circular economy. A growing focus on local sourcing, insights into life cycle impacts, and cascading opportunities for resource re-circulation have been motivational factors. Further, digital technologies and advancement in alternate materials have opened new opportunities. Above all, policy frameworks consisting of regulations and economic instruments play an important role to "demand" or "push" innovations in a circular economy. Strategic partnerships between entrepreneurs and financing institutions are however, needed to translate innovations into viable and inspiring business models, Innovation supported through partnerships, including Public-Private Partnerships (PPPs), is perhaps the key.

Cities and regions can provide venues for experimenting with different partnerships and solutions and have the flexibility and scope for policy experimentation. Their high business and consumer density, presence of universities and research institutes and connectivity, make them ideal locations for innovation hubs and incubator spaces[1]. For example, Finland is one of the pioneers of innovative partnerships for a circular economy and has adopted an ambitious national roadmap on a circular economy based on innovation partnerships.

Initiatives in Birmingham based on industrial symbiosis, demonstrates how city-level innovative partnerships can drive the transition toward a circular economy. Box 9.1 describes the Tyseley project launched by International Synergies in partnership with Birmingham City Council in the United Kingdom.

There are numerous examples of new technologies, processes, services, and business models that are re-shaping product life cycles from design through production and paths for resource recirculation such as recycling. New materials and technologies are making products and assets more durable and resilient, while digital platforms, data, and analytics are transforming how the products and assets are used for fullest utilization and enjoy extended life. Businesses across the world have invested extensively in this direction.

The Bitz and Prezel innovation festival – convened by founders for founders – is now giving a focus on the need for "impact" driven innovation to help achieve the Sustainable Development Goals (SDGs). The combination of entrepreneurial passion and energy with emerging technologies from the Industry 4.0 revolution, are starting to generate incredible outcomes[3].

Horizon 2020 is the biggest European Union (EU) Research and Innovation programme ever with nearly EUR 80 billion of funding available over 7 years (2014–2020). The program attracts investments from the private sector. Seen as a means to drive economic growth and create jobs, Horizon 2020 has the political backing of Europe's leaders and the Members of the European Parliament. These leaders endorsed that research is an investment in the future. Horizon 2020 has therefore been at the heart of the EU's blueprint for smart, sustainable, and inclusive growth and jobs[4].

Eco-innovation offers a huge market for enterprises and has become one of the cornerstones of the EU strategy in response to the global environmental and economic challenges being faced. Eco-innovation is the introduction of any new or significantly improved product (good or service), process, organizational change or marketing solution that reduces the use of natural resources (including materials, energy, water, and land) and decreases the

BOX 9.1 EXPLOITING INDUSTRIAL SYMBIOSIS IN BIRMINGHAM FOR INNOVATIONS – CASE OF TYSELEY PROJECT[2]

Birmingham was one of the first cities to adopt a pro-active industrial symbiosis approach to develop a medium and long-term strategy for sustainable economic development for the Tyseley Environmental Enterprise Zone (TEEZ). Birmingham City Council commissioned International Synergies to undertake an analysis of infrastructure and resource flows in TEEZ applying industrial symbiosis methodology to devise a short, medium, and long-term plan for economic regeneration.

The potential economic, social, and environmental impact of implementing the industrial symbiosis opportunities identified for today, tomorrow, and the future resulted in:

- 400–500 direct jobs (and further jobs related to investment).
- 55,000 tons per annum of Greenhouse Gas (GHG) emission reduction.
- Cost savings for existing companies more than GBP 1.9 million per annum.
- Additional revenue for Birmingham-based businesses of GBP 8–10 million per annum.
- Total Gross Value-Added impact GBP 12–15 million per annum.

Birmingham is widely acknowledged for its vibrant metals industry. The study identified opportunities for the recovery of precious metals, rare earth elements, and other critical materials from local resource flows in TEEZ, which would yield immediate benefits and help reduce future UK dependence on imports.

Examples of this included the potential to recover platinum and palladium from municipal road sweepings collected in the area using extraction and processing equipment already available. The study also identified opportunities for the recovery of silver from dental amalgam and X-rays.

release of harmful substances across the whole life cycle[5]. Box 9.2 describes two initiatives on eco-innovation in the EU.

A.SPIRE is the European Association that is committed to manage and implement the SPIRE PPP. It represents innovative process industries, and more than 150 industrial and research process stakeholders from over a dozen countries spread throughout Europe. SPIRE brings together cement, ceramics, chemicals, engineering, minerals and ores, non-ferrous metals, pulp and

BOX 9.2 ECO-INNOVATION OBSERVATORY AND EIT RAWMATERIALS IN THE EU

ECO-INNOVATION OBSERVATORY

Eco-Innovation Observatory was established as 3 year initiative financed by the European Commission's Directorate-General for the Environment from the Competitiveness and Innovation framework Programme (CIP)[5]. The website has several useful resources to download such as eco-innovation Small and Medium Enterprise (SME) guides[6]. These guides provide an overview of emerging business opportunities in eco-innovation and circular economy and how to reconsider business models, develop new products, technologies, or services, or improve production processes.

EIT RAWMATERIALS[7]

EIT RawMaterials is the world's largest community dedicated to innovation and education in the raw materials domain. EIT RawMaterials is initiated and funded by the European Institute of Innovation and Technology (EIT), a body of the European Union. EIT RawMaterials connects roughly 400 organizations from leading industries, universities, research, and technology organizations (the so-called knowledge triangle) from more than 20 EU countries.

One of the core thematic areas covered by EIT RawMaterials is the "Design of Products and Services for the Circular Economy". The partners include organizations that are active across the entire raw materials value chain of industrial minerals and metals (with emphasis on, e.g., mining and recycling) and include end-user industries (e.g., car components manufacturers and automotive industries). Within this large Innovation Community, they collaborate in implementing innovation, and education projects.

As of 2019, the project portfolio of EIT RawMaterials includes almost 240 projects that are expected to result in the introduction of innovative and sustainable products, processes, and services, as well as talented people who will deliver increased economic, environmental, and social sustainable impacts to the European society.

paper, refining, steel, and water sectors, several being world-leading sectors operating from Europe. The mission of A.SPIRE is to ensure the development of enabling technologies and best practices along all the stages of large scale existing value chain productions that will contribute to a resource-efficient process industry[8].

Advance London is a 3 year program which places London's small and medium businesses at the heart of the circular economy. It is jointly funded by the European Regional Development Fund and London Waste and Recycling Board. Advance London business supports SMEs with existing circular economy offerings, as well as SMEs that want to transition to a circular economy business model. With their support, businesses can improve and scale, as they change their processes to become greener and more circular. Since its inception, it has provided over 150 SMEs with customized support[9].

Sitra is a public fund in Finland set up to promote innovation to improve Finland's competitiveness and the well-being of the Finnish people. As an independent organization, Sitra carries out practical experiments, compiles cross-boundary networks and develops and finances business operations. Sitra has compiled 39 inspiring innovations in circular economy across the world[10].

The Geological Survey of Finland, VTT Research and Aalto University work together with industry to develop solutions for the needs of a carbon-neutral and resource-efficient society. A joint laboratory of circular economy was opened in Otaniemi. The Circular Raw Materials Hub or Circular Raw Material Infrastructure (RAMI) is a nationally significant strategic infrastructure assessed by the Academy of Finland. Various components within RAMI provide the foundations for a holistic approach toward the whole material value chain that encompasses primary production and new materials through to product design and reuse/recycling. Their key focus areas include[11]:

- Knowledge-driven circular economy.
- Toward sustainable processing.
- Materials for renewable energy.

Five million euros have been invested in the infrastructure of the joint laboratory. It will serve research and industry as well as educate future experts in the circular economy.

Second Muse[12] is a global innovation company that builds inclusive economies to benefit people and protect the planet. They work with visionary organizations all over the world (i.e. The Rockefeller Foundation, the National Aeronautics and Space Administration [NASA], World Bank) to tackle some of the world's biggest challenges. Some of the current work includes catalyzing solutions to end plastic pollution (the Incubation Network), improving youth wellbeing through digital experiences (Headstream) and designing the future of the global food system (Food System Vision Prize).

SecondMuse started 10 years ago as a radical experiment to see if they could build new economic relationships without employing competitive frameworks. They had a theory of change rooted in collaboration and inclusion, and a vision of a prosperous future that was just and equitable for all. Programs include implementing incubators and accelerators, running global

BOX 9.3 RABO BANK'S CIRCULAR BUSINESS CHALLENGE[13]

Rabo Bank's Circular Business Challenge consists of a region scan, company scan, various workshops, and designing a circular action plan.

The results of the Circular Business Challenge at Rabo Bank have been very encouraging. The circular businesses that participated could reduce their "footprints" and make their reputation grow. They could establish new contacts, form new partnerships, venture into new markets and the sales rose accordingly.

One-third of the participating businesses achieved a direct financial benefit from circularity. Another third expects to benefit financially from circular economy in the near future. Three-quarters of participants have made circularity as a part of their business strategy.

The businesses that participated in the challenge were diverse, proving that any business can help make the economy more circular. More importantly, the business models proposed were sustainable. Because of the cross-sector discussions facilitated by the dialogue sessions, the participants could access new circular networks and interact with some of their regions' role models.

To make the Circular Business Challenge more effective, Rabo Bank partnered with CSR Netherlands and KPMG in 2017. CSR Netherlands deploys its network, knowledge, and active matchmaking capabilities to guide entrepreneurs during the challenges. KPMG is involved in working on the analyses performed in the context of the region and conducting company scans.

hackathons, managing innovation challenges, and designing and deploying investment funds.

More recently, innovations in a circular economy are encouraged by holding challenges and hackathons.

As a part of its commitment to promote innovations in the business of circular economy, Rabo Bank organized a Circular Business Challenge. Box 9.3 shows some of the findings.

EU-Resource Efficiency Initiative and Ekonnect Knowledge Foundation in India launched a Circular Economy Challenge in September 2020 reflecting on the COVID-19 pandemic situation. Proposals were invited from India and the EU to respond to the following three challenges:

- Avoiding Single Use Plastic and Managing Safe Disposal in the wake of COVID-19.
- Extending Product Life – Innovative Repair and Refurbishing Services, Smart Waste Sorting and Cleaning Systems.
- Business models on Extended Producer Responsibility.

BOX 9.4 CIRCULARS ACCELERATOR PROGRAM TO PROMOTE INNOVATIONS IN CIRCULAR ECONOMY[15]

Circulars ran a highly successful Circulars awards program that received 1500+ applications from over 70 countries during the 5-year program. The Circulars has transitioned from an awards program recognizing individuals and organizations championing the circular economy agenda to an accelerator program. The 2021 6-month program will connect innovators and entrepreneurs with industry leaders and circular experts for tailored mentorship. The program also unlocks opportunities for accelerator participants to work directly with cross-industry leaders to scale their circular solutions.

The program will consist of:

- Strategic, business, and circular expertise delivered in a series of workshops and coaching modules.
- A select number of 1:1 session with a variety of experts to explore key barriers to scale.
- Invitation to key circular economy events, facilitating networking opportunities.
- Touchpoints with investor community during the program and at the Innovation Showcase, and
- Ongoing support from Accelerator Team to guide overall engagement and help shape opportunities.

In this Challenge, the shortlisted teams were coached through three mentoring workshops on preparing a business plan from the concept proposal, ending with pitching the business plan to the jury. Post the announcements of the awards, a compendium was published that includes the business plans prepared by the participants[14].

Initially, the idea of the sponsor of Challenges or Hackathons has been to give awards and provide for shortlisted teams mentorship support prior to final evaluation, but now the trend is to continue the support further by expanding to an accelerator program. Box 9.4 describes Circulars Accelerator Program to promote innovations in a circular economy.

More recently, the World Economic Forum (WEF) has created the Scale360° Playbook[16], an initiative to build ecosystems for the circular economy and help solutions scale. Launched in September 2020, Scale360°'s diverse global community is expected to create much-needed connectivity across a wide range of stakeholders, including financiers, technologists, activists, public servants and others in need of circular innovation solutions. Brought together by Scale360°, the members can share challenges, innovations, lessons learned, and opportunities.

Operating platforms with stakeholders in a circular economy can facilitate financing. To accelerate transition toward circular economy, WEF collaborated with the Ellen MacArthur Foundation for a number of projects to scale business driven circular economy innovations. Building on this work, the Platform for Accelerating the Circular Economy (PACE) [17] was launched in 2017 as a public-private collaboration. The Platform will bring key stakeholders to the table, to collaboratively design public policy reforms, from both a global perspective through the leaders' network and a local view through the projects[18].

PACE aims to create systems change at speed and scale by enabling partners to[20]:

- Develop blended financing models for circular economy projects, especially in developing and emerging economies. Applying mixed funding approaches to de-risk private investments would help identify strategies for national governments to scale private sector Circular Economy activities such as secondary materials recovery systems and reorient their larger-scale investments to integrate circular economy design/principles. The Platform will help scale existing activities by brokering partnerships and test collaborative funding approaches through the network of the private, public sector and institutional partners.
- Help create and adjust enabling policy frameworks to address specific barriers to advancing the circular economy. Policies and regulations often surfaced as both key barriers and enablers for scaling up circular economic efforts, including trade policies, waste regulations, public procurement policies, resource pricing, etc. These cross-cutting systemic policy issues need to be addressed from both global and national perspectives and through innovations. Further solutions need to be designed collaboratively by integrating government, business, and civil society views.

ACTIVITY 9.1 DISCUSSION ON INNOVATION SUPPORT AND INTELLECTUAL PROPERTY RIGHTS

Financing innovation and operation on innovation platforms for knowledge exchange play an important role. While a partnership or collaborative approach to support innovation is recommended, given the circular economy's complex and multidisciplinary canvas, protecting intellectual property rights could be challenging. Research on how to address this challenge while co-creating and nurturing circular innovations.

9.2 Financing Circular Business

In the last three decades, financing has been provided through grants (especially for demonstration of innovative projects) and concessional loans (mainly for full-scale implementation) targeted to MSMEs to support cleaner production projects. The money is provided through special funds or lines of credit. The finance has been in some cases conditional, e.g., limited to 50% in the case of high-risk demonstration projects or performance-based, e.g., 20% of the loan is reimbursed as a grant if the project meets the target on completion. As PPP models have become popular, especially in water, energy, and waste sectors, municipal governments have resorted to models such as Design, Build and Operate where the private sector operator does part sharing of the capital.

It is hard to distinguish the projects that strictly follow the investment space defined by circular economy as the business canvas of circular economy is rather large and difficult to scope as evident from the 8 business models described in Chapter 8. There are also questions whether dedicated financing for circular economy should be developed as an investment class and rather operate the funds on a broad thematic basis such as wastewater recycling, energy conservation, waste recycling, climate change mitigation, or conservation finance. Hence in theory, a large number of financing opportunities exist under project finance for circular economy.

Financing in circular economy should support start-ups in incubation and intermediary phases of the business. While the start-ups need to put their own funding, government grants and subsidies can add up. Sometimes funding can be made available from programs that support GHG mitigation or renewable energy. Funding from Corporate Social Responsibility (CSR) budgets of philanthropic corporate could also be sought. Funds set up for Socially Responsible Investments and Impact Funds have started to play an increasing role in the start-up and intermediary stages.

As the start-up establishes and reaches an intermediary stage demonstrating commercial viability, angel investors and Venture Capitalists (VCs) step in. Angel investors are high net worth individuals, often successful business leaders, who use their own funds to invest that help seed the start-ups to grow to intermediary stage. Angel investors like to see that the business blends professionalism with a deep personal commitment. The amount they invest varies from 10,000 USD and 100,000 USD or more when angels group together.

The VCs manage pooled money sourced from others by setting a professionally managed fund. A VC is essentially an investor who provides capital to firms that exhibit high growth potential in exchange for an equity stake. VCs target firms that are at the stage where they are looking to commercialize their idea, so VCs are helpful both at start up and more so in the intermediary stage. Capital provided by venture capital funds often starts from USD 1 million and up to USD 25 million and hence VC funding is more applicable for business in the intermediary stage. VC funding is useful in the case when the business does not have access to capital markets, bank loans, or other debt instruments. Typically, VCs follow a staged funding like in start-ups as seed and later depending on performance of the company, higher investments called as series A, B, and C in the intermediate and maturity stages[19].

Private Equity (PE) firms mostly buy mature companies that are already established. PE firms buy these companies and streamline operations to increase revenues. PE firms therefore buy more than 50% of the ownership of the companies in which they invest. Typically, VC firms invest much less in the equity. PE firms usually invest starting from 25 million USD to USD 100 million in a single company.

Businesses that are successful to demonstrate successful innovation can access several windows of project finance, e.g., loans from commercial banks and applying for larger funds that have been explicitly set up for investing in circular economy. Fund raising from the market (e.g., through green bonds and climate bonds) may bring in additional inflows as the business reaches a maturity stage. A typical progression of mobilizing finance is illustrated in Figure 9.1.

Box 9.5 presents the example of MMC Ventures and its Enterprise Investment Scheme (EIS).

Impact investing, a sub-set of socially responsible investments, aims to use funds for positive social and environmental outcomes and particularly to support social enterprises. Impact investing attracts high net worth individuals as well as institutional investors including hedge funds, private foundations, banks, pension funds, and other fund managers. Corporates and socially conscious financial service companies also contribute to the impact funds. Women are often the beneficiaries of impact funds. Impact funds are expected to play a greater role to support circular economy business.

In 2010, JP Morgan and Rockefeller Foundation estimated that the impact investment sector could reach USD 400 billion to USD 1 trillion by 2020[22]. Visit the Global Impact Investor Network for more information on impact funds.[23].

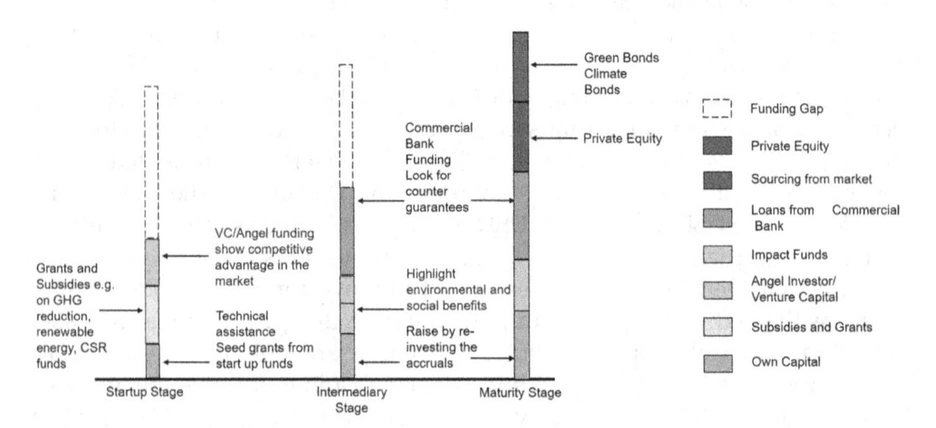

FIGURE 9.1
Different stages of business and sources used for financing.

BOX 9.5 ENTERPRISE INVESTMENT SCHEME AT MMC VENTURES[20]

MMC Ventures, the London-based VC that typically invests at seed and Series A. MMC's Enterprise Investment Scheme (EIS) invests in transformative companies that are changing the face of society: challenging and replacing cultural and business norms, popularizing new customer behaviors, inventing new industries and creating new jobs. Many companies venturing into circular economy therefore qualify to see venture capital from MMC's EIS fund. These are the businesses MMC identifies and scales from Seed and Series A. The objective of the MMC EIS Fund is significant capital growth, targeting a return of 2–3 times on each investor's portfolio. Some of the companies invested by EIS are listed below that are very much in the space of circular economy[21].

- Gousto, an AI-powered recipe kit company — and now one of the UK's fastest-growing scale-ups — that is eliminating food waste.
- Senseye, a predictive maintenance company that uses cutting-edge technology to extend the life of industrial manufacturing equipment.
- QFlow, an environmental data management platform that is reducing the environmental impact of construction.
- Unmade, a pioneering fashion-tech company that's reducing waste in the fashion supply chain by enabling "demand-driven" creation.

In 2019, MMC launched a new GBP 100 million "Scale Up" fund to provide expansion capital to its later-stage portfolio companies. This fund will provide primary capital to current portfolio companies that have grown beyond the mandate of MMC Ventures' existing funds such as EIS and also open for secondary transactions. In this way, capital can be recycled within the early-stage funding ecosystem, whereby, for example, angel investors can go again by backing newly formed companies, while MMC maintains a longer-term outlook. The Scale Up Fund rounds off a number of new funds managed by MMC. The firm recently floated a new GBP 52 million seed fund in partnership with the mayor of London. And combined with its annual EIS fundraise, the VC has added GBP 200 million to its coffers in the last 12 months. Over the last year, MMC has invested more than GBP 85 million across the pre-seed, seed, series A and later stages in amounts ranging from GBP 100,000 to GBP 25 million.

BOX 9.6 IMPACT FUNDS IN INDIA THAT SUPPORT INNOVATIVE BUSINESSES IN CIRCULAR ECONOMY

T-HUB[24]

T- Hub leads India's pioneering innovation ecosystem that powers next-generation products and new business models. Since its incorporation in 2015, it has provided 1100+ national and international startups access to better technology, talent, mentors, customers, corporations, investors, and government agencies. T-Hub incubated startups Banyan Nation (described in Chapter 6) and Sanshodhan E Waste Exchange and signed MoUs with the Telangana Government for e-waste management.

AAVISHKAAR GROUP[25]

Aavishkaar group focuses on developing the impact ecosystem in the continents of Asia and Africa. The group manages assets more than USD 1 Billion across Equity and Credit, with 6000+ employees present across India, Indonesia, Bangladesh, and Kenya. The group's ecosystem includes Aavishkaar Capital – Pioneer in equity led impact investing, Arohan – One of India's largest Technology led Financial inclusion platform, IntelleGrow – specialized lender to small and growing businesses, Tribe – a fintech player, Intellecap – Thought Leader and Advisory business with a focus on sustainability and Sankalp Forum – one of the world's largest inclusive development led platform. Company NEPRA described in Chapter 8 received two stage funding from Aavishkaar Capital. The Group's shareholders are e TIAA-Nuveen, Triodos Bank, Shell Foundation and Dutch Entrepreneurial Development Bank FMO.

Some of the prominent impact funds in India have started including in their portfolio start-ups who are looking for support in setting circular business models.

Box 9.6 presents examples of two such impact funds.

There have been several funds set up to support resource efficiency. The European Regional Development Fund (ERDF) has supported resource efficiency in European countries to promote a circular economy. Box 9.7 shows examples of two such funds in Leeds, UK, and Scotland. The Resource Efficiency Fund in Leeds is unique as it offers free resource efficiency audits as technical assistance. Many public and private sector financing institutions have followed this blending technical and financial assistance strategy to ensure quality investments.

The Government of South Australia set up a Circular Economy Investment Fund called Resource Efficiency and Productivity (REAP) Grants[28]. This fund is more of technical assistance to help businesses make an action plan toward circular economy by carrying out resource efficiency and productivity

BOX 9.7 EXAMPLES OF FINANCING RESOURCE EFFICIENCY SUPPORTED BY THE EUROPEAN REGIONAL DEVELOPMENT FUND

RESOURCE EFFICIENCY FUND IN LEEDS, UK[26]

Between November 2016 and December 2019, the Resource Efficiency Fund operated by Leeds City Region Partnership supported by ERDF financed over 700 small or medium-sized enterprises to help them lower energy usage and reduce water and waste costs, and minimize their impact on the environment. The support available included:

- Free resource or energy efficiency audit worth up to GBP 1200.
- 40% funding toward capital investment projects to a maximum GBP 40,000 grant.
- Free circular economy consultancy of up to 30 days intensive support.

The program provided over 370 free resource efficiency audits and funded 160 projects to save businesses GBP 663,000, 7.7 million kWh of energy, and 2422 tCO_2 every year.

THE CIRCULAR ECONOMY INVESTMENT FUND IN SCOTLAND[27]

The GBP 18 million (approximately EUR 20 million) funds, administered by Zero Waste Scotland, offer SMEs investment based in Scotland and support work that will deliver circular economy growth. The European Regional Development Fund supports it through the GBP 73 million (approximately EUR 82 million) Resource Efficiency Circular Economy Accelerator Programme. The fund is designed to fund organizations wishing to explore and pioneer circular approaches; identify, develop, and bring new business models and innovative technologies to market. The fund is available for:

- Small to medium-sized enterprises (less than 250 employees/ turnover less than EUR 50 million).
- Non-profit organizations (charities and social enterprises).

assessments. These assessments are provided as a grant (maximum of AUD 10,000), similar to the Leeds model. An additional AUD 10,000 is made available to kick-start implementation of the recommendations.

The number of private-sector funds with a focus on circular economy has grown tenfold since 2016[29]. These funds include venture capital, private equity and private debt funds. Not all funds are dedicated to a circular economy and the scope in some cases is broader such as environmental protection and conservation. Funds available under the theme of Cleantech target on renewable energy and climate change. Box 9.8 illustrates three funds that are focused on circular economy.

Closed Loop Fund was set up by a New York-based investment firm, which comprised of venture capital, growth equity, private equity, project finance, and an innovation center focused on building the circular economy. More details about this fund are provided in Box 9.9.

Example of Closed Loop Infrastructure Fund shows a case of increased interest in financing circular infrastructure. In specific, opportunities have come up in financing common wastewater and recycling projects, waste to energy plants, common heat and power plants and material recovery facilities. In some countries, national schemes have been set up to provide some conditional grants (typically 25%). Soft loans are made available from commercial banks either through development finance (e.g., multi-lateral and bilateral funds), or by providing counter guarantees (e.g., by United States Agency for International Development (USAID)). Financial closure is done by contribution from the beneficiaries (e.g., industries) and capital provided by the technology provider or private sector partner. A wide range of PPP models have been implemented however with mixed results. Challenges have been due to not well drafted concession contracts, poor risk identification and allocation, changes in inputs (e.g., waste quantities and composition) and market distortion due to policy change.

One must remember that financing sought for or provided depends a lot on type of circular economy business model. In some cases, the companies look for short term debt for working capital (e.g., for sharing business models) or a long-term equity (e.g., in resource recovery projects) or mezzanine financing (e.g., series B) in business of providing facilitation (e.g., digital services).

Financing can be a challenge for newly evolved circular businesses such as product as a service or PaaS. Companies who operate such a business need to invest significantly upfront, with returns spaced over time as the service is used. This constraints a company's growth and limits range of pricing schemes that can be offered to its customers and sharing risks and returns with value chain partners. Here, financial products that are performance based, providing flexible, time spaced cash flows can help. Rabo Bank used such an approach for supporting Bundles a hybrid PaaS based company providing washing services. To address high administrative costs of the service, Rabo Bank supported a pilot project to develop and test a circular service network that could be shared by a number of service providers[37].

Public sector organizations, typically urban local bodies look for PPP based models seeking part financing for wastewater treatment and recycling projects from private sector partner. It is recommended that you refer to the knowledge paper "Closing the water loop: Reuse of treated wastewater in urban India" by PwC. In India, under the National Mission for Clean Ganga,

BOX 9.8 FUNDS WITH A FOCUS ON CIRCULAR ECONOMY

CIRCULATE CAPITAL OCEAN FUND

Circulate Capital is an investment management firm dedicated to financing innovation, companies, and infrastructure that prevents the flow of plastic waste into the world's ocean while advancing the circular economy. Circulate Capital established the world's first investment fund dedicated to preventing ocean plastic: Circulate Capital Ocean Fund (CCOF). CCOF is a new, blended financing mechanism, bringing together the public and private sectors. It was created in 2018 with a fund amount of USD 106 million[30]. Circulate Capital established this fund in partnership with leading corporations viz. PepsiCo, the first investor, Coca-Cola, Danone, Dow, Procter & Gamble, Unilever, Chevron, Phillips Chemical and backed by USAID[31]. The CCOF provides both debt and equity financing to waste management, recycling, and circular economy start-ups and SMEs in India, Indonesia, Thailand, Vietnam, and the Philippines. CCOF helps fund solutions to scale and replicate by connecting them to the supply chains of the world's leading companies. Companies include start-ups such as PET recycling companies Tridis Oasis and Lucro[32].

DECALIA CIRCULAR ECONOMY FUND

Decalia Circular Economy Fund was set up in 2018 and is an actively managed, first equity investment fund dedicated to promoting circular economy. It covers 8 proprietary sectors that include sharing economy and platform-as-a-service, prevention and diagnostics, nutrition, renewable energy and smart grid, waste and recycling, water management, smart and green materials, and Industry 4.0. The investment universe thus comprises about 580 companies worldwide, including 215 in the USA, 165 in Europe, 90 in Japan and 110 in other countries[33]. As of October 31, 2020, the fund size was GBP 62.73 million[34].

BGF CIRCULAR ECONOMY FUND

Blackrock set up BGF Circular Economy Fund in 2019. The fund aims to provide a return on investment through a combination of capital growth and income on the fund's assets. The fund invests globally at least 80% of its total assets in the equity securities (i.e., shares) of companies that benefit from, or contribute to, the advancement of circular economy, excluding investments in the coal sector and oil and gas producers. As of November 2020, the fund size was USD 1077.75 million[35].

BOX 9.9 THE CLOSED LOOP FUND[36]

Closed Loop Fund was set up by a New York based investment firm, and comprised of venture capital, growth equity, private equity, and project finance as well as an innovation center focused on building the circular economy.

The Fund has built an ecosystem that connects entrepreneurs, industry experts, global consumer goods companies, retailers, financial institutions, and municipalities. The funds available are:

- Closed Loop Infrastructure Fund focusing on recycling and circular economy infrastructure across North America.
- Closed Loop Beverage Fund set up in partnership with the American Beverage Association, with an objective to improve the collection of the industry's valuable plastic bottles so they can be made into new bottles in the United States.
- Closed Loop Venture Fund where early-stage capital is deployed into companies developing breakthrough solutions for the circular economy. The portfolio includes companies developing leading innovations in material science, robotics, agri-tech, sustainable consumer products and advanced technologies that further the circular economy.
- Growth Equity Closed Loop Fashion Fund that focuses on the global fashion industry.
- Private Equity Closed Loop Leadership Fund that focuses on acquiring companies along the value chain to build circular supply chains.

tenders have been issued based on Design, Build, Operate (DBO) model for financing wastewater recycling projects in the Ganga river basin. Securing market for treated wastewater however remains a challenge as water price in most countries is highly subsidized.

In the energy efficiency sector, financing Energy Service Companies (ESCO) has been a strategy and a variety of financing options are blended depending on the need, context, and interest of the co-financiers. Commercial banks and the Non-Banking Financing Corporations (NBFCs) play an intermediary role while disbursing central funds often pooled from the government, contributed by Multilateral funds (like Global Environment Facility) or private sector development institutions such as International Finance Corporation.

Refer to the "Guidelines for Financing Energy Efficiency Projects" produced by Bureau of Energy Efficiency in India[38]. Report on "Financing Energy Efficiency - Lessons from Brazil, China, India, and Beyond"[39] presents interesting case studies that provide insight to different financing models followed to support the ESCO. A recent article on "Unlocking green financing for building energy retrofit: A survey in the western China" suggests innovative models for financing ESCOs[40]. Table 9.1 shows different types of financing energy efficiency presented by Harris Williams.

TABLE 9.1

Different Types of Financing for Energy Efficiency[41]

Type	Description
Energy Service Agreements	Energy Service Agreements are agreements between a customer and the ESA provider that provides financing for the project and delivers energy savings (i.e., megawatt hours) at a negotiated price (less than retail rates for energy services).
Energy Savings Performance Contract	Under an Energy Savings Performance Contract (ESPC), ESCO coordinates installation and maintenance of efficiency equipment in a customer's facilities and is paid from the associated energy savings. The ESCO typically provides a savings guarantee.
Commercial PACE Financing	Commercial property-assessed clean energy is a financing structure in which building owners borrow money for energy efficiency, renewable energy or other projects and make repayments via an assessment on their property tax bill. The financing arrangement then remains with the property even if it is sold, facilitating long-term investments in building performance.
On-Bill Financing/ Repayment	On-Bill Financing/Repayment are financing options in which a utility or private lender supplies capital to a customer to fund energy efficiency, renewable energy or other generation projects and is repaid through regular payments on an existing utility bill.
Power Purchase Agreement	A Power Purchase Agreement (PPA) is an arrangement in which a third-party developer installs, owns, and operates an energy system on a customer's property. The customer then purchases the system's electric output for a predetermined period. A PPA allows the customer to receive stable and often low-cost electricity with no up-front cost.
Efficiency-as-a-Service	Efficiency-as-a-service is a pay-for-performance, off-balance-sheet financing solution that allows customers to implement energy and water efficiency projects with no up-front capital expenditure. The provider pays for project development, construction, and maintenance costs.
Loan or Debt Financing	Customers can borrow money directly from banks or other lenders to pay for energy efficiency, renewable energy, and other generation projects. The customer must then arrange the purchase, installation, and management of equipment by a third-party contractor or in-house staff.
Lease Financing	A lease is a simple financing structure that allows a customer to use energy efficiency, renewable energy, or other generation equipment without purchasing it outright. The two most common types are on-balance-sheet capital leases and off-balance-sheet operating leases.
Internal Funding	Internal funding refers to the use of an organization's existing financial resources to pay for energy efficiency, renewable energy or other generation projects, rather than seeking external financing. This is often the most simple and direct method for funding projects.

Projects developed for wastewater recycling, energy efficiency and for waste recovery align toward circular economy and hence financing them certainly helps to push circularity in the business as well as governance. Ideally, however, projects identified for financing should consider resource efficiency and resource circulation as the basis, and covering sectors such as water, energy and waste, to come up with robust and truly circular interventions. There are several other financing instruments that could be tapped. For example, energy efficiency projects help reduce GHG emissions and hence can qualify for applying to CleanTech and climate funds. Instruments such as Green bonds could be used for fund raising. The project identification and preparatory tools, described in Chapter 3, such as waste minimization audits, cleaner production opportunity assessments, resource efficiency audits and GHG inventorization should be therefore integrated and align with the business models and various financing options.

Green bonds are designated bonds intended to encourage sustainability and to support climate-related or other types of special environmental projects. More specifically, green bonds finance projects are aimed at energy efficiency, pollution prevention, sustainable agriculture, fishery and forestry, the protection of aquatic and terrestrial ecosystems, clean transportation, clean water, and sustainable water management etc. They also finance the development of environmentally friendly technologies and the mitigation of climate change.

Box 9.10 presents the example of Climate Bonds Initiative to encourage waste reduction and sound waste management.

Finally, some of the corporate CSR funds have started supporting social enterprises that are engaged in circular economy. This trend is laudable as it provides grant funds that are useful in the early stage of operations. When these grants are provided on a continued basis, these enterprises grow and lead to similar initiatives that operate on a scale and attract investors for expansion and replication. Box 9.11 gives an example of Greensole in India.

Sustainable Salons is a unique social enterprise that helps its salon members, and their clients reduce their impact on the planet, while investing in local communities. Paul and Ewelina officially launched Sustainable Salons in 2015. Box 9.12 provides more details on Sustainable Salons program.

Given emergence of such social enterprises, and especially those following inclusive business models as described in Chapter 8, these investments need to be assessed based on Social Rate of Return (SRR) apart from financial returns. The mindset of investment committees managing funds in circular economy have to look at these new paradigms.

Given such needs, in the last decade, particularly, the circular economy's opportunities have reshaped the financial sector itself, geared toward long term value creation and the enablement of an economy that is distributed, diverse, and inclusive[21]. Funds like the Osmosis Investment Management[47] use their proprietary "resource efficiency factor" in investing with the belief that companies that are resource efficient will be rewarded, and those who monetize sustainability in the balance sheet are more likely to outperform than their peers.

BOX 9.10 CLIMATE BONDS INITIATIVE INCLUDES WASTE MANAGEMENT CRITERIA[42]

The waste sector has the potential to contribute a 10–15% reduction in global greenhouse gas emissions. Opportunities in prevention, reuse, recycling, and energy recovery can achieve significant mitigation by reducing landfill emissions, reducing emissions linked to resource extraction and production using virgin materials, and providing an alternative energy source that substitutes fossil fuels. According to UBS research carried out in early 2020 only 4% of green bonds issued are related to waste[43].

Recently, Climate Bonds Initiative published updated waste management bond criteria and included these criteria in the climate bonds standard. These Criteria apply to assets and projects relating to the following aspects of waste management:

- Collection (including collection infrastructure, containers).
- Sorting to separate recyclables.
- Reuse and recycling (including processing into secondary raw materials and repair).
- Composting & anaerobic digestion of green/garden/yard and food waste.
- Thermal treatment with energy recovery of residual waste (outside the EU only).
- Installation of gas recovery systems for landfill site.

The Climate Bonds Initiative sees a role for the USD 100 trillion bond market in addressing the huge funding needs of the global waste sector, estimated to be hundreds of billions of dollars. The new Criteria has now been launched (as on December 11, 2019) and investment is open for certification.

ABN AMRO, ING, and Rabo Bank launched joint circular-economy finance guidelines internationally in July 2018, inspired by the ambition to create a common framework for financing the circular economy worldwide. The three banks lay the groundwork for accelerating investment in and funding of circular business models by introducing these guidelines[48].

Policy frameworks consisting of strategies, legislation, and economic instruments play a critical role in stimulating, expanding, and sustaining circular businesses. Chapter 10 provides an overview of the policy frameworks in a circular economy, illustrating initiatives taken by various countries across the world.

BOX 9.11 GREENSOLE – A SUCCESSFUL SOCIAL ENTERPRISE PROMOTING CIRCULAR ECONOMY[44]

Worldwide every year more than 350,000,000 pairs of shoes are discarded, while as per the report by World Health Organisation (WHO), 1.5 billion people are infected by diseases that could be prevented by wearing proper footwear[45]. Greensole in India recycles discarded shoes to comfortable footwear, keeping them away from landfills and provide them to children in need. They also retail upcycled footwear toward building a self-sustaining social venture.

Greensole collects the old footwear and reuses the soles to make new shoes. The Insole are used to make Uppers. Uppers are used to make pouches and the laces are used for packing. The entire discharged shoes get a second life, especially for those who need.

Greensole funds itself by selling the refurbished footwear through retail. Donation drives are held in the form of buying refurbished pair of footwear. Donors buy pair of refurbished footwear at a price of INR 199 (2.8 USD) and Greensole donates a particular number of pairs for people in need. Minimum number to be donated or purchased is 2000 pairs, so that an entire locality can be benefited. Many corporates support this drive through contributions made by employees. Greensole footwear are provided as per requirement for the employees of the company. Company in addition can provide matching funds as a donation.

With help of corporate CSR funds, Greensole has now expanded its operations by building skill centers in collaboration with Tata Steel and Lion's Club, at the skill center. Greesole trains, skills and employs underprivileged people. Trainers at the center teach people on how to use the machinery and the complete refurbishment procedure. Greensole plans to open more than 30 skill centers across India with the help of corporate CSR support, thus promoting more social enterprises in circular economy.

BOX 9.12 SUSTAINABLE SALONS[46]

The Sustainable Salons program has more than 908 salon members across Australia and New Zealand. As of 2019, 188,600 kgs of foil was diverted from landfill and made into new aluminum packaging, 2250 kgs of e-waste was repurposed, 192,700 kgs of paper was recycled, among other environmental and social impacts.

An additional AUD 2 is charged on the final bill of customers visiting the members of the Sustainable Salons network. The AUD 2 Sustainable Salons Fee supports a range of important sustainability and community initiatives. The program began in the hairdressing industry and now creates waste solutions for barber shops, beauty salons, dermal clinics, and pet-grooming salons as well. Sustainable Salons provides each member the in-salon separation bins and outside collection bins to suit the salon's needs. It also offers a fortnightly waste collection service from the salon.

Following are the ways that the Sustainable Salons Program diverts waste from the landfills:

- Plastic packaging is sent to specialist local plastics recyclers where it's cleaned and made into outdoor furniture, landscape supplies and new product packaging. All aluminum, foil, color tubes, cardboard, paper, magazines, razor blades, unwanted tools and select disposables are sold for recycling, and the proceeds are donated to OzHarvest and KiwiHarvest to provide meals for hungry people.
- Hair clippings are collected from the salon floor and stuffed into stockings to make Hair Booms that can one day help clean up oil spills along coastlines. Hair is also repurposed in local community gardens for composting or used in sustainable art installations to educate the community.
- All collected ponytails 20cm or longer are distributed to charitable organizations to create wigs for those suffering from cancer or alopecia. Sustainable Salons is the largest donor of ponytails in the Southern Hemisphere.
- By collecting small amounts from a large salon network, Sustainable Salons is the first company to recycle excess chemicals in the salon industry. These are pooled and sent to chemical recycling plants where they're neutralized and turned into recycled water used in roadworks and construction.

Sustainable solutions are also promoted to help the salon make greener choices. Via Rewards Points, salons can access environmentally friendly options for everyday salon products, such as eco cleaning products, recycled toilet paper, biodegradable coffee pods, recycled foil, biodegradable towels, and biodegradable gloves.

Sustainable Salons also promotes ethical and community inclusive practices by encouraging salon professionals in their network to volunteer their time and skills for those who don't have access to salon services. Alongside, in partnership with Endeavour Foundation and Mambourin, Sustainable Salons is providing employment to people with disabilities within their material collection and processing streams.

ACTIVITY 9.2 DISCUSSION ON DEDICATED FUNDS AND METRICS FOR FINANCING CIRCULAR ECONOMY

While setting up funds with a focus on a circular economy could be a good idea, will this be a good business to the fund managers? Isn't it hard to limit the scope of circular financing? Won't the general funds on environment and those aligned to energy, climate finance, CleanTech etc. be more convenient and competitive? Conversely, how can circular principles be integrated in the thematic funds?

Investors in their metrics need to consider both Internal rate of Return and SRR to ensure that the circular financing leads to meeting of the SDGs. Will this be possible in the case of impact investors? Can you cite some examples? What are the other indicators that should be considered in circular financing?

9.3 Key Takeaways

- Innovation plays an important role in circular economy, especially in fostering circular businesses.
- Creation of innovation platforms in partnership is an effective strategy, given the canvas of life cycle in the circular ecosystems.
- Financing should be a part of the innovation platforms. Here impact funds, angel investors and venture capitalists can play an important role.
- Models that are based on PPP, help in raising finance especially in circular infrastructure projects.
- Circular economy has influenced the financing community in the philosophy of investing.
- Setting up of dedicated funds in circular economy has been a recent trend. Funds focusing on CleanTech and climate change have also started expanding to cover opportunities in circular economy.

ADDITIONAL READING

1. **Green financing: Financing circular economy companies:** Transitioning from linear economy to the CE requires changes in four areas: material and product design, business models, global reverse networks and enabling business environments.

This study considers the financing needs of CE companies as a result of business model changes. Through the case studies of Ragn-Sellsföretagen AB and Inrego AB, analysed with secondary data from ING Bank and primary data collected through semi-structured interviews with the case companies, this research sheds more light on the financing needs of circular economy companies and how they are actually financed. Findings from this research suggest that the financing needs of circular economy companies depend on the value proposition of the company. In accordance with the pecking order of capital structure, all financing needs of the companies studied are financed from internal sources, particularly retained earnings before external debt financing is accessed. Findings indicate the willingness of banks to finance circular economy companies.

The results of this research suggest that the circular economy companies studied do not need financial support from the government or its agencies to succeed even though favorable laws are welcomed. They report that their long-term success depends on their ability to remain innovative in their business models, aligning with Schumpeter's creative destruction model.

Source: Acheampong, J., 2016. *Green Financing: Financing Circular Economy Companies*. KTH Industrial Engineering and Management. <https://www.diva-portal.org/smash/get/diva2:938393/FULLTEXT01.pdf> [Accessed 9 November 2020].

2. **Circular economy finance guidelines:**
Financial institutions have a major role in the transition from linear to the circular economy, as the latter sees the advent of a multitude of new business models that require different types of finance, where knowledge of business models is key. The Circular Economy Finance Guidelines aim to promote and develop the role that finance can play in this transition. In 2018, ABN AMRO, ING and Rabobank launched the joint circular-economy finance guidelines internationally. The Guidelines are a voluntary process guideline that recommend transparency and disclosure and promote integrity in the debt and equity market for the circular economy. The Guidelines propose to consider both the business model and the socio-economic impacts (includes environmental impact) in the circular assessment of the company or project. Only propositions based on a circular business model that generate long-term positive impact should be considered as circular propositions.

Source: Rabo Bank. 2018. ABN AMRO, ING And Rabobank Launch Finance Guidelines For Circular Economy. <https://www.rabobank.com/en/press/search/2018/20180702-abn-amro-ing-and-rabobank-launch-finance-guidelines-for-circular-economy.html> [Accessed 9 November 2020].

3. **Financing the circular economy:**
   Since the beginning of 2020, assets managed through public equity funds with the circular economy as the sole or partial investment focus have increased 6-fold, from USD 0.3 billion to over USD 2 billion. This paper focuses primarily on private sector finance and explores the circular economy's value creation potential for investors, banks, and other financial services firms. First, this paper shows how the circular economy can help achieve climate and other Environmental & Social Governance (ESG) goals while creating opportunities for new forms of better economic growth, effectively moving beyond the initial progress and focus that ESG investment has achieved over recent years. Second, it highlights how investors, banks, and insurers are already capturing these opportunities, showing that the market for financing the circular economy is rapidly taking off across asset classes and sectors. Third, it provides a direction of travel for finance to fully capitalize on the opportunity, by helping to rapidly scale the circular economy.

   This paper is intended as an initial exploration rather than a detailed analysis of any individual aspect of finance. Its purpose is to stimulate discussion about how the financial services sector can help scale the circular economy to drive new and better growth that is more distributed, diverse, and inclusive, and help build an economy that is restorative and regenerative by design.

Source: Ellen Macarthur Foundation. 2020. *Financing The Circular Economy*. <https://www.ellenmacarthurfoundation.org/assets/downloads/Financing-the-circular-economy.pdf> [Accessed 8 November 2020].

4. **Financing circularity: Demystifying finance for circular economies:**
   The COVID-19 pandemic crisis in 2020 has created mixed signals regarding the shift to sustainable consumption and production and circularity. It has temporarily disturbed waste

collection based on fears of collecting contaminated materials. The need for personal protection equipment has created unexpected volumes of single-use wastes and plastic products. These short-term pandemic related upsets can impose barriers, hurdles or even temporary downturns for the growth toward circularity in favor of the more wasteful linear take-make-waste principle.

The report provides insights into practical approaches to financing circularity, such as the application of sectoral metrics in decision-making, and encourages financial institutions to formalize industry-wide support programs and commitments for the transition to a circular economy and more sustainable patterns of consumption and production.

Source: UN Environment Program. 2020. *Financing Circularity: October 2020 Demystifying Finance for Circular Economies.* <https://www.unepfi.org/wordpress/wp-content/uploads/2020/10/UNEPFI_DemystfyingFinanceCircularity-2020.pdf> [Accessed 17 November 2020].

Notes

1. UNECE. 2018. *How Can Innovation and Public–Private Partnerships Support Sustainable Production and Consumption and The Shift To A Circular Economy?.* <https://www.unece.org/info/media/news/innovation/2018/how-can-innovation-and-public-private-partnerships-support-sustainable-production-and-consumption-and-the-shift-to-a-circular-economy/doc.html> [Accessed 8 November 2020].
2. International Synergies. n.d. *Exploiting Industrial Symbiosis In Birmingham.* <https://www.international-synergies.com/projects/exploiting-industrial-symbiosis-in-birmingham-2/> [Accessed 8 November 2020].
3. Gawel, A., 2019. *The Economy Of The Future Is Circular. Here's How Entrepreneurship Can Help.* World Economic Forum. <https://www.weforum.org/agenda/2019/10/innovation-entrepreneurship-waste-circular-economy/> [Accessed 8 November 2020].
4. European Commission. n.d. *What Is Horizon 2020?.* <https://ec.europa.eu/programmes/horizon2020/en/what-horizon-2020> [Accessed 8 November 2020].
5. Eco-innovation Observatory. n.d. *About Us.* <https://www.eco-innovation.eu/index.php/about-us> [Accessed 8 November 2020].
6. Eco-Innovation Observatory. n.d. *Eco-Innovation SME Guides.* <https://www.eco-innovation.eu/index.php/guide-for-smes> [Accessed 8 November 2020].

7. EIT RawMaterials. 2020. *Circular Economy And Innovation.* <https://eitrawmaterials.eu/circular-economy-and-innovation/> [Accessed 8 November 2020].
8. A.SPIRE. n.d. *The Association.* <https://www.spire2030.eu/spire/the-association> [Accessed 8 November 2020].
9. Crisp, B., 2019. *Innovation In The Circular Economy.* Medium. <https://medium.com/design-thinkers-academy-london/innovation-in-the-circular-economy-e8172034fd36> [Accessed 8 November 2020].
10. Lehtinen, A. and Ekengren, A., 2020. *Introducing 39 Circular Economy Solutions To Inspire The World.* Sitra. <https://www.sitra.fi/en/articles/introducing-39-circular-economy-solutions-to-inspire-the-world/> [Accessed 8 November 2020].
11. Aalto University. 2020. *A Joint Laboratory Of Circular Economy Was Opened In Otaniemi.* <https://www.aalto.fi/en/news/a-joint-laboratory-of-circular-economy-was-opened-in-otaniemi> [Accessed 8 November 2020].
12. SecondMuse. n.d. *About.* <https://www.secondmuse.com/about/> [Accessed 8 November 2020].
13. Rabobank. 2019. *Rabobank To Help Businesses With Circular Enterprise.* <https://www.rabobank.com/en/about-rabobank/in-society/sustainability/articles/2019/20190508-rabobank-to-help-businesses-with-circular-enterprise.html> [Accessed 8 November 2020].
14. Ekonnect Knowledge Foundation - https://ekonnect.net/
15. The Circulars Accelerator. n.d. *About The Circulars.* <https://thecirculars.org/> [Accessed 8 November 2020].
16. World Economic Forum. 2019. *Scale360° Circular Innovation.* <https://www.weforum.org/scale360-circular-innovation/> [Accessed 8 November 2020].
17. Platform for Accelerating the Circular Economy. n.d. *Home.* <https://pacecircular.org/> [Accessed 8 November 2020].
18. TERI. 2018. *Circular Economy: A Business Imperative For India.* <https://www.teriin.org/project/circular-economy-business-imperative-india> [Accessed 9 November 2020].
19. Investopedia. n.d. *Series A, B, C Funding: How It Works.* <https://www.investopedia.com/articles/personal-finance/102015/series-b-c-funding-what-it-all-means-and-how-it-works.asp> [Accessed 17 November 2020].
20. MMC Ventures. n.d. *Funds.* <https://mmc.vc/> [Accessed 17 November 2020].
21. Kelnar, D., 2019. *The 'Circular Economy': We'Re Investing In A Sustainable Future.* Medium. <https://medium.com/mmc-writes/the-circular-economy-we-re-investing-in-a-sustainable-future-3aeacb840cd2> [Accessed 17 November 2020].
22. O'donohoe, N., Leijonhufvud, C., Saltuk, Y., Bugg-levine, A. and Brandenburg, M., 2010. *Impact Investments: An Emerging Asset Class.* The GIIN. <https://thegiin.org/research/publication/impact-investments-an-emerging-asset-class> [Accessed 17 November 2020].
23. The GIIN. n.d. *About Impact Investment.* <https://thegiin.org/> [Accessed 17 November 2020].
24. T-Hub. n.d. *About.* <https://t-hub.co/> [Accessed 8 November 2020].
25. Intellecap. n.d. *About Us.* <https://www.intellecap.com/group/> [Accessed 8 November 2020].

26. Leed's City Region Enterprise Partnership. n.d. *Resource Efficiency and The Circular Economy.* <https://www.the-lep.com/business-support/growth-support/resource-efficiency-and-the-circular-economy/> [Accessed 9 November 2020].
27. Zero Waste Scotland. n.d. *Circular Economy Investment Fund.* <https://www.zerowastescotland.org.uk/circular-economy/investment-fund> [Accessed 8 November 2020].
28. Government of South Australia. n.d. *Resource Efficiency and Productivity- REAP Grants.* <https://www.greenindustries.sa.gov.au/reap> [Accessed 8 November 2020].
29. Ellen Macarthur Foundation. 2020. *Financing The Circular Economy.* <https://www.ellenmacarthurfoundation.org/assets/downloads/Financing-the-circular-economy.pdf> [Accessed 8 November 2020].
30. Toto, D., 2019. *Circulate Capital Closes $106M Circulate Capital Ocean Fund.* Recycling Today. <https://www.recyclingtoday.com/article/circulate-capital-ocean-fund-announces-first-close/#:~:text=Circulate%20Capital%2C%20the%20New%20York,Capital%20Ocean%20Fund%20(CCOF).> [Accessed 9 November 2020].
31. Circulate Capital. n.d. *Home.* <https://www.circulatecapital.com/> [Accessed 8 November 2020].
32. Circulate Capital. 2020. *Our Investment Portfolio.* <https://www.circulatecapital.com/investments> [Accessed 9 November 2020].
33. DECALIA. 2018. *DECALIA Launches The First Equity Fund Dedicated To The Circular Economy - DECALIA.* <https://www.decaliagroup.com/en/decalia-launches-the-first-equity-fund-dedicated-to-the-circular-economy/> [Accessed 8 November 2020].
34. Financial Times. 2020. *Decalia Circular Economy.* <https://markets.ft.com/data/funds/tearsheet/summary?s=LU1787060471:USD> [Accessed 9 November 2020].
35. BlackRock. 2020. *BGF Circular Economy.* <https://www.blackrock.com/ch/individual/en/products/310165/blackrock-circular-economy-fund> [Accessed 8 November 2020].
36. Closed Loop Partners. n.d. *Our Funds.* <https://www.closedlooppartners.com/funds/> [Accessed 8 November 2020].
37. Sustainable Finance Lab. 2019. *The Circular Service Platform.* <https://sustainablefinancelab.nl/wp-content/uploads/sites/334/2019/04/The-Circular-Service-Platform-White-Paper.pdf> [Accessed 17 November 2020].
38. Bureau of Energy Efficiency. 2017. *Guidelines For Financing Energy Efficiency Projects In India.* <https://beeindia.gov.in/sites/default/files/EE-Finance-Guidelines_0.pdf> [Accessed 17 November 2020].
39. Taylor, R., Govindarajalu, C., Levin, J., Meyer, A. and Ward, W., n.d. *FINANCING ENERGY EFFICIENCY.* The World Bank. <https://www.esmap.org/sites/default/files/esmap-files/financing_energy_efficiency.pdf> [Accessed 17 November 2020].
40. Zhang, M., Lian, Y., Zhao, H. and Xia-Bauer, C., 2020. Unlocking green financing for building energy retrofit: A survey in the western China. *Energy Strategy Reviews*, 30, p.100520.
41. Harris Williams. 2020. *Opportunities In Energy Transition – Energy Service Companies.* <https://www.harriswilliams.com/sites/default/files/industry_reports/opportunities_in_the_energy_transition_escos_1.22.20_v2_0.pdf> [Accessed 17 November 2020].

42. Climate Bonds Initiative. n.d. *Waste Management*. <https://www.climatebonds.net/standard/waste> [Accessed 17 November 2020].
43. UBS. 2020. *Future Of Waste*. <https://bit.ly/3pE6OUS> [Accessed 17 November 2020].
44. About Greensole. n.d. *Greensole*. <https://www.greensole.com/> [Accessed 17 November 2020].
45. Bhattacharjee, S., 2018. *How This Mumbai-Based Start-Up Refurbishes Old Shoes Into Trendy Footwear*. Business-standard.com. <https://www.business-standard.com/article/companies/how-this-mumbai-based-start-up-refurbishes-old-shoes-into-trendy-footwear-118121500573_1.html#:~:text=Worldwide%20every%20year%2C%20more%20than,prevented%20by%20wearing%20proper%20footwear.> [Accessed 17 November 2020].
46. Sustainable Salons. n.d. *About Us*. <https://sustainablesalons.org/about-us/> [Accessed 20 November 2020].
47. Osmosis Investment Management. n.d. *Investment Philosophy*. <https://www.osmosisim.com/rw/philosophy/> [Accessed 8 November 2020].
48. Rabo Bank. 2018. *ABN AMRO, ING And Rabobank Launch Finance Guidelines for Circular Economy*. <https://www.rabobank.com/en/press/search/2018/20180702-abn-amro-ing-and-rabobank-launch-finance-guidelines-for-circular-economy.html> [Accessed 9 November 2020].

10

Governance in CE

Policies and regulations have played a key role in environmental governance. Circular economy is not an exception. In this chapter, a global overview on governance in circular economy is presented along with some of the noteworthy initiatives.

Policies provide the guiding and operational principles that enshrine the intent and commitment of the governments. Regulations help to put these principles into practice and assign roles to institutions for enforcement. Economic, market, and information-based instruments are used to support the implementation. These instruments stimulate and promote increased participation of stakeholders to ensure the adoption of the circular principles to a mutual advantage.

In some countries, apex or coordinating institutions have been created to implement the agenda on circular economy with clearly defined mandates. Generally, ministries of environment have taken charge. Strategic action plans are launched that are multi-sectoral, and in some cases, focus on specific themes, e.g., resource efficiency, water conservation, phaseouts of toxic chemicals or financing technology upgradation. Here, coordination between various line ministries becomes critical. Cross-thematic and cross-sectoral interventions help in leveraging.

There is no silver bullet solution however and for each situation, a customized approach is necessary to build the action plan on circular economy. An adaptive approach is also necessary as the success largely depends on innovation and behavioral change that takes much longer a time to support and align.

Sections below present examples of initiatives that have played an important role in the governance on circular economy.

10.1 Resource Efficiency as the Driver

Resource efficiency has been one of the principal drivers of circular economy. Box 10.1 illustrates the German Resource Efficiency Programme.

BOX 10.1 RESOURCE EFFICIENCY DRIVEN CIRCULAR ECONOMY – CASE OF GERMANY[1]

With the adoption of the German Resource Efficiency Programme (ProgRess) in February 2012, Germany was among the first countries to determine targets, guiding principles and approaches to the conservation of natural resources. The German government is obligated to submit a report to the Bundestag on developments in resource efficiency in Germany every four years and to update the program. The first update report, ProgRess II, was adopted by the Federal Cabinet on March 2, 2016, the second, ProgRess III, on June 17, 2020. The German Resource Efficiency Programme has led to the development of a broad-based political and social process to implement resource efficiency measures[2].

ProgRess describes measures for increasing energy efficiency along the entire value chain – from raw material extraction and product design to production, use and circular economy. One aspect of updating the program is to address current challenges in order to further improve the program's effectiveness.

ProgRess focuses only on abiotic raw materials, excluding fossil fuels used for energy production and including biomass used as a material. It covers the entire value chain and defines ten action areas with corresponding policy approaches (123 in total). Germany has set itself ambitious targets in ProgRess: To double raw material productivity of domestic production by 2020, compared to 1994. Germany seeks to continue the trend in total raw material productivity from 2000 to 2010 until 2030[3].

Focus in ProgRess I: 2012–2015

- 20 strategic approaches – e.g. consumption – public awareness, resource efficiency as criterion for trade and consumer, certification systems for raw materials, public procurement.

Focus in ProgRess II: 2016–2019

- 10 action approaches – Supply of raw materials, resource-efficient production, resource-efficient products and consumption, resource-efficient circular economy, sustainable construction and sustainable urban development, resource-efficient ICT, overarching instruments, synergies with other policy areas, support of resource policy at local and regional level, strengthening of resource policy at European Union (EU) and international level.

The ProgRess III report incorporates current environmental policy challenges into the program[4].

New elements of the update:

- Stresses the importance of resource efficiency for achieving Germany's climate targets.
- Analyses the potential and risks of digital transformation in the field of resource efficiency.
- Takes a look at mobility from the point of view of resource efficiency.
- Identifies priority actions.

Adopting these principles has helped overcome the problem encountered by policymakers when drawing up the program of making clear distinctions between the various concepts that exist in the area. Agreeing on a definition of resource efficiency and the metrics to measure it presented similar problems. Officials also found that the effort to standardize guidelines to industry was much more complicated than for instance in the field of energy efficiency. In addition to achieving such common understandings, policymakers found that to anchor resource-efficiency in the mindsets of stakeholders is a long process that policy must aim to support and sustain[2].

One of the key outcomes of the first ProgRess action plan was the building of networks to amplify policy measures. These networks have proved successful in accelerating the exchange of knowledge among stakeholders. The principle cross industry example, the 31 member Resource Efficiency Network, is managed by the Association of German Engineers and brings together industry associations, special-interest groups, chambers of commerce, research institutes, and Federal and Länder (regional) bodies to share best practice, expertise and experience in resource-efficient production, products, and management. There are also sector specific organizations, for example the Round Table on Resource Efficiency in Buildings[2].

In 2015, leaders of the G7 nations held their annual Summit in June 2015, in Germany and established G7 Alliance on Resource Efficiency. This Alliance serves as a forum to share knowledge, create information networks across G7 countries, and encourage collaboration with large and small businesses and relevant stakeholders to advance resource efficiency, promote voluntary best practices, and foster innovation[5].

The G20 countries decided at their summit in Hamburg in July 2017 to establish a "G20 Resource Efficiency Dialogue". The Dialogue is expected to exchange good practices and national experiences to improve the efficiency and sustainability of natural resource use across the entire life cycle, and to promote sustainable consumption and production patterns[6].

EU has been engaged in bilateral collaborations to promote resource efficiency. In July 2020, EU and Government of India signed a Joint Declaration to establish an India-EU Resource Efficiency and Circular Economy Partnership. This partnership will bring together representatives of relevant stakeholders from both sides, including governments, businesses (including start-ups), academia, and research institutes. The objectives of this partnership is to enhance resource efficiency and move towards a more circular economic model that reduces primary resource consumption, strives towards non-toxic material cycles, and enhances the use of secondary raw materials. This partnership is considered relevant and important, in the context of the post COVID-19 pandemic economic recovery efforts. See Box 10.2 on the steps taken by the Government of India on resource efficiency and circular economy.

BOX 10.2 INITIATIVES TAKEN BY GOVERNMENT OF INDIA ON RESOURCE EFFICIENCY AND CIRCULAR ECONOMY

In 2017, India's National Institution for Transforming India (NITI) Aayog prepared a strategy paper on Resource Efficiency[7]. The focus of this paper was to recommend a broad strategy for enhancing resource-use efficiency in Indian economy. Government of India established the Indian Resource Panel in 2016 – an advisory body under the Ministry of Environment, Forest and Climate Change (MoEFCC). The strategy paper on resource efficiency is borne out of the work of the Indian Resource Panel.

Later, NITI Aayog with support of EU-REI through GIZ updated the strategy paper and published in 2019 Status Paper on Resource Efficiency and Circular Economy – Current Status and Way Forward[8]. This paper was an outcome of yearlong consultations with 9 key ministries, taking their inputs. This led to a harmonized plan with the ongoing initiatives as well as those from the private sector. In addition to the strategy paper, four sectoral reports were prepared on resource efficiency on steel, aluminum, construction and demolition waste and electronic waste. These papers following a life cycle approach.

In 2019, MoEFCC formulated a draft policy on resource efficiency and formed a resource efficiency cell at the ministry. With the help of EU-REI and guidance from NITI Aayog, the State of Goa published Strategy for Resource Efficiency and Circular Economy in Goa[9].

Visit https://www.eu-rei.com/publications.html to access all the important policy papers produced.

BOX 10.3 RESOURCE SUSTAINABILITY ACT (2019) IN SINGAPORE[10]

The Ministry of Environment and Water Resources in Singapore designated 2019 as the "Year Towards Zero Waste". The idea was to rally Singaporeans to conserve and protect its resources and build a strong 3R culture and pave the way towards a circular economy and a zero-waste nation.

The government passed the Resource Sustainability Bill in 2019 and subsequently the act in September 2019. The purpose of this Act is to impose obligations relating to the collection and treatment of electrical and electronic waste and food waste, to require reporting of packaging imported into or used in Singapore, to regulate persons operating producer responsibility schemes, and to promote resource sustainability. The Act also has a separate "purposes clause" that sets out its purposes as:

- Implement a framework where persons who profit from the supply of products bear the cost of collecting and treating these products when they become waste.
- Encourage producers of packaging to reduce, re-use or recycle packaging.
- Enable proper segregation and treatment of food waste.

In some countries, the agenda on resource efficiency has been pushed through legislation. Box 10.3 cites example of Singapore.

Resource efficiency should be complemented by strategies towards resource-recirculation. Given the material flows across the sectors and at various scales, a framework helps to guide and involve key stakeholders and assign the roles. South Korea is an interesting example where a framework was developed for resource-recirculation. Box 10.4 presents such a framework.

BOX 10.4 FRAMEWORK FOR RESOURCE-RECIRCULATION IN SOUTH KOREA[11]

In South Korea, in the 2000s, a framework for a resource-recirculation society was pursued by developing a master plan. Accordingly, waste was not simply treated and disposed but recycled as a resource. Currently, Korea follows a "Zero Waste" policy, which seeks to utilize wastes as a source of resources, in addition to minimizing waste generation.

In 2002, the "Second National Master Plan for Waste Management" (2002–2011) was established, and in 2007 this plan was revised. In 2003, the "Deposit System" was expanded to an "Extended Producer Responsibility" system. The legislative framework was extended by the "Construction Waste Recycling Promotion Act", the prohibition on direct landfill burial of food waste (2005) and the "Act on Resource Circulation of Electrical and Electronic Equipment and Vehicles" (2007).

From late 2000s onwards, the government emphasized the need for resource and energy recovery from waste. In September 2011, the Korean Ministry of Environment established the "First Framework Plan (2011–2015) for Resource Circulation" to form a foundation for upcycling waste resources and thereby promote a resource circulating (zero-waste) society. Besides, the "Measure for Promotion of Transition to a Resource Circulation Society" (2013) were introduced that included:

- Collection and transportation of recyclable resources through a free-of-charge collection of large-sized domestic electronic equipment waste.
- Consolidation of the sorting system.
- Increased installation of facilities for the energy utilization of waste resources. As such, foundations of recycling were also announced.
- Creation of a market for recycled products and support for the industries.

The "Framework Act on Resource Circulation" (FARC) was decreed in 2016 to form a basis for the implementation of these measures and associated policies. FARC was enforced since 2018. Using FARC, Korea intends to transform the mass production-oriented and mass waste-producing economic structure into a much more sustainable and efficient resource-circulating one at a fundamental level. The provisions of the framework address three categories such as "Recyclable Resource Recognition Program", "Resource Circulation Performance Management Program", and "Waste Disposal Fees'.

For more information, read: Korean Environment Institute. 2016. Introduction of the Framework Act on Resource Circulation Toward Establishing a Resource Circulating Society in Korea.[12]

Resource efficiency plays an important role to combat climate change by reducing Greenhouse Gas (GHG) emissions across the life cycle of materials. In response to a request by leaders of the Group of 7 nations International Resources Panel (IRP) conducted a rigorous assessment of the contribution

BOX 10.5 MATERIAL EFFICIENCY STRATEGIES FOR A LOW-CARBON FUTURE[13]

According to the IRP report, GHG emissions from the material cycle of residential buildings in the G7 countries and China could be reduced by at least 80% by 2050. This could be possible through more intensive use of homes, design with less materials, improved recycling of construction materials, and other strategies.

Significant reductions of GHG emissions could also be achieved in the production, use and disposal of cars. IRP modeling showed that GHG emissions from the material cycle of passenger cars in 2050 could be reduced by up to 70% in G7 countries and 60% in China and India through ridesharing, car-sharing, and a shift towards trip-appropriate smaller cars, among others.

Cross-cutting Policy interventions such as revision of building standards and codes, use of building certification systems by governments, green public procurement, virgin material taxation, removal of virgin resource subsidies, and recycled content mandates are required if material efficiency benefits are to be achieved.

Policy instruments such as taxation, zoning, and land use regulation play a role, but so do consumer preferences and behavior.

of material efficiency to GHG abatement strategies. Its focus was reduction potential of GHG emissions in residential buildings and light duty vehicles, and review policies that address these strategies. Box 10.5 shows the major findings of this report.

The IRP report cited above argues the importance of material efficiency in meeting the Nationally Determined Contributions (NDCs) of the Paris agreement. NDCs currently include limited commitments to material efficiency, exception being NDCs of Japan, India, China, and Turkey. Material efficiency may gain more importance if it is reflected in the scope of targets in the NDCs.

10.2 Zero Waste based Roadmaps

Many countries have prepared national waste policies and created a framework for action towards minimizing waste generation and circulation of wastes as materials to close the loop. Targets on Zero Waste have been the driver. Box 10.6 shows example of Australia.

BOX 10.6 AUSTRALIA'S NATIONAL WASTE POLICY (2018)[14]

Australia developed a National Waste Policy (Less Waste, more resources) in 2018 building on 2009 waste policy. This policy provides a framework for collective action by businesses, governments, communities, and individuals until 2030. The policy is based on following key principles:

- Avoid waste.
- Prioritise waste avoidance, encourage efficient use, reuse and repair.
- Design products so waste is minimized.
- Products are made to last and easy recovery of materials should be possible.
- Improve resource recovery.
- Improve material collection systems and processes for recycling.
- Improve the quality of recycled material that is produced.
- Increase use of recycled material and build demand and markets for recycled products.
- Better manage material flows to benefit human health, the environment, and the economy.
- Improve information to support innovation, guide investment and enable informed consumer decisions.

This policy responds to the challenges facing waste management and resource recovery in Australia – excluding radioactive waste. It also reflects the global shift towards a circular economy. It also acknowledges the need to improve capacity to better design, reuse, repair and recycle the goods that are used. Finally, the policy provides a framework for businesses to embrace innovation and develop technologies that create new opportunities.

There are also examples where, while the focus is on waste, extended producer responsibility (EPR) is also explicitly considered in preparing the circular economy strategy. Box 10.7 describes the case of Welsh Government's new circular economy strategy (2020) as an example.

Road maps help to ensure that actions towards circular economy are well coordinated and are synergized. Road maps include targets that are measurable and can be used to report the progress made on the outcomes. Importantly, action plans made through road maps help in mobilizing the finances and provides all the key stakeholders a sense of direction so that

BOX 10.7 WALES, 2020 – CIRCULAR ECONOMY STRATEGY THAT INTEGRATES EPR[15]

The Welsh Government's new circular economy strategy has set out the nation's ambitious goal to become "the world leader in recycling". Published ten years after Wales' original "Towards Zero Waste" strategy, the government's new waste strategy, entitled "Beyond Recycling", outlines plan for Wales to become a zero-waste, net-zero emissions nation by 2050, making savings of up to GBP 2 billion for the Welsh economy.

The Welsh Government set an interim target of a 70% household, municipal and industrial recycling rate by 2025, alongside a 50% reduction in avoidable food waste and a limit of five per cent waste to landfill[16].

As part of its aim to phase out single-use plastic, the Welsh Government plans to work with the other governments of the UK to accelerate the introduction of EPR for packaging and a deposit return scheme (DRS) for drinks containers, with an EPR regime and a ban on highly littered single-use plastics to be implemented by 2021.

In 2019, the Government sought consultation on EPR strategies[17]. There was a strong support for the use of EPR, with respondents viewing it as in line with the polluter pays principle, enforcing environmental externalities and prompting more responsible changes at design level. EPR was generally preferable to a direct tax on use of certain materials. Larger manufacturers preferred the idea of EPR being applied along the supply chain, whereas retailers believed there should be a single point of compliance further up the supply chain. On DRS a concern was expressed that it could undermine existing collections and reduce overall recycling performance.

they are better prepared. Box 10.8 provides examples of road maps in Finland, the Netherlands, and Sweden.

EU prepared a circular economy action plan in 2015 that consisted of 54 actions. Implementation of this action plan has accelerated the transition towards a circular economy. In 2016, sectors relevant to the circular economy employed more than four million workers, a 6% increase compared to 2012. Circularity also opened up new business opportunities, giving rise to new business models and developed new markets, domestically and outside the EU. In 2016, circular activities such as repair, reuse or recycling generated almost EUR 147 billion in value added while accounting for around EUR 17.5 billion worth of investments. In 2020, EU updated its circular economy action plan as support to the European Green Deal to boost efficient use of resources by moving to a clean, circular economy, restore biodiversity and reduce pollution. Box 10.9 provides some of the highlights.

BOX 10.8 EXAMPLES OF ROADMAPS TOWARDS CIRCULAR ECONOMY

FINLAND'S ROAD MAP[18]

Finland was the first country in the world to prepare a national road map to a circular economy in 2016, under the leadership of the research organization Sitra. In March 2019, an updated version was published that includes descriptions of the essential circular economy measures to which Finnish stakeholders have already committed themselves. The publication includes the most effective circular economy measures and solutions that Finland proposes in order to tackle the challenges of climate change, depletion of natural resources and urbanization. Road map 2.0 updates the solutions and hones the vision and strategic objectives.

NETHERLANDS, ROAD MAP TOWARD CIRCULAR ECONOMY GOAL 2050[19]

In 2016, The Netherlands set a goal to be fully circular by 2050. The cabinet formulated three objectives to make the Dutch economy circular:

1. Existing production processes should make more efficient use of raw materials, so that fewer raw materials are needed.
2. When new raw materials are needed, use is made as much as possible of sustainably produced, renewable and generally available raw materials, like biomass, which is raw material from plants, trees, and food residues. This will make the Netherlands less dependent on fossil sources.
3. Develop new production methods and design new products that are circular.

A Circular Economy Implementation Program was developed to meet the goal for the years between 2019 and 2023. The program will be updated every year until 2023 and new initiatives will be added as relevant.

SWEDEN'S NATIONAL ACTION PLAN 2020[20]

In 2020, Sweden adopted a national strategy for a circular economy that sets out the direction and ambition for a long-term and sustainable transition of Swedish society. This was an important step towards Sweden becoming the world's first fossil-free welfare nation.

The core of the strategy is a vision that states "A society in which resources are used efficiently in toxin-free circular flows, replacing new materials".

The national action plan focuses on sustainable production and product design; sustainable consumption of materials, products, and sustainable services; toxin-free and circular eco cycles. Circular economy is considered as a driving force for the business and other actors by promoting innovation and circular business models.

BOX 10.9 EU'S CIRCULAR ECONOMY ACTION PLAN (2020)[21]

On 11 March 2020, the European Commission adopted a new Circular Economy Action Plan. It presents new initiatives along the entire life cycle of products in order to modernize and transform our economy while protecting the environment. It aims at accelerating the transformational change required by the European Green Deal, while building on circular economy actions implemented since 2015.

The action plan is driven by the ambition to make sustainable products that last and to enable our citizens to take full part in the circular economy initiatives and benefit from the positive change that it brings about. The action plan announces initiatives along the entire life cycle of products, targeting for example their design, promoting circular economy processes, fostering sustainable consumption, and aiming to ensure that the resources used are kept in the EU economy for as long as possible. The action plan also introduces legislative as well as non-legislative measures targeting areas where action at the EU level brings real added value.

ACTIVITY 10.1 DISCUSSION ON NATIONAL POLICIES ON CIRCULAR ECONOMY

Research on examples where policies of different ministries have been in conflict posing challenges towards transition to circular economy. Make a comparison between strategies adopted on circular economy between Finland, China, and Australia. What could be the lessons learnt?

10.3 Legislating Circular Economy

There has been a debate on whether circular economy should be legislated. Many of the existing regulations on resource use and management of residues can be strengthened and integrated in a life cycle perspective and as such there may not be a need for a separate legislation.

Box 10.10 describes the Closed Substance Cycle and Waste Management Act (1996) of Germany and Box 10.11 describes the Sound Material Society Act of Japan (2000) and the supporting legislations respectively. These legislations evolved further. The Closed Substance Cycle and Waste Management Act (1996) of Germany was transformed to Circular Economy Act in 2012 with extended producer responsibility as the focus. The scope and intent of the Sound Material Society Act of Japan (2000) was broadened to address the vision of sustainable society.

While to some, legislating circular economy is a panacea, enforcement of "framework legislation" is not easy as multiple ministries are involved. There is also a role played by regional and local authorities. It often takes several years for the legislation on circular economy to mature and given the

BOX 10.10 THE CLOSED SUBSTANCE CYCLE AND WASTE MANAGEMENT ACT IN GERMANY[22]

In Germany, the Closed Substance Cycle and Waste Management Act of 1996 was enhanced and transformed into the Circular Economy Act in 2012.

The Act aims to ensure the complete prevention and recovery of waste, including hazardous waste. Thus, prevention takes precedence over recovery, which in turn comes before disposal.

Waste prevention is implemented inter-alia, through EPR, which on the one hand involves developing products and substances with the longest possible service life and, on the other, introducing production techniques that generate the minimum possible volume of waste through Best Available Technologies requirements as part of a permitting system for industrial installations.

Under the EPR, producers of a commodity are required to consider the environmental impacts and possible risks of a product during its entire life cycle as a precautionary measure.

In collaboration with the other parties involved – producers, distributors, consumers, disposal and recycling companies, government departments – the producer is required to create a system that minimizes the adverse environmental impacts and maximizes the recovery of resources through reuse and recycling.

BOX 10.11 SUPPORTING LEGISLATIONS TO THE SOUND MATERIAL SOCIETY ACT OF JAPAN (2000)[23]

The Sound Material Society Act was aimed to create a society where the consumption of natural resources will be conserved and the adverse environmental impact will be reduced to the extent possible, by preventing or reducing the generation of wastes, by promoting cyclical use of products and by ensuring proper disposal of circulative resources not put into cyclical use.

The Sound Material-Cycle Society is in conformity with the philosophy of the Environment Basic Act (Act No. 91 of 1993), clarifying the responsibilities of the State, local governments, business operators and citizens, and articulating fundamental matters for making policies and plan for the formation of a Sound Material-Cycle society.

The creation of a sound material-cycle society was accelerated significantly for about 10 years after 2000. Three indicators such as "inlet", "in circulation", and "outlet" were developed to measure the progress on circularity. These indicators were later expanded as the goal was broadened towards creation of sustainable society. In 2018, a Fourth Fundamental Plan was developed that will be reviewed in 2023.

In recent years, in Japan, all the indicators of resource productivity, cyclic use rate at the inlet, and final disposal amount are leveling off. While there are some items of waste, such as industrial scrap metal waste, that are recycled almost entirely, there are also some items, including plastics and food waste, that require further 3R efforts.

Following legislations have supported implementation of the sound material society act:

- Waste Management Act.
- Act on the Promotion of Effective Utilization of Resources.
- Container and Packaging Recycling Act.
- Home Appliance Recycling Act.
- Small Home Appliance Recycling Act and Construction Recycling Act.
- Food Recycling Act and End-of-life Vehicle Recycling Act.
- Act on Special Measures against Industrial Waste (Temporary legislation until 2022).
- Act on Special Measures concerning Promotion of Proper Treatment of PCB Wastes.
- Act on Promotion of Procurement of Eco-Friendly Goods and Services by the State and Other Entities.

complexity and the changing perspectives of the society, it often provides mixed results.

In 2008, China passed Circular Economy Promotion Law. This law refers to the reduction, reuse, and recycling (3R) activities in the production, circulation, and consumption of products. Circular economy in China has two purposes. Firstly, scarcity of resources should be partly solved by improving energy and material use efficiency and reducing the consumption of resources and energy. Secondly, emissions of pollutants and greenhouse gas should be reduced by mitigating pollution caused by rapid industrial development[24]. More about the implementation of this Law will be discussed later in this chapter.

Sometimes, existing legislations can be used as a basis to push circularity of materials. The Resource Conservation & Recovery Act (RCRA) of 1976 in the United States of America (USA) provided a legislative basis for environment Protection Program's (EPA) Sustainable Materials Management Program (SMM). This program set a strong preference for resource conservation over disposal. US EPA's report, Beyond RCRA: Waste and Materials Management in 2020 (released in 2002) made the argument for focusing efforts on materials management, and their report, SMM: The Road Ahead (released in 2009) provides recommendations and an analytical framework for moving toward sustainable materials management. In SMM, EPA's waste hierarchy continues to provide guidance, highlighting source reduction/waste prevention & reuse over recycling and composting, energy recovery, and treatment & disposal[25].

The Sustainable Materials Management Program's Strategic Plan for Fiscal Years 2017–2022[26] has identified three strategic focus areas:

- The Built Environment – conserve materials and develop community resiliency to climate change through improvements to construction, maintenance, and end-of-life management of our nation's roads, buildings, and infrastructure.
- Sustainable Food Management – focus on reducing food loss & waste.
- Sustainable Packaging – increase the quantity and quality of materials recovered from municipal solid waste and develop critically important collection and processing infrastructure.

Another example in the USA is about various legislations introduced in thematic areas or sectors e.g., reducing life cycle impacts of built environment. Box 10.12 describes USA laws on Sustainable Design that show such a progression.

Trend on legislative circular economy seems to be on the rise. The Australian Senate passed Recycling Waste Reduction Bill 2020. This legislation will permanently ban the export of waste including plastic, paper, and tyres, beginning from January 1, 2021.

BOX 10.12 USA LAWS ON SUSTAINABLE DESIGN[27]

The Energy Policy Act (EPAct) of 2005 in the US addressed energy production and included building-related provisions to "design new federal buildings to achieve energy efficiency at least 30% better than ASHRAE 90.1 standards, where life-cycle cost is effective".

In 2006, 19 federal agencies signed a Memorandum of Understanding committing to "federal leadership in the design, construction, and operation of High-Performance Sustainable Buildings". This inter-agency memo yielded what is now called the Guiding Principles for Sustainable Federal Buildings and charged agencies to optimize buildings' performance while maximizing assets' life-cycle value. Executive Orders, including 2007's E.O. 13423 Strengthening Federal Environmental, Energy and Transportation Management, have required Federal agencies to make annual progress toward 100% portfolio compliance with the Guiding Principles.

The Energy Independence and Security Act of 2007 established additional environmental management goals. Accordingly, new General Services Administration buildings and major renovations must meet requirements including: reducing fossil-fuel-generated energy consumption by 65% by 2015 and by 100% by 2030; managing water from 95th percentile rain events onsite; and applying sustainable design principles to siting, design, and construction.

In 2018, Executive Order 13834 regarding Efficient Federal Operations superseded the Executive Orders 13423, 13514, and 13693. It directs federal agencies to manage their buildings, vehicles, and overall operations to optimize energy and environmental performance, reduce waste, and cut costs. It calls for agencies to cost-effectively meet goals including:

- Achieve and maintain annual reductions in building energy use, and implement energy efficiency measures that reduce costs.
- Reduce potable and non-potable water consumption, and comply with storm water management requirements.
- Ensure that new construction and major renovations conform to applicable building energy efficiency requirements and sustainable design principles. Revised Guiding Principles for Sustainable Federal Buildings were issued by the Council on Environmental Quality in 2016.

ACTIVITY 10.2 DEBATE ON CHINA'S CIRCULAR ECONOMY PROMOTION LAW

Hold a debate on "Is legislation and enforcement a solution for adoption of circular economy?". Research on China's Circular Economy Promotion Law, its evolution, and results.

10.4 Sustainable Public Procurement

Sustainable Public Procurement (SPP) has its roots in the term Green Public Procurement (GPP). SPP expanded the scope of GPP to include social and ethical considerations. SPP may simply be defined as "public procurement for a better environment and socio-economic development". SPP is a process whereby government and its agencies (such as public sector organizations) seek to procure goods, services and works with reduced environmental and social impact throughout their life cycle. SPP promotes innovation, resource efficiency and resource recirculation and create green jobs. More importantly, SPP influences the consumption and production patterns. By promoting and using SPP, public authorities can provide industry with real incentives for developing green materials, technologies, and products. SPP is therefore a strong stimulus for eco-innovation. SPP is fundamentally a voluntary instrument, but it can be legislated.

Japan already has a law on GPP. In 2000, in South Africa, Department of Environment Affairs adopted a Preferential Procurement Policy under the 'Preferential Procurement Policy Framework Act, 2000'. In China, from January 2007, provincial and central governments have made a list of environment friendly products certified by China Certification Committee for Environmental Labeling and these products must mandatorily meet environmental protection and energy saving standards. In Mexico, the 2007–2012 National Development Plan brought in sustainability criteria in the procurement policy followed by a procurement law. The law recognized that all wood and furniture procurement by public agencies requires a certificate highlighting its legal origin and paper procured by public agencies will need to have 50% recycled content[28].

EU adopted two directives on February 26, 2014. Today many of the EU countries have transposed these directives or rules into national laws. These directives support Innovation partnerships where a contracting authority wishes to purchase goods or services, which are not currently available on the market. The procedure for establishing an innovation partnership is set out in Article 31 of Directive 2014/24/EU[28].

To be effective, SPP requires the inclusion of clear and verifiable environmental criteria for products and services in the public procurement process. Several countries in the world have developed guidance in this area, by defining general sourcing and product specific sustainability criteria or by prescribing eco-labels and by asking Environmental Products Declarations (EPD).

A new standard, ISO 20400 is developed on Sustainable procurement to provide guidelines for organizations wanting to integrate sustainability into their procurement processes. This standard was launched in April 2017. Since sustainable procurement is a key aspect of social responsibility, ISO 20400 complements ISO 26000, Guidance on social responsibility[29].

SPP has been introduced in many countries through policies, legislations and eco-labeling programs that are considered as information-based instruments. The approach over the years has been progressive. Tables 10.1 and 10.2 show such a progression on GPP in Japan and Korea.

Experience has shown that SPP requires change management and this is possible only when the capacity of the procurement officers is built-in public-sector organizations. Box 10.13 describes some of the efforts taken in this direction.

UNEP carried out a survey in 2017 on SPP and published a global review. Box 10.14 highlights some of the key findings[41].

TABLE 10.1

Progression of SPP in Japan

Year	Action	Details
1989	Eco Mark Program[30]	Launched by The Japan Environment Association to encourage suppliers and consumers to use Eco Mark certified products and services (ISO14024 and ISO14020).
1996	Establishment of the Green Purchasing Network[31]	A non-profit organization with 2,876 member organizations (as on 2019) from businesses, local governments, and NGOs, is an important agency supporting the government with the implementation and promotion of green public procurement, particularly in the areas of training and awareness-raising.
2000	Act on Promoting Green Purchasing[32]	The Act requires all government bodies including federal, cities, prefectures, towns, and villages to develop a green purchasing policy, implement a green purchasing system and publish a summary of green purchasing records on an annual basis. The environmental criteria and specifications for each of the items targeted for green public procurement under the act are published and updated by the MOE, Japan.
2007	Green Contract Law[33]	The law concerning the promotion of contracts considering reduction of greenhouse gases and other emissions by the state and other entities.

TABLE 10.2

Progression of SPP in Korea

Year	Action	Details
1992	Korean Eco-label Program[34]	Launched by the Ministry of Environment (MOE) for four selected product groups. However, as on 2019, The MOE provides different standards and requirements of eco-labeling which applies to 161 categories of products.
1994	Act on Development and Support of Environmental Technology[35]	The Act recommends the preferential purchase of products with the Korea Eco-label or Good Recycled Mark certification to public institutions.
2005	Act of Promotion of Purchase of Green Products[36]	The Act serves as the basis for the implementation of Green Public Procurement (GPP) in Korea and requires all government agencies, including central and local governments and public corporations, institutes and education institutions, to submit to Korea Environmental Industry and Technology Institute (KEITI) an annual GPP implementation plan in which each organization sets its own voluntary target and performance report on the amount, in expenditure and number, of green products purchased.
2010	Framework Act on Low Carbon, Green Growth[37]	The purpose of the Act is to achieve a national economy based on low-carbon green growth through green technology and industry development. GPP is one of the instruments for the facilitation of green technology and industry development. Hence, the Act led to the promotion of GPP on a large scale.

BOX 10.13 HIGHLIGHTS OF SOME OF THE TRAINING AND CAPACITY BUILDING EFFORTS ON SPP

EU supported creation of GPP Training Toolkit (consisting of six independent modules and ten operational modules, with PowerPoint presentations and accompanying guidance) for use by public purchasers and by GPP trainers[38].

Ministry of Finance in China provides training for procurement staff to help them become more confident in applying green public procurement policies, new procedures and developing criteria that include environmental considerations[39].

Briefing sessions are carried out by Japanese Government on GPP in eight prefectures every year. Seminars cover information about "Law on Promoting Green Purchasing". To encourage the enforcement on green purchasing law by local public institutions, a workshop targeted to local procurers is held 3 times every year. The workshops aim to engage those public bodies that are not implementing GPP[40].

In Korea, Public Procurement Service (PSS) has developed the "Green Purchasing Educational Course" in the Public Procurement Human

Resources Development Center. The program is oriented to procurement officers of public organizations. The training is undertaken twice a year with a total of 21 h. Nationwide GPP training is also offered to public officials every year. GPP guidelines developed by KEITI are distributed before the training session. Annual nationwide training is provided from November to December for the following year's GPP implementation, and additional training is provided for newly appointed GPP staff.

BOX 10.14 GLOBAL REVIEW ON SPP BY UNEP

Nearly all European and Asian national governments participating in the survey adopted their first SPP policies between 2001 and 2009. Most of the countries in Europe approved their National Green Public Procurement Action Plans after 2006, following policy recommendations from the European Commission. Since 2012, many European countries have revised and renewed their National Green Public Procurement Action Plans.

SPP policies vary widely across national governments. Countries are using different policy vehicles to drive SPP, ranging from single-aspect regulations, such as focusing on procurement from army veterans (e.g., the Republic of Korea) or buying recycled-content products, often characterizing early efforts, to comprehensive action plans. Most national governments participating in the survey included SPP provisions in overarching or thematic policies and strategies, while a smaller proportion included them in procurement regulations or in policies specifically dedicated to the promotion of SPP. Some governments, particularly in Asia, focus exclusively on environmental issues, and are not yet considering the socio-economic dimension. However, countries in EU prioritize on both socio-economic and ethical issues in addition to focusing on the environment.

Countries also show variation in the level of enforcement prescribed by SPP policies. Mandatory policy frameworks for SPP were found to be typically more effective in driving implementation since they do not depend as heavily on the initiative of individual ministries, departments, or procurers. However, one common feature across governments was that the ministries or agencies involved in the design of SPP policies are predominantly those associated with environmental, economic, and financial affairs, i.e., procurement agencies and ministries of environment, economy, and/or finance.

Public and private procurers increasingly support the idea of sustainable procurement. Few however realize that SPP is more than just purchasing more environmentally friendly products and could potentially be a strategy towards circular economy.

Circular public procurement is now considered as an expansion of SPP. Circular procurement sets out an approach to SPP focusing on "purchase of works, goods or services that seek to contribute to the closed energy and material loops within supply chains, whilst minimizing, and in the best case avoiding, negative environmental impacts and waste creation across the whole life-cycle". More detail can be obtained from the guidebook published by EU named Public Procurement for A Circular Economy[42].

ACTIVITY 10.3 DISCUSSION ON SPP

What may be the challenges in introducing SPP? How could these challenges be addressed? Look for examples where corporates have followed sustainable procurement influencing the markets as well as the supply chains.

10.5 Economic, Market Based and Information Driven Instruments

Economic, market-based and information driven instruments are important in the governance. These instruments help in the enforcement of legislation (or the "push") and provide a win-win situation (or the "pull") to the stakeholders, especially consumers and business organizations.

These instruments often demonstrate that meeting compliance can be a preferred option. They essentially make a business case for compliance through cost-savings, competitiveness, innovation, and branding by providing positive and negative incentives.

Positive incentives are tax reduction for refurbished/recycled products and negative incentives for example could be penalties for not complying to waste segregation, high charges, and restrictions for the landfilling and incineration of waste. Some incentives include both positive and negative aspects, e.g., deposit refund schemes can be voluntary and generating income but opting out would lead to tax increase.

France has EPR schemes for different product types such as electrical or electronic waste, packaging, furniture, and paper. This means it integrates a fee, which covers the cost of disposing a product, into the purchase price paid by the user. The schemes are adjusted based on how recyclable the product is: the more recyclable the material, the lower the fee. This is, essentially, an incentive for manufacturers to design their products so that they are more easily recyclable. For example, manufacturers who place vacuum cleaners with brominated flame-retardant plastics on the market pay 20% more than manufacturers who do not use this hazardous substance[43].

Whalen, Milios, and Nussholz (2018)[44] categorized the instruments into technological, educational, social, regulatory, institutional, market based, fiscal and industrial arrangements. It would be good to read this resource. Some of the important instruments are:

- Take-back incentives.
- Lowering labor taxes. Tax incentives on meeting EPR. Skills development (training and educational activities) – e.g., training for refurbishers.
- Obligations to provide spare parts.
- Obligations to provide product information to repairers, refurbishers, remanufacturers.
- Enforcement of longer warranty periods for consumers.
- Support to innovative, circular economy-focused business models.
- Development of infrastructure for consumers to return used products.
- Introduction of material efficiency and durability in product design regulation.
- Legal framework to facilitate trade of repaired and refurbished goods.
- Reduction of value-added tax (VAT) for refurbished products.
- Creation of subsidies for reuse that could help reduce operational costs and assist reuse operations.

The subject of policy instruments is vast. At the end of this chapter few key references as additional reading have been provided. It may be observed that much research on policy instruments in circular economy has been carried out in the EU with publications that report the impact or outcomes. Most countries have not yet made use of these instruments and relied on legislation and enforcement.

10.6 Need for Policy Impact Assessments

Developing policy frameworks to transition to circular economy is a complex task. It is a strategic process that needs to prioritize and gradually evolve by integrating and harmonizing existing regulations and instruments. In most instances it is a learning process, requiring impact assessment involving consultation with stakeholders for realigning and adaptation. Participation of key ministries is necessary for coordination and mobilization of resources.

While experience from other countries does help, each situation is unique and needs to be contextualized considering economic, social, and environmental considerations. Given the global trade of material flows and markets, policies on circular economy cannot anymore limit to national boundaries.

Introducing sudden bans, issuing directions to phase out of hazardous substances or imposing ambitious targets on resource efficiency often seem like appealing options. But such interventions if proposed on ad-hoc basis, without promising alternatives and in the absence of a framework of incentives, may hit more obstacles. A "systems thinking" is therefore necessary.

It is important that the policy toolbox for circular economy is used very carefully to "design" a "balanced" governance recognizing the capacities and the readiness. Introducing too many interventions at a time can be disastrous. Political will is necessary to take the leadership.

Strategic Environmental Impact Assessment is perhaps one of the most powerful tools that can help in the inquiry, assessment of alternatives and holding a dialogue with stakeholders. Science based participatory methods should be used to build scenarios. Here, application of modeling approaches such as System Dynamics coupled with Life Cycle Sustainability Assessment (LCSA), described in Chapter 3, should be encouraged.

Box 10.15 presents case of ban of importation of used computers in Uganda and in Box 10.16, case of banning E-waste to landfills in Victoria in Australia that used policy impact assessment.

BOX 10.15 IMPACT ASSESSMENT ON THE BAN ON IMPORTATION OF USED COMPUTERS IN UGANDA[45]

During the reading of the 2009/2010 financial budget, Government of Uganda imposed a ban on importation of used computers with a view of combating the accumulation of electronic waste in the country. Whilst this was for good intention, there was a general outcry that this ban stifled economic activities. As a result, traders and other stakeholders vehemently resisted the ban and petitioned against it. There was need to review the ban on used computers because it lacked clear specifications of old, used, new, assembled, and refurbished computers. For example, a computer used for only two weeks was considered as used computer hence banned. Based on their petitions and other considerations Cabinet on November 2, 2011 directed the Ministry of Finance Planning and Economic Development in consultation with the Ministry of ICT to review the importation of used computers. The need to spread use of computers through provision of affordable computers was recognized. The study assessed the impact of the ban on importation of used computers in Uganda and used the outcomes and recommendations of the assignment to form a basis for reviewing and lifting of the ban.

BOX 10.16 BANNING E-WASTE FROM VICTORIAN LANDFILLS (CASE OF POLICY IMPACT ASSESSMENT)[46]

E-waste is growing up to three times faster than general municipal waste in Australia and covers a range of items we use and discard from our daily working and home lives, including televisions, computers, mobile phones, kitchen appliances and white goods. These items contain both hazardous materials, which can harm the environment and human health, and valuable materials which are scarce and worth recovering.

The Victorian Government in Australia sought views from the community and industry on the proposed approach to managing "e-waste" in Victoria. A package of proposed measures was developed to reduce e-waste from landfill, increase resource recovery and support jobs and investment in the recycling sector.

Since it was considered that all Victorians have a say on the details of the proposed changes the Government proposed to consult the people and use feedback to refine the arrangements for the ban on e-waste from landfill. The timeline for the policy package was from October 4, 2017 to June 25, 2018, followed by implementation till June 2019. To guide this process, the Victorian Government prepared a 200+ pages exhaustive document on Policy Impact Assessment.

Sustainability Victoria launched a new campaign, implementing an AUD 1.5 million community education program on July 4, 2018 to educate Victorians about the value of e-waste and how it can be recycled. The campaign featured a new website, ewaste.vic.gov.au, which includes an animated video showcasing the valuable materials inside our electronics and social media and digital advertising.

To support the rollout of the ban of e-waste from landfills, the Victorian Government is investing AUD 16.5 million to upgrade e-waste collection and storage facilities across the state and to deliver an education campaign to support the ban.

With a ban proposed to start in July 2019, a package of AUD 16.5 million was mobilized to encourage safe management of hazardous materials found in e-waste and to enable greater recovery of the valuable materials. The objective of this package was to lead to a more stable industry and more jobs for Victoria.

Another area that deserves attention is the need for impact assessments of disruptive technology innovations. Disruptive technologies play an important role in a circular economy. Today, disruptive technologies include areas such as autonomous vehicles, The Internet of Things, 3D printing, energy storage, advanced robotics with senses, blockchain, artificial intelligence and machine learning, advanced materials, gene editing, spatial computing or augmented reality, hydrogen economy, synthetic biology, etc. Many of these technologies help in improving resource efficiency and resource re-circulation.

Ideally, the technology developers and innovation sponsors should be required to anticipate the potential risks and side-effects of disruptive technologies. Unfortunately, right now, concerns about the potential adverse impacts of disruptive technologies, especially in circular economy are not discussed and seem to be left to the critics.

Take an example of 3-D printing. Many of you know that additive manufacturing may provide custom-made and highly functional products, reduce the material intensity and weight of traditionally made products, and simplify logistical requirements through decentralized manufacturing that takes place closer to customers. But at the same time this type of production may require more energy than conventional manufacturing. It may also potentially ramp-up resource consumption by providing affordable objects that are difficult to fix or even dismantle for waste-recovery purposes.

Though the term 'disruptive technologies' is relatively new, the competitive effect of such innovations is much older. Take an example. Until the 1970s most of the world's steel was made by large, integrated steel companies that served all types of customers from users of high-end sheet steel to low-end "rebar" steel for reinforcing cement. However, other manufacturers began to use *minimill* technology to enter the steel market, melting scrap metal in electric furnaces to produce steel more cheaply than the integrated companies. At first, minimills could only sell to the rebar market, and the integrated steel mills let rebar customers go in order to concentrate on the more profitable high-quality steel. But as their technology advanced, minimills were able to progress producing higher quality steel, encroaching further on the market served by integrated steel mills. Several integrated steel mills went bankrupt as their customer base was reduced to the high-end, low volume segments. So, while the industry got restructured, there was a significant economic and social impact[47].

Not too far away in the future, we will face a significant number of disruptive technologies that will, in accordance with their nature, disrupt life and the economy and social systems. While the driver could be to accelerate towards circular economy, technology and humans will become closer and humans will probably get far away from the nature as a compromise. Today, we know about technology assessment and impact assessment. What is missing is a form of midstream governance, which Hasselbalch labels as "innovation assessment"[48]. One needs to look at circular economy from such perspectives, given both the opportunities and challenges. Chapter 11 discusses the future of circular economy.

10.7 Key Takeaways

- Developing policy frameworks to transition to circular economy is a complex task. It is a strategic process that needs to prioritize and gradually evolve by integrating and harmonizing existing regulations and instruments. In most instances it is a learning process, requiring impact assessment involving consultation with stakeholders for realigning and adaptation. Participation of key ministries is necessary for coordination and mobilization of resources.
- While experience from other countries does help, each situation is unique and needs to be contextualized considering economic, social, and environmental considerations. Given the global trade of material flows and markets, policies on circular economy cannot anymore limit to national boundaries.
- Mainstreaming resource efficiency in nationally determined commitments to climate change could be strategic. Response to climate change through nationally determined commitments will widen the canvas of circular economy.
- Introducing sudden bans, issuing directions to phase out of hazardous substances or imposing ambitious targets on resource efficiency often seem like appealing options. But such interventions if proposed on ad-hoc basis, without promising alternatives and in the absence of a framework of incentives, may hit more obstacles. A "systems thinking" is therefore necessary.
- It is important that the policy toolbox for circular economy is used very carefully to "design" a "balanced" governance, recognizing the capacities and the readiness. Introducing too many interventions at a time can be disastrous. Political will is necessary to take the leadership.
- Strategic Environmental Impact Assessment is perhaps one of the most powerful tools that can help in the inquiry, assessment of alternatives and holding a dialogue with stakeholders. Science based participatory methods should be used to build scenarios. Here, application of modeling approaches such as System Dynamics coupled with Life Cycle Sustainability Assessment (LCSA), described in Chapter 3, should be encouraged.

ADDITIONAL READING

1. **Delivering the circular economy: A toolkit for policymakers** – Sector-by-sector analysis can be a valuable approach to address the variety of opportunities and challenges involved in transitioning towards the circular economy. Within each sector,

effective circular economy policymaking requires the combination of many policy interventions and does not rely on a "silver bullet" or blanket solutions. Policymakers can address market and regulatory failures to create the right enabling conditions for circular economy initiatives to reach scale. They can also more actively steer and stimulate market activity by setting targets, implementing circular and total cost of ownership-oriented public procurement, and investing in innovative pilots and R&D.

Source: Ellen Macarthur Foundation. n.d. *Delivering the circular economy: A toolkit for policymaker.* <https://www.ellenmacarthurfoundation.org/assets/downloads/government/EMF_TFPM_FullReportEnhanced_11-9-15.pdf> [Accessed 10 November 2020]

2. **Advancing to a circular economy: Three essential ingredients for a comprehensive policy mix** – Three policy areas that can contribute to closing material loops and increasing resource efficiency are thoroughly discussed and their application challenges are highlighted. The three policy areas are: (1) policies for reuse, repair and remanufacturing; (2) green public procurement and innovation procurement; and (3) policies for improving secondary materials markets. Finally, a potential policy mix, including policy instruments from the three mentioned policy areas – together with policy mixing principles – is presented to outline a possible pathway for transitioning to Circular Economy policy making.

Source: Milios, L., 2017. Advancing to a Circular Economy: Three essential ingredients for a comprehensive policy mix. *Sustainability Science*, 13(3), pp. 861–878.

3. **POLICE project** – The aim of the POLICE project supported by EIT RawMaterials and led by VTT was to identify and describe different types of incentives to promote and boost the implementation of the circular economy concept, considering relevant policy instruments and incentives and their mixes. Another key objective of the project was to compare positive versus negative policy instruments and incentives. Furthermore, the third target of the project was to evaluate and analyse the effectiveness of incentives based on the information on circular economy barriers and drivers.

Source: EIT Raw Materials. 2020. *Policy Instruments and Incentives For Circular Economy - Final Report*. <https://eitrawmaterials.eu/wp-content/uploads/2020/07/EIT-RawMaterials-project-POLICE-Final-report.pdf> [Accessed 10 November 2020].

4. **SITRA project on economic instruments for CE** – During 2018, the project developed by SITRA, explored and compiled the best European economic policy instruments and practices for promoting a circular economy. The project also analysed and modelled the effects and opportunities brought to Finland by an ecological tax shift, as well as produced a scenario of how a shift in the focus of tax and aid policy could best promote a circular economy in Finland. Green Budget Europe (the consortium leader) was in charge of implementing this project together with the Ex'Tax project and the Institute for European Environmental Policy (IEEP). Cambridge Econometrics created a model for assessing the effects of the shift in the focus of taxation. The project identified economic policy instruments that support a circular economy and opportunities for changes in the focus of tax and aid policy.

Source: Sitra. n.d. *Economic Policy Instruments for a Circular Economy*. <https://www.sitra.fi/en/projects/economic-policy-instruments-circular-economy/#what-is-it-about> [Accessed 10 November 2020].

5. **Financial instruments for a circular economy** – The use of taxes, subsidies, and public procurement can accelerate the transition to a circular economy. Smart use of these financial instruments will reduce environmental damage and uncertainty about the supply of raw materials in the future, as well as creating economic opportunities through more efficient use of raw and other materials, and new business models. Moreover, reducing consumption of raw and other materials, combined with emphasizing recycling and reuse, will lead in many cases to a decrease in environmental damage in the form of CO_2 emissions, because less energy will be needed for the production process. This is the key message of the study "Financial instruments for a circular economy", which the Social and Economic Council of the Netherlands (SER) published in May 2018. The report identifies the criteria for use of financial instruments that can stimulate the circular economy and details areas of concern and research questions.

Source: SER. 2018. *Financial Instruments for a Circular Economy.* <https://www.ser.nl/-/media/ser/downloads/engels/2018/financial-instruments-circular-economy.pdf> [Accessed 10 November 2020].

6. **The role of market-based instruments in achieving a resource efficient economy** – This study investigated how market-based instruments (MBIs), can support and drive the move towards resource efficiency. These tools are designed to "get the prices right", meaning that markets better reflect environmental impacts (externalities) and resource scarcity in prices so that producers and consumers can respond appropriately. This is widely understood to be more economically efficient than directly legislating or regulating for similar goals. The objective of this study was to identify the market based instruments being used, particularly those that demonstrate best practice in promoting resource efficiency, and examine how they can be improved, what lessons can be drawn and the recommendations for the future, taking into account the cost, competitiveness and other impacts.

Source: European Commission. 2011. *The Role of Market-Based Instruments in Achieving a Resource Efficient Economy.* <https://ec.europa.eu/environment/enveco/mbi/pdf/studies/role_marketbased.pdf> [Accessed 10 November 2020]

Notes

1. Federal Ministry for the Environment, Nature Conservation and Nuclear Safety. n.d. *German Resource Efficiency Programme (Progress).* <https://www.bmu.de/en/topics/economy-products-resources-tourism/resource-efficiency/overview-of-german-resource-efficiency-programme-progress/#:~:text=Achieving%20more%20with%20less%3A%20The,the%20conservation%20of%20natural%20resources.> [Accessed 4 November 2020].
2. Ellen Macarthur Foundation. n.d. *German Resource Efficiency Programme (Progress II).* <https://www.ellenmacarthurfoundation.org/case-studies/german-resource-efficiency-programme-progress-ii> [Accessed 10 November 2020].
3. International Climate Initiative. n.d. *Resource Efficiency.* <https://www.international-climate-initiative.com/fileadmin/Dokumente/2018/20181011_Policy-Brief_Resource_Efficiency.pdf> [Accessed 10 November 2020].
4. Federal Ministry for the Environment, Nature Conservation and Nuclear Safety. n.d. *German Resource Efficiency Programme (Progress) – An Overview.* <https://www.bmu.de/en/topics/economy-products-resources-tourism/resource-efficiency/overview-of-german-resource-efficiency-programme-progress/> [Accessed 10 November 2020].

5. US EPA. n.d. *G7 Alliance On Resource Efficiency.* <https://www.epa.gov/smm/g7-alliance-resource-efficiency-us-hosted-workshop-use-life-cycle-concepts-supply-chain> [Accessed 21 October 2020].
6. GIZ. n.d. *Initiative Resource Efficiency And Climate Action* <https://www.giz.de/en/downloads/giz2019-en-ressourceneffizienz.pdf> [Accessed 21 October 2020].
7. Niti Aayog. 2017. *Strategy Paper On Resource Efficiency.* <https://niti.gov.in/writereaddata/files/document_publication/Strategy%20Paper%20on%20Resource%20Efficiency.pdf> [Accessed 4 November 2020].
8. Niti Aayog. 2019. *Status Paper On Resource Efficiency and Circular Economy – Current Status and Way Forward.* <https://www.eu-rei.com/pdf/publication/NA_EU_Status%20Paper%20&%20Way%20Forward_Jan%202019.pdf> [Accessed 4 November 2020].
9. TERI. 2020. *Strategy for Fostering Resource Efficiency and Circular Economy In Goa.* [online] Available at: <https://www.eu-rei.com/pdf/publication/GOA-EU%20Report_Mail.pdf> [Accessed 4 November 2020].
10. Allen & Gledhill. 2019. *Resource Sustainability Bill Passed In Furtherance of Zero Waste Goal.* <https://www.allenandgledhill.com/sg/publication/articles/13337/resource-sustainability-bill-passed-in-furtherance-of-zero-waste-goal> [Accessed 4 November 2020].
11. Global-recycling. 2020. *South Korea: The Aim is a Resource-Circulating Society.* <https://global-recycling.info/archives/3205> [Accessed 4 November 2020].
12. Korea Environment Institute. 2016. Introduction of the Framework Act on Resource Circulation toward Establishing a Resource circulating Society in Korea <https://www.greengrowthknowledge.org/sites/default/files/downloads/policy-database/Introduction%20of%20the%20Framework%20Act%20on%20Resource%20Circulation%20toward%20Establishing%20a%20Resource-Circulating%20Society%20in%20Korea.pdf> [Accessed 4 November 2020].
13. IRP. 2020. Resource Efficiency and Climate Change: Material Efficiency Strategies for a Low-Carbon Future. Hertwich, E., Lifset, R., Pauliuk, S., Heeren, N. A report of the International Resource Panel. United Nations Environment Programme, Nairobi, Kenya. https://www.resourcepanel.org/reports/resource-efficiency-and-climate-change
14. Department of Agriculture, Water and the Environment. 2018. *2018 National Waste Policy: Less Waste, More Resources.* <https://www.environment.gov.au/protection/waste-resource-recovery/publications/national-waste-policy-2018> [Accessed 4 November 2020].
15. Government of Wales. n.d. *Circular Economy Strategy.* <https://gov.wales/circular-economy-strategy> [Accessed 4 November 2020].
16. Government of Wales. 2019. *Options For Extended Producer Responsibility: Food And Drink Packaging Waste.* <https://gov.wales/options-extended-producer-responsibility-food-and-drink-packaging-waste> [Accessed 10 November 2020].
17. Government of Wales. 2020. *Consultation – Summary Of Responses: Beyond Recycling.* <https://gov.wales/sites/default/files/consultations/2020-09/beyond-recycling-summary-of-responses_1.pdf> [Accessed 10 November 2020].
18. Sitra. n.d. *Finnish Road Map To A Circular Economy 2016-2025.* <https://www.sitra.fi/en/projects/leading-the-cycle-finnish-road-map-to-a-circular-economy-2016-2025/#what-is-it-about> [Accessed 4 November 2020].
19. Rijksoverheid. n.d. *Netherlands Circular In 2050.* <https://www.rijksoverheid.nl/onderwerpen/circulaire-economie/nederland-circulair-in-2050> [Accessed 4 November 2020].

20. Government Offices of Sweden. n.d. *Sweden Transitioning To A Circular Economy* <https://www.government.se/press-releases/2020/07/sweden-transitioning-to-a-circular-economy/#:~:text=The%20Government%20has%20adopted%20a,first%20fossil%2Dfree%20welfare%20nation.> [Accessed 4 November 2020].
21. European Commission. n.d. *First Circular Economy Action Plan.* <https://ec.europa.eu/environment/circular-economy/first_circular_economy_action_plan.html> [Accessed 4 November 2020]
22. Federal Ministry for the Environment, Nature Conservation and Nuclear Safety. n.d. *Waste Policy.* <https://www.bmu.de/en/topics/water-waste-soil/waste-management/waste-policy/> [Accessed 4 November 2020]
23. Ministry of Environment. 2018. *Fundamental Plan For Establishing A Sound Material-Cycle Society.* <https://www.env.go.jp/en/recycle/smcs/4th-f_Plan.pdf> [Accessed 4 November 2020].
24. Li, W. and Lin, W., 2016. Circular Economy Policies in China. In Anbumozhi, V. and Kim, J. (eds.), Towards a Circular Economy: Corporate Management and Policy Pathways. ERIA Research Project Report 2014-44, Jakarta: ERIA, pp. 95–111.
25. US EPA. 2015. *U.S. EPA Sustainable Materials Management Program Strategic Plan.* <https://www.epa.gov/sites/production/files/2016-03/documents/smm_strategic_plan_october_2015.pdf> [Accessed 4 November 2020].
26. US EPA. n.d. *Sustainable Materials Management Basics.* <https://www.epa.gov/smm/sustainable-materials-management-basics#smm%20stategic%20plan> [Accessed 4 November 2020].
27. US General Services Administration. n.d. *Sustainable Design.* <https://bit.ly/30VwRgi> [Accessed 4 November 2020].
28. TERI. 2013. *Engagement With Sustainability Concerns In Public Procurement In India: Why And How*: <http://www.teriin.org/policybrief/docs/spp_2013.pdf> [Accessed 10 November 2020].
29. ISO. 2015. *First International Standard for Sustainable Procurement Nears Publication.* <http://www.iso.org/iso/home/news_index/news_archive/news.htm?refid=Ref2105> [Accessed 10 November 2020].
30. Japan Environment Association. n.d. *The Eco Mark Program.* <https://www.ecomark.jp/english/> [Accessed 10 November 2020].
31. Green Purchasing Network. n.d. *About GPN.* <https://www.gpn.jp/english/> [Accessed 10 November 2020].
32. Japanese Law Translation. 2003. *Act on Promotion Of Procurement Of Eco-Friendly Goods And Services By The State And Other Entities.* <http://www.japaneselawtranslation.go.jp/law/detail/?ft=2&yo=&ia=03&kn%5b%5d=%E3%81%8F&_x=17&_y=13&ky=&page=2&re=02> [Accessed 10 November 2020].
33. Ministry of Environment of Japan. 2007. *Act On Promotion Of Contracts Of The State And Other Entities, Which Show Consideration For Reduction Of Emissions Of Greenhouse Gases, Etc.* <https://www.env.go.jp/en/policy/economy/pdf/contract1.pdf> [Accessed 10 November 2020].
34. Korea Environmental Industry and Technology Institute. n.d. *Eco-Label Use.* <http://el.keiti.re.kr/enservice/enindex.do> [Accessed 10 November 2020].
35. Ecolex. 2011. *Environmental Technology And Industry Support Act.* <https://www.ecolex.org/details/legislation/environmental-technology-and-industry-support-act-lex-faoc168076/> [Accessed 10 November 2020].

36. One Planet Network. 2011. *ACT ON PROMOTION OF PURCHASE OF GREEN PRODUCTS*. <https://www.oneplanetnetwork.org/sites/default/files/korea_act_on_promotion_of_purchase_of_green_products.pdf> [Accessed 10 November 2020].
37. UNESCAP. n.d. *Republic Of Korea'S Framework Act On Low Carbon, Green Growth*. <https://www.unescap.org/sites/default/files/33.%20CS-Republic-of-Korea-Framework-Act-on-Low-CarbonGreen-Growth.pdf> [Accessed 10 November 2020].
38. European Commission. 2019. *GPP Training Toolkit*. <https://ec.europa.eu/environment/gpp/toolkit_en.htm> [Accessed 10 November 2020].
39. Green Growth Knowledge. 2017. *Comparative Analysis Of Green Public Procurement And Ecolabelling Programmes In China, Japan, Thailand And The Republic Of Korea: Lessons Learned And Common Success Factors*. <https://www.greengrowthknowledge.org/sites/default/files/downloads/resource/UNEP_green_public_procurement_ecolabelling_China_Japan_Korea_Thailand_report.pdf> [Accessed 10 November 2020].
40. Asia-Pacific Economic Cooperation. 2013. *Green Public Procurement In Asia And Pacific Region: Challenges And Opportunities For Green Growth*. <https://apec.org/Publications/2013/06/Green-Public-Procurement-in-the-Asia-Pacific-Region-Challenges-and-Opportunities-for-Green-Growth-an> [Accessed 10 November 2020].
41. One Planet Network. 2017. *Global Review Of Sustainable Public Procurement*. <https://www.oneplanetnetwork.org/resource/2017-global-review-sustainable-public-procurement> [Accessed 10 November 2020].
42. European Commission. 2017. *Public Procurement for a Circular Economy*. <https://ec.europa.eu/environment/gpp/pdf/CP_European_Commission_Brochure_webversion_small.pdf> [Accessed 4 November 2020].
43. European Environmental Bereau. n.d. *Economic Instruments for a Circular Economy*. <https://euagenda.eu/upload/publications/untitled-88054-ea.pdf> [Accessed 10 November 2020].
44. Whalen, K. A., Milios, L. and Nussholz, J., 2018. Bridging the gap: Barriers and potential for scaling reuse practices in the Swedish ICT sector, Resources, Conservation and Recycling. Elsevier, 135, pp. 123–131. doi: 10.1016/J.RESCONREC.2017.07.029.
45. NITA Uganda. n.d. *Impact Assessment On The Ban On Importation Of Used Computers*. <https://www.nita.go.ug/publication/impact-assessment-ban-importation-used-computers> [Accessed 10 November 2020].
46. Engage Victoria. n.d. *E-Waste Landfill Ban*. <https://engage.vic.gov.au/waste/e-waste> [Accessed 10 November 2020].
47. Diplock, T. and Wheatland, J., 2016. *Why Disruptive Technologies Matter*. L.E.K. <https://www.lek.com/sites/default/files/insights/pdf-attachments/Disruptive_Technologies_Part1_Tom_Diplock_Jeremy_Wheatland.pdf> [Accessed 17 November 2020].
48. Hasselbalch, J., 2018. Innovation assessment: Governing through periods of disruptive technological change. *Journal of European Public Policy*, 25(12), pp. 1855–1873, DOI: 10.1080/13501763.2017.1363805

11

More Insights and Way Forward

In this chapter some of the important aspects that need to be considered while planning, strategizing, and implementing a circular economy will be addressed. Metrics is the first step to understand the status on circularity in our economy and accordingly set targets to develop a roadmap. Such an assessment is required at the product and company/organizational levels to understand the return on investments or make a business case. Assessments at regional, national, and global levels help in understanding the impact of policy measures on a circular economy.

11.1 Measuring Circular Economy

Measuring the circularity of a product or service can be a challenge due to the complexity and variety of actions, activities, and projects that could be called circular. Unfortunately, no single accepted framework exists to enable organizations to assess and report their circularity. This absence of a universally agreed framework is one of the greatest needs.

There is a compelling need to develop a framework to measure the Returns on Investment (RoI) for circular economy business models. This would provide evidence and encourage more businesses to adopt circular economy practices. Several organizations are developing metrics that should help sustainability professionals and other executives develop business cases for why transitioning to circularity can provide strong RoIs[1]. Companies often set corporate sustainability goals, such as reducing their carbon footprint or increasing their use of recycled content. However, they do not create actionable steps or have any way of measuring their progress towards their goals.

Box 11.1 shows an example of Circulytics™ that is one of the tools developed for measuring circularity for companies/organizations.

Many tools for metrics on circularity build on and expand the results of Life Cycle Assessment (LCA) and Material Flow Analysis (MFA). MFA is also

BOX 11.1 CIRCULYTICS™ – COMPREHENSIVE CIRCULARITY MEASUREMENT TOOL FOR COMPANIES[2]

The Ellen MacArthur Foundation launched Circulytics, a tool to support a company's transition towards the circular economy, regardless of industry, complexity, and size. Going beyond assessing products and material flows, this company-level measuring tool reveals the extent to which a company has achieved circularity across its entire operations. The tool enables following:

- Measures a company's entire circularity, not just products and material flows.
- Supports decision making and strategic development for circular economy adoption.
- Demonstrates strengths and highlights the areas for improvement.
- Provides optional transparency to investors and customers about a company's circular economy adoption, and
- Delivers unprecedented clarity about circular economy performance, opening up new opportunities to generate brand value with key stakeholders.

The Circulytics indicators have been developed by the Foundation in collaboration with 13 Global Partners and CE100 member companies and have been tested by over 30 companies during 2019. Version 2.0 is now available.

used to assess the "before" and "after" situation. Box 11.2 presents examples of COMPASS and SCORE that help companies to develop metrics on circularity at company or product or portfolio of products level. More importantly, the tools help in generating and assessing alternatives in product and packaging design, allowing assessment of both functional and circularity-related performance, to make robust decisions.

The US Chamber of Commerce Foundation has identified metrics that can help measure different aspects of circularity of a project. Here, the metrics associated with each of the projects is summarized, quantifying how material is flowing through a supply chain and back into new products. For example, a company recycling a product may look at how much material they recover, but a company manufacturing bottles with recycled content would look at how much of that same material they used, and a company that has a closed-loop material system can look at both the numbers. "Double counting" occurs when different actors can use the same material and legitimately

BOX 11.2 COMPASS AND SCORE – TOOLS FOR DECISION MAKING BASED ON CIRCULARITY METRICS

COMPASS was conceived, funded, and launched by the Sustainable Packaging Coalition (SPC) in the United States in 2006. A cross functional team from brand manufacturers, packaging suppliers, retailers, and LCA professionals defined and gathered requirements. It was implemented by Trayak[3].

COMPASS lets user "define" the product and packaging and computes the LCA metrics such as fossil fuel use, water use, mineral resource use, greenhouse gas (GHG) emissions and carbon uptake, human impact in disability-adjusted life year, freshwater eutrophication and freshwater ecotoxicity.

However, LCA does not tell you if a product is recyclable, or if it includes bio-renewable or Post-Consumer Recycled (PCR) content. SCORE allows to visualize trade-offs between LCA metrics and packaging metrics based on company priorities by setting weights. Here "functional" parameters such as damage rate, packaging-product weight ratio, cube efficiency, shelf life, packaging weight reduction, total cost of packaging etc. are considered along with circularity indicators such as packaging recovery rate, package reuse rate, bio-renewable content, PCR content, certified content, material wasted and Material Circularity Index (MCI)[2]. SCORE provides a framework to assess circular performance at the product portfolio as well to allow prioritization in decision making on circular interventions. So, metrics is not just for reporting or tracking performance but also for decision making. SCORE thus gives a strategic direction towards product and package development processes at the operational level to facilitate company's sustainability strategy and achieve product and brand goals.

SCORE can be fully integrated into your processes by embedding in your enterprise IT systems. Once fully embedded, SCORE's assessment can be automated making the analysis a natural part of design process. These vital integrations yield quantifiable results for material optimization, waste reduction, manufacturing efficiency and profitability in business.

claim the benefits of the activity that produced the material. This is a significant concern when trying to quantify a circular system and one of the challenges facing any standard framework for measuring circularity[4].

Common global indicators to measure circular economy for businesses are still being developed by different institutions. It would be good to read the report "Circular Economy Indicators for Businesses" by Institut National de l'Economie Circulaire and Entreprises pour l'Environnement[5]. The report identifies indicators that mainly focus on information related to waste and, to a lesser degree, products, and product use (eco-design, lifespan and share of sales). It also showcases several case studies of businesses using various circularity indicators. Table 11.1 lists the indicators under each category against

TABLE 11.1

Example of Company-Level Material Circularity Framework[6]

Objectives	Indicators	Sub-weights	Category Score	Weights	Total Score
Making Supply chain Circular	Use of recycled material vis-à-vis virgin material	0.2		0.25	
	Elimination of hazardous materials and greater use of biodegradable materials	0.2			
	Efforts to reduce Supply Chain GHG Emissions and Transition to Renewable Energy	0.2			
	Greater Transparency to disclose the list of suppliers making their products, including what the supplier does and where it is located.	0.2			
	Natural capital valuation (monetary valuation of company's environmental impact, e.g., placing a price on carbon, water, etc.)	0.2			
Product life extension	Design for reusability, reparability, and recycling	0.33		0.25	
	Manuals and Spare parts available for repairs and upgrades	0.33			
	Certification of refurbished and recycled products	0.33			
Products as a service, asset sharing and use of sharing platforms	Consumer demand for company's product utility rather than ownership	0.25		0.25	
	Intra and inter organization sharing of idle or underutilized assets (shared services availed vis-à-vis owning the equipment)	0.25			
	Internet of things facilitating sustained utility through monitoring and predictive maintenance	0.25			
	Product Standardization-improving the viability of extended service program	0.25			
Recovery and Recycling	Take back systems	0.33		0.25	
	Recovery and Recycling rates	0.33			
	How much material goes into the landfill?	0.33			

which company's performance can be assessed in India. This has been developed by The Energy Research Institute (TERI) and YES Bank and is based on weights and scoring system.

Circular economy indicators are used by countries such as Japan, Korea, China, and in the European Union (EU) to report progress. The World Business Council for Sustainable Development (WBCSD) has prepared a landscape analysis on circular metrics for business, NGOs & Academia, and governments[7].

The Global Circularity Gap Report[8] is an annual report measuring the state of circularity. Its goal is to measure, assess the progress and come up with recommendations to governments and businesses for a transition to a global circular economy. The report is published by Circle Economy[9] on an annual basis since 2018. Circle Economy has introduced a metric to measure global circularity to track and target performance of circular economy initiatives.

The first Circularity Gap Assessment report released in 2018, established that the world is only 9.1% circular, leaving a massive gap in circularity. In 2020, the third report stated that circularity has gone down from 9.1% to 8.6% over the course of just 2 years. This report was based on the status of circularity practices adopted by 176 countries. The results of the Circularity Gap reports must be interpreted with caution as the methodology as well as the datasets on resource extraction and consumption have changed. Changes to the report are expected in the 2021 report because of the COVID-19 pandemic. Box 11.3 shows some of the important observations of the 2020 report.

BOX 11.3 IMPORTANT OBSERVATIONS FROM THE 2020 CIRCULARITY GAP ASSESSMENT REPORT

- **High rates of extraction:** "Rate of extraction of resources increases, has outpaced improvements in the resource recirculation by a factor of two to three. Over the last five decades, the global use of materials has more than tripled."

  Here, practicing circular supplies or preferring secondary resource materials will help in reducing the extraction of virgin resource. Strategy of using more renewable energy and stepping up the resource efficiency will also be useful. Sustainability considerations translated in the form of regulations will also help in limiting the resource extraction.
- **Stock build-up:** "More materials are added to build up our global housing stock, infrastructure and heavy machinery – to meet the needs of a growing global population. Urbanization,

as a global phenomenon, is increasing an accelerating demand for housing, so driving the stock build-up dynamic worldwide".

Chapter 1 described issue of concern. Perhaps, move towards extending product life, offering product as a service, and promoting shared economy will reduce the pace of the stock build up.

- **Low levels of end-of-use processing and cycling:** "Poor design of products (in terms of reparability and recyclability) and deficiencies in reverse logistics and the recycling infrastructure has led to a continued linearity in our production and consumption systems. This has exacerbated demand for virgin materials, especially on non-renewable resources. On a positive note, however, it is seen in parts of the world, recovery rates are on the rise."

  Strengthening the informal sector, enforcing EPR, using advanced technologies for sorting, cleaning, and recycling and restricting hazardous substances in the products could be effective strategies.

ACTIVITY 11.1 DISCUSSION ON METRICS FOR CIRCULAR ECONOMY

While having metrics for circular economy is desirable, it is fraught with following challenges:

- Absence of a uniform framework (organizations like Ellen Macarthur Foundation and Global Reporting Initiative should support here. Global Reporting Initiative (GRI) has started working on GRI 306: Waste 2020 standard[10]).
- Challenges in estimation of impacts, scoring and aggregation into indices. This is perhaps one of the greatest challenges and experience on the practice level application of tools like LCA has repeatedly brought out this concern.
- Difficulties in acquiring data and poor quality. This limitation is particularly true in developing economies where informal sector plays an important role in a circular economy.

Discuss how these challenges could be addressed.

11.2 Importance of Regional Circular Economy

Governance in circular economy could be viewed at various scales such as national, sectoral, industrial clusters or parks, cities, corporates together with supply chains and also communities. See Figure 11.1.

Complexity in implementing circular economy increases at a macro scale (e.g., national, state, and sector levels). Efforts for coordination also increase to reduce ambiguities, overlaps, conflicts and redundancy in the policy design, regulations, and responsible institutions. In countries, where a federal structure is followed for governance, an integrated approach is required addressing all the scales, especially States or regional authorities. A mechanism needs to be created for the purpose of integration and harmonization.

In transitioning to circular economy, leading countries like Germany, Japan, and China followed a top-down approach. In China, the circular economy plan was implemented at micro level (such as consumer or company) to meso (e.g., eco-industrial parks) and macro levels (e.g., provinces, regions, and cities). Corporates are implementing circular economy across their business ecosystems close to meso scales, covering facilities as well as supply chains. Here tools like organizational- life cycle assessment (o-LCA), described in Chapter 3 become useful to strategize circular economy.

As shown in Figure 11.1, life cycle thinking plays a very important role to close the technical and biological flows. Experience has shown that one of the critical pillars is stakeholder consultation and involvement. All stages of life cycle, e.g., sourcing of materials, design, production, distribution, consumption, and waste management need to be addressed for a balanced governance. Involving communities and bringing in a behavior change is extremely

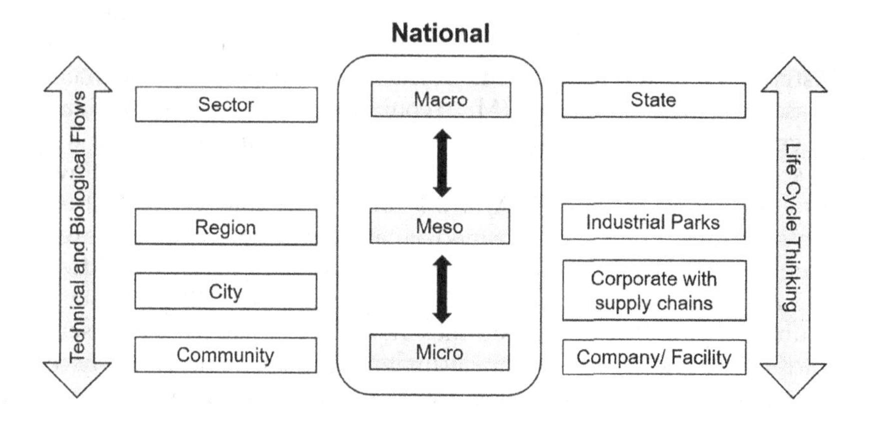

FIGURE 11.1
Circular economy at various scales.

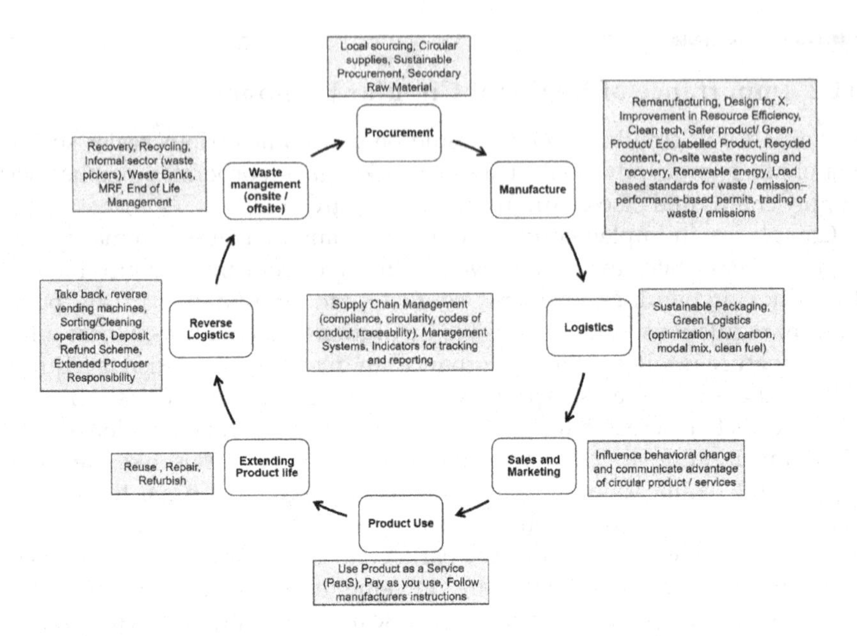

FIGURE 11.2
Circular strategies across life cycle.

critical. Figure 11.2 presents strategies that can be considered for transitioning to circular economy at various stages of the life cycle. Many of these strategies with examples have already been discussed in earlier chapters.

Amongst these different scales of circular economy, regional circular economy plays an important role. Regional circular economy has the potential to influence both macro and micro scales – i.e., both "upstream" and "downstream". Introducing circular economy at regional scale allows closer involvement of key stakeholders consisting of local government, cities, industry and industry associations, investors, research, and academia. Tools such as Material Flow Analysis (MFA) coupled with Life Cycle Sustainability Assessment (LCSA) and modeling with System Dynamics can help to generate scenarios. These scenarios could be then assessed following Strategic Environmental Assessment (SEA) as discussed in Chapter 10, to identify preferred strategies. Implementation models can then be developed that are partnership based and continue over long term guided by a vision. Addressing Sustainable Development Goals (SDG), especially SDG 12 on Sustainable Consumption and Production becomes more effective on a regional scale.

At a regional scale, closing technical (material) and biological loops becomes eminent, particularly latter, as bioeconomy plays an important role in several regions of the world. Regional circular economy essentially provides a better opportunity for "systems thinking". Adopting a functional approach and going beyond the administrative boundaries of cities helps in better resource management and socio-economic development.

Cities and industrial parks are not isolated ecosystems but spaces for inflows and outflows of materials, resources, and products, in connection with surrounding areas and beyond. Therefore, linkages across urban and rural areas (e.g., related to agriculture and forestry) are key to promote local production and recycling of wastes. At the regional level, loops related to a series of economic activities (e.g., the bioeconomy) can thus be closed with longer time for resource recirculation. Boxes 11.4 and 11.5 present case studies of Kawasaki

BOX 11.4 KAWASAKI ECO-TOWN[12]

Eco-Towns in Japan originated through a subsidy system established by Ministry of Economy, Trade, and Industry (METI) in Japan and Ministry of Environment in Japan in 1997. Around that time, Japan was confronted by a serious shortage of dump yards and the necessity to revive local economy. The national government established Eco-Towns to solve garbage problems and assist companies in declining industries such as steel, cement by the Zero-Emission concept.

Kawasaki Eco-Town was approved in 1997 as one of the first Eco-towns in Japan. Kawasaki City is in the Tokyo Metropolitan Area, it is the ninth most populated city in Japan. Kawasaki Eco-Town Plan targets an area broadly defined to include almost the entire stretch of the Kawasaki coastal area, and aims at creating a resource-recycling society, and revitalizing the coastal area. This concept envisions that the industrial companies that will be located in the area will minimize their operations' impact on the environment.

Kawasaki Eco-Town employed four developmental steps:

- Promote environmental measures at company level.
- Promote environmental measures at the recycling zone level.
- Research to realize sustained growth.
- Contribute results to society and developing nations.

Facilities were approved for sustainable businesses that included Waste Plastics Recycling System for Use as Raw Material for Blast Furnace, NF Board for Concrete Forms Manufactured from Recycled Plastic, paper recycling facility, PET-to-PET recycling facility and Waste Plastics Recycling System for use as material for ammonia.

Several other recycling facilities also like home appliance recycling & cement manufacturing with recycling process were also established. 15 enterprises were established in a complex to carry out metal-processing, paper, plating, forging, and stamping, building them around the industrial park to promote cooperation in terms of resource recycling with the existing companies. Around JPY 25 Billion in total were provided as a subsidy by the National Government and the Kawasaki City.

Between 1997 and 2006, METI rolled-out Eco-town programs in 26 major cities. In all, 95 waste-facilities have been developed as a result of the Eco-Town Programme and the resource productivity rose by 20% between 2000 and 2006[13].

Eco-Town and Eco-Industrial parks in Vietnam as examples where circular systems can be promoted on a regional or meso scale. Recent work on circular cities in another example[11]. While preparing a regional or national level circular economy action plan, meso scale plans should be prepared for cities and industrial parks by involving the concerned authorities.

Capacity building on circular economy of regional authorities e.g., city managers and heads of the industrial parks, however, becomes very important. Role of coalitions, faciliatory organizations, knowledge and innovation platforms becomes critical. Box 11.6 shows results of one of the recently conducted

BOX 11.5 VIETNAM ECO-INDUSTRIAL PARK[14]

To facilitate the establishment of new industries, the Government of Vietnam created industrial zones, which account for a significant portion of the country's GDP. The transformation of conventional industrial parks into Eco-Industrial Parks (EIP) presented an effective opportunity to attain inclusive and sustainable industrial development and to meet the objectives of the 2030 Agenda and the SDGs.

The Project "Implementation of eco-industrial park initiative for sustainable industrial zones in Viet Nam" was jointly established by the Ministry of Planning and Investment (MPI) and UNIDO. The project started in Vietnam, in the following regions: Khanh Phu – Ninh Binh, Hoa Khanh – Da Nang and Tra Noc – Can Tho. National Green Growth Strategy was adopted in 2014 that assigned 66 different tasks to ministries, sectors and localities, along with it the Ministry of Planning and Investment conducted studies to create a new developmental model for industrial zones. The Idea initiated with the implementation of the National Green Growth Strategy by Prime Minister Nguyễn Tấn Dũng in 2012.

There were 24 enterprises in 3 pilot industrial zones participating in the project. The project received donations from the Global Environment Facility (GEF), State Secretariat for Economic Affairs (SECO) & United Nations Industrial Development Organisation (UNIDO). The total budget sanctioned was USD 4.5 million.

BOX 11.6 RESULTS OF OECD SURVEY ON THE CIRCULAR ECONOMY IN CITIES AND REGIONS[15]

According to the results of the OECD Survey on the Circular Economy in Cities and Regions, it was found that the respondents face five major categories in terms of gaps:

- Financial gaps: A vast majority of the 51 surveyed cities and regions reported challenges related to insufficient funding (73%), as well as financial risks (69%), lack of critical scale for business and investments (59%), and lack of private sector engagement (43%).
- Regulatory gaps: Regulatory barriers can inhibit the development and implementation of circular economy strategies. Inadequate regulatory frameworks and incoherent regulation across levels of government represent a challenge for respectively 73% and 55% of respondents.
- Policy gaps: Several local policies and strategies share objectives with the circular economy at large. However, the lack of a holistic vision is a major obstacle for 67% of respondents, often due to poor leadership and co-ordination, and/or the lack of political will.
- Awareness gaps: Cultural barriers represent a challenge for 67% of surveyed cities and regions along with lack of awareness (63%) and inadequate information (55%) for policymakers to take decisions, businesses to innovate and residents to embrace sustainable consumption patterns.
- Capacity gaps: The lack of human resources is a challenge for 61% of surveyed cities and regions. Technical capacities should not just aim for optimizing linear systems but strive towards changing relations across value chains and preventing resource waste.

The report suggests cities and regions act simultaneously as promoters, facilitators, and enablers of the circular economy, in a shared responsibility with national governments and stakeholders.

OECD came out with a 3Ps (People, Policies, and Places) Framework, which provides a conceptual framework to make circular economy happen in cities and regions.

surveys by Organisation for Economic Co-operation and Development (OECD) that emphasize the need for capacity building.

According to OECD, the potential of the circular economy to support sustainable cities, regions and countries still needs to be unlocked. Firstly, the approach should be *people* oriented and inclusive. Further, it requires a holistic and systemic approach that cuts across sectoral policies. Finally, a functional approach is necessary going beyond the administrative boundaries of cities and linking them to their hinterland and rural areas to close and ensure an efficient and longer resource recirculation and at the right places and scale. Regional approach to circular economy also allows to consider bioeconomy which links people, policies, and practice for resource conservation, in the agricultural and food sector.

11.3 Developing Circular Economy Action Plans

Developing circular action plan at regional and national level is a complex process. These action plans need to be based on guiding and operational principles of circular economy. A strategic framework such as CIRCULAR described in Box 11.7 can help in developing an overarching framework for the regional or national action plans. Priorities on the strategies proposed will depend on the socio-economic and political situation, resource availability and security. The CIRCULAR model is developed based on ReSOLVE framework of six actions for circular economy in business[16] and the 7-element based DISRUPT model[17].

One of the first steps to start the action planning process is the identification of priority sectors. Priority sectors help in scoping the plan. However, given the cross-sector material flows, focusing on priority sectors does not restrict involvement of other sectors and limit the "impact" of introducing circularity.

Figure 11.3 shows an example of cross sectoral material flows across two "production" sectors (steel and aluminum) and two "waste management" sectors (e-waste and construction and demolition waste) as an illustration. Clearly, the policy and regulatory reforms in one sector will influence material flows in other related sectors. It is important that these interlinkages are recognized by building a cross-sectoral enabling framework and positioned in the action plan.

Sectors are generally prioritized based on economic, environmental, and social considerations. Box 11.8 shows an example of such considerations that were used to identify steel sector as a priority for preparing a circular economy action plan in India.

Priority sectors need to be identified by examining economic, environmental, and social perspectives holding inter-ministerial consultation. The next

BOX 11.7 A "CIRCULAR" MODEL FOR PREPARING ACTION PLAN ON CIRCULAR ECONOMY

Close the Loops – Keep components and materials in closed loops and prioritize the inner circles. Track and optimize resource recirculation and strengthen connections between supply-chain actors through digital, online platforms and technologies.

Integrate with Informal Sector or be Inclusive – Integrate informal sector into waste management systems, to create new jobs and opportunities for businesses as well as accelerate circularity efforts. Address concerns regarding health, safety and dignity at work and include skilling or capacity building.

Regenerate resources and ecosystems – Shift to renewable energy and replace used materials with renewable, reusable, non-toxic resource materials, by using safe materials (chemicals) and resource efficient technology. Reclaim, retain, and regenerate the health of ecosystems; and return recovered biological resources to the biosphere.

Coordinate, catalyze and leverage partnerships – Work together throughout the supply chain. Collaborate within organizations engaged with outreach, education & training, research & innovation, and financing institutions, both in private and public sector.

Utilize waste as a resource – Utilize waste streams as a source of secondary resources and recover waste for reuse and recycling. Support remanufacturing.

Lessen consumption of virgin resources – Deliver product as a service, virtualize utilities, practice collaborative consumption. Maintain, repair, and refurbish products in use to maximize their lifetime and give them a second life through take-back strategies, where applicable.

Augment product lifespans – Adopt a systemic perspective during product design such as Design for Durability, Reparability and Disassembly to promote repairs, refurbish and remanufacturing for extended use and optimal recovery.

Rethink – Stay within needs and not driven by wants, innovate policies, governance, and business models, while thinking out the box, remember to address the "spill-over effects", keep the SDGs in focus.

steps in the action planning process are policy mapping (showing the trail of development of policies and regulations in each priority sector), policy and program review.

Circular economy action plans are generally built on the existing policy and regulatory frameworks (see case studies in Chapter 10). Lessons from the existing and past programs also help. It is important therefore that an exercise on policy and program review is undertaken.

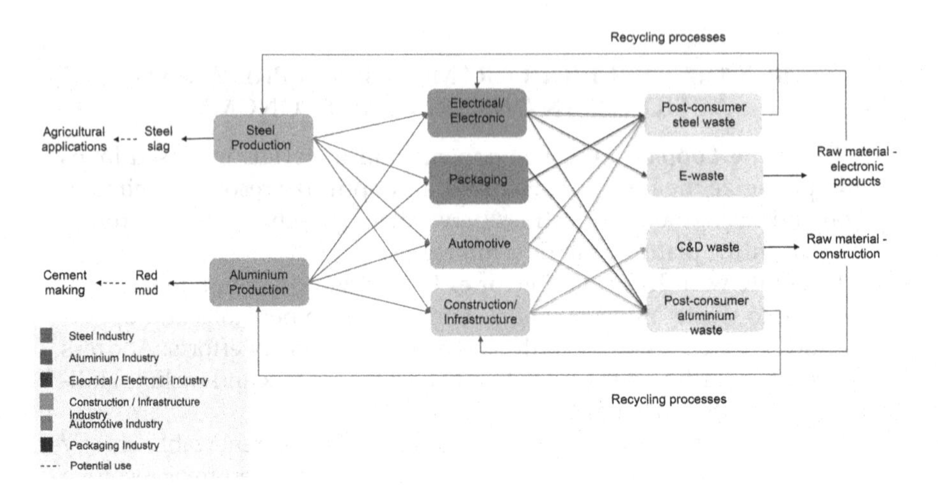

FIGURE 11.3
Illustration of material flows across sectors[18].

BOX 11.8 ECONOMIC, ENVIRONMENTAL, AND SOCIAL PERSPECTIVES OF STEEL SECTOR IN INDIA[18]

Economic: In FY18, India produced 104.98 million metric tons (MMT) of finished steel. Crude steel production during 2017–2018 stood at 102.34 MMT.[19] The National Steel Policy 2017 projects crude steel production to reach 300 MMT by 2030–2031, nearly three times the production in 2017–2018.

According to the Ministry of Steel, in 2011 there are 13 major integrated steel plants[20], both in the public and private sector in India. In addition, there are several medium and small steel units in India which comprise of Mini Blast Furnace units, Sponge Iron units, Induction Furnace units and Rolling Mills. The total number of such units was estimated to be 3647 based on the Joint Plant Committee survey conducted in 2009–2010.[21] There are approximately 313 sponge iron producers, 42. Electric Arc Furnaces (EAF), 1126. Induction Furnace (IF), and around 1157 small and medium sized steel rerolling mills scattered over the country. They are usually found in clusters, with each cluster having about 50–400 units.[22] Presently, India imports some steel scrap for secondary steel production. The domestic steel scrap industry needs to be formalized to ensure higher rates of domestic scrap recovery so that the import dependency is reduced.

Environmental[23]: The Iron and Steel Industry (involving processing from ore/scrap/integrated steel plants)[24] is included in the 17 major polluting industries in India and is classified as a "RED" Category industry (i.e., heavily polluting and covered under the Central Action Plan) by MoEFCC. Steel production has several impacts on the environment, including air emissions (CO, CO_2, SO_x, NO_x, $PM_{2.5}$, and PM_{10}), wastewater release, and generation of solid and hazardous wastes. The major environmental impacts from integrated steel mills are from coking and ironmaking. Most of the GHG emissions associated with steel production are from the emissions related to energy consumption. According to the Steel Authority of India[25], the steel industry accounts for about 6–7% of the total GHG emissions in India.

Slag is a waste which is generated during manufacturing of pig iron & steel and is classified as a potential by-product by the Indian Bureau of Mines[26]. There are four types of slag which are generated from primary and secondary steel production[40]:

1. Blast Furnace Slag.
2. Steel Slag (also known as Linz-Donawitz or LD slag).
3. EAF Slag.
4. Induction Furnace (IF) Slag.

Social: The Indian steel industry provides employment to over 2.5 million people directly and indirectly. The potential for revenue generation in steel scrap industry is of the order of 20,000 million INR/million ton per annum of steel scrap processed. This will require skilling people in new trades as well as bringing focus on new innovative ideas/research in the MSME sector[40].

Figure 11.4 shows an illustration of such policy mapping and policy trail for Steel sector in India. It should be noted that such an exercise is to be carried out by considering other policies of relevant ministries and not limiting to the ministry in charge of the priority sector alone. For example, policies of Ministries of Environment & Forests, Mining, Housing, etc. had to be reviewed along with those of Ministry of Steel.

Policy review helps in identifying concerns, ambiguities and overlaps that need to be addressed. Achieving harmonization and coordination are also the important objectives of policy review. More importantly, a discussion is also required to know the efficacy or challenges faced in implementing policies that push circular economy. Box 11.9 presents an example of how key triggers were identified in the national steel policy in India.

Lessons from past and existing programs (e.g., Cleaner Production, 10YFP under Sustainable Consumption and Production, Green Economy) greatly help to understand what has worked and what has not. Bringing all the program

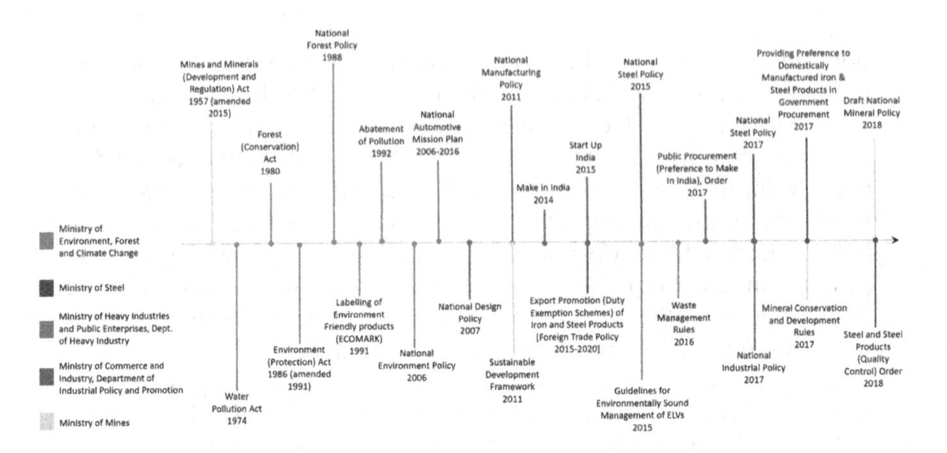

FIGURE 11.4
Timeline of various policies and program related to steel[18].

BOX 11.9 IDENTIFICATION OF TRIGGERS TOWARDS CIRCULAR ECONOMY IN NATIONAL STEEL POLICY OF MINISTRY OF STEEL, GOVERNMENT OF INDIA[15]

NATIONAL STEEL POLICY, 2017

- "The Ministry will promote cluster-based consortium approach for optimum land use, easy availability of raw materials and economies of scale" – **Chapter 4.2 Steel capacity, Point 4.2.7, Pg. 10**
- "Necessary policy environment will also be provided to promote gas-based steel plants, electric/induction furnaces and other technologies which will bring down usage of coking coal in blast furnaces" – **Chapter 4.2 Steel capacity, Point 4.2.8, Pg. 10**
- "...actions will be initiated to increase availability of ferrous scrap. Options will also be evaluated in coordination with other concerned ministries to develop a scrap segregation (quality-wise), collection, processing, and recycling policy" – **Chapter 4.3.9, Ferrous Scrap, Point 4.3.9.1, Pg. 13**
- "...there has been a major thrust by the Government on reduction of discharge from the steel plants which will require innovative solutions and techniques to effectively recycle treated waste water" – **Chapter 4.4 Land, Water & Power, Point 4.4.4, Pg. 14**

- "Ministry of Steel will facilitate the use of waste heat recovery in Steel plants in consultation with other ministries" – **Chapter 4.4 Land, Water & Power, Point 4.4.6, Pg. 15**
- "...are conducive to effective & efficient utilization of domestic resources". – **Chapter 4.7 Technological Efficiency, Point 4.7.1.1, Pg. 17**
- "Optimize resource utilization and facilitate modernization of the steel industry". – **Chapter 4.7 Technological Efficiency, Point 4.7.1.3, Pg. 17**
- "Adoption of energy efficient technologies in the MSME steel sector will be encouraged". – **Chapter 4.8. MSME Steel Sector, Point 4.8.3.3, Pg. 18**
- "...and clusters for optimal utilization of land and reach economies of scale". – **Chapter 4.8. MSME Steel Sector, Point 4.8.3.3, Pg. 18**
- "...such as NEDO model projects in CPSEs and UNDP-AUSAID-MOS steel project in steel re-rolling mills to facilitate improvement in energy efficiency". – **Chapter 4.10. Environment Management, Point 4.10.3, Pg. 19**
- "Considering all waste materials as an economic asset, Ministry will encourage the steel companies to develop a Waste Management Plan for additional impetus on zero-waste or complete waste recycling. Concrete efforts will further be made by Ministry to promote use of iron & steel slag in alternate uses like road making, rail ballast, construction material, soil conditioner etc. Simultaneously, steel plants will be pursued to implement SMS slag weathering/steam ageing plants to enable them to supply processed/sized SMS slag for road making, rail ballast etc." – **Chapter 4.10. Environment Management, Point 4.10.4, Pg. 19**

lead agencies or the actors together in the form of coalition is important right in the early stages of action planning process. Refer to Chapter 2 where several programs and initiatives that intersect with circular economy are described.

Life Cycle Thinking (LCT) should form a basis for identifying the interventions. Here the strategies outlined in Figure 11.2 for each stage of life cycle may help. The CIRCULAR framework described in Box 11.7 could also be referred. All interventions should be identified in a 4P format consisting Policy, Plan, Program, and Projects. Mere identification of projects will not make the circular economy action plan impacting and sustainable.

In the action identification exercise, consultations need to be held with the ministries concerned and the key stakeholders. Note that for each priority sector, the relevant ministries and key stakeholders could be different for a life cycle stage. Actions are to be devised to respond to current and future scenarios of economic growth and environmental and social challenges. Importantly, alternatives are to be assessed to come up with preferred actions. Here some of the tools described in Chapter 3 could be used. A LCT based approach ensures that objectives of reduced resource consumption, enhanced resource efficiency and longer resource recirculation are achieved at each stage of the value chain.

At the end of this process, a sectoral action plan for circular economy gets prepared for the ministry concerned. Combining the sectoral plans of the ministries, an integrated and harmonized plan is developed at a regional or national scale. In this plan, cross-sectoral interventions are to be included with involvement of faciliatory organizations. It would be beneficial to refer to the circular economy action plan developed on this basis, consisting of 30 interventions involving 9 major ministries in India[18].

ACTIVITY 11.2 USING THE CIRCULAR FRAMEWORK

Process of developing circular economy action plans at regional and national scale is a complex process. In Chapter 10, some of the recently formulated circular economy action plans have been described. Research on the processes followed in countries such as Germany, China, Korea, Japan, EU, and Australia, and hold a debate on comparison.

Map the CIRCULAR framework and the circular strategies across life cycle (Figure 11.2) to make such a comparison.

Form groups and allocate a country for an in-depth analysis to make the debate productive.

Do you think the process and methodology for developing a circular economy action plan can be templated as a guidebook?

11.4 Circular Economy in Supply Chains – Codes of Conduct and Traceability

Many corporates operate on a scale and have extensive supply chains operating across the regions. The circular interventions taken up by such corporates have created both "enveloping" as well as "ripple effects" on meso and macro scales. Circular economy across supply chains helps the corporate to be competitive, resilient and create a brand. In some countries, corporates active with supply chains have played a significant role in the national circular economies. Box 11.10 illustrates an example of H&M.

BOX 11.10 APPLICATION OF PRINCIPLES OF CIRCULAR ECONOMY ACROSS SUPPLY CHAIN OPERATIONS – CASE OF H&M[27]

Hennes & Mauritz AB (H&M) is a Swedish clothing-retail company working in fast-fashion for men, women, teenagers & children. As of 2019, H&M operates in 74 countries with over 5000 stores, it is the second-largest global clothing retailer.

To reduce the impact of raw materials such as cotton, H&M decided to use recycled or sustainably sourced raw materials such as sustainably sourced cotton, recycled polyester or lyocell by at least 2030. H&M is working with organizations like World Wide Fund (WWF) & Swedish Textile Waste Initiative to improve the sustainability performance of fabric & yarn production by addressing the concerns regarding water, chemical, carbon emission & working conditions. Further, the company is working with the suppliers to reduce the water consumption by 25% per unit produced by 2022.

At H&M, upstream transport represents only 3% of the total GHG emissions in a garment's lifecycle. This was possible because 90% of transport carried out using ships and trains and solutions such as Electrical trucks and bikes are used for the last mile delivery.

Having a vast network of 5076 stores, H&M has managed to supply all the offices, stores and warehouses with renewable energy. In fact, 96% of electricity comes from renewable sources.

In 2019, H&M collected 29,005 tons (equivalent to 145 million t-shirts) of garments for recycling and reuse. About 97% of cotton used is either organic, recycled or sustainably sourced. H&M is also working on innovating ways to prolong the life of garments by reuse, by investing in re-sale, rental servicing & restoring damaged clothing.

In the supply chain management, codes of conduct play an important role to address the social dimension of circular economy. Given the situation of pandemics like COVID-19, adhering to the codes of conduct becomes very important.

The most comprehensive and universally applicable standard directly addressing the responsibilities of businesses operating internationally is the International Labur Organisation (ILO) Tripartite Declaration of Principles concerning multinational enterprises and social policy. The Ethical Trading Initiative Base Code[28] was founded on the conventions of the ILO and is an internationally recognized code of labor practice.

SA 8000 is an international certification standard that encourages organizations to develop, maintain and apply socially acceptable practices in the workplace. SA 8000 certification addresses issues including forced and child labor, occupational health and safety, freedom of association and collective bargaining, discrimination, disciplinary practices, working hours, compensation, and management systems[29].

Adopting SA 8000 certification means an organization must consider the social impact of their operations in addition to the conditions under which their employees, partners and suppliers operate. Systems like SA8000 are useful when tools like Social-LCA (s-LCA) are applied[25].

Responsible Business Alliance (RBA)[30] is the world's largest industry coalition dedicated to corporate social responsibility in global supply chains. RBA is a multi-industry, multi-stakeholder initiative focused on ensuring that the rights of workers vulnerable to forced labor in global supply chains are consistently respected and promoted. The RBA Code of Conduct is a set of social, environmental, and ethical industry standards. The standards set out in the Code of Conduct reference international norms and standards including the Universal Declaration of Human Rights, ILO International Labor Standards, OECD Guidelines for Multinational Enterprises, International Organization for Standardization (ISO), and SA standards, and many more. While the Code of Conduct originated with the electronics industry in mind, it is applicable to and used by many industries beyond electronics. This reference code becomes relevant during the application of LCSA across the supply chains.

Traceability is understood as "the ability to trace the history, application or location of an object" in a supply chain[31]. It a process by which enterprises track materials and products and the conditions in which they were produced through the supply chain"[32].Traceability in the supply chain is relevant when business operations are engaged in recycling or when recycled content is included in the products.

There is now a trend to include traceability as a requirement in ecolabel certifications and Environmental Product Declarations (EPD). The EU Ecolabel, the Global Organic Textiles Standard (GOTS) and the Fairtrade Textiles Standards contain elements of traceability implementation for textiles. BASF is launching the pilot platform reciChain in the province of British Columbia (BC), Canada, as a response to plastic waste. A scalable blockchain solution will support the track, trace, and monetization of plastics within the value chain, through innovative Digital Badge and Loop Count technology[33].

Everledger developed a mechanism that supports businesses who actively choose supply chains that enable a circular economy model. In early 2020, Everledger completed a pilot project with a team of partner companies in Queensland, Australia. The partnership was facilitated by Circular Economy Lab, an initiative involving 27 industry partners supported by the Australian government. A team of five companies set out to successfully demonstrate a validated circular material flow and connect material throughput to authenticated recycling outcomes[34].

OPTEL[35], a company operating at five locations across four continents offers a traceability and consumer engagement platform called Intelligent Supply Chain developed for food and beverage industry. The platform can manage all types of marking technologies, including QR Codes, RFID, NFC, laser engraving, GTIN and more. More importantly, the platform can help the industry to ensure and prove responsible sourcing and fair trade, reduce waste, and increase recycling and engage with consumers to encourage and reward environmentally responsible and ethical consumption.

In 2013, the European Resource Efficiency Platform (EREP) was set up to guide European policymakers towards a resource efficient society. One of their proposals was to create "product passports". Box 11.11 provides a description of Product Passport.

BOX 11.11 PRODUCT PASSPORT

Product Passports are a set of information about the components and materials that a product contains, and how they can be disassembled and recycled at the end of the product's useful life. This would encourage a move to a circular economy in which the material content of obsolete products re-enters the production cycle.

Product passports have potential to tackle the problem of inadequate information on what resources a product contains. Recycling and reuse as an economic activity could be greatly boosted if recyclers and re-users can obtain clear information on the resources contained in a product, and any risks that might be posed, for example by hazardous chemicals in electronics.

There are several concerns related to product passports. For instance, concerns about data confidentiality arises. Manufacturers request to protect the data related to products in line with their intellectual property rights. Another, trickier issue is if a business other than the original manufacturer changes the properties of the product, it can pose safety issues for the consumers[36].

A form of product passport already exists: Environmental Product Declaration (EPD). This is a certified environmental declaration developed in accordance with the standard ISO 14025 on environmental labels and declarations. EPDs include a description of product components and the materials they are made from[37]. However, the EPD website lists only 1100 EPDs, from 47 countries – a tiny fraction of the number of products on the market[38].

At a company level, Maersk Line, that provides trade and transport solutions, developed a Cradle-to-Cradle Passport in June 2013. The Passport, a first for the shipping industry, comprises an online database to create a detailed inventory that can be used to identify and recycle the components to a higher quality than is currently possible. By creating a resource that is flexible, manageable, and can be maintained throughout the 30-year lifetime of a ship, Maersk Line gains an improved understanding of the composition of the vessel that enters the recycling yard. As a result, the materials – including the 60,000 tons of steel per ship – can be sorted and processed more effectively, maintaining their inherent properties, and hopefully commanding a better price when re-sold[39].

At a regional level, the implementation of product passports is being implemented by Germany. In 2020 the Government of Germany launched the environmental policy digital agenda[40]. One key measure is the idea of developing a "digital product passport" for more transparency about the environmental impacts of different products[41]. Concepts for digital product passport do exist, however, they have not yet been institutionalized by mandatory standard data sets or central databases[42].

ACTIVITY 11.3 DISCUSSION ON o-LCA AND s-LCA

Discuss how tools such as o-LCA and s-LCA could be potentially used in managing circular supply chains.

How should circularity be promoted and ensured in the supply chains? By elimination or by strengthening or capacity building of the suppliers? What is the approach that you would recommend?

11.5 Education, Training, and Knowledge Networking

Education, training, and knowledge networking are critical in circular economy where transition requires behavioral change, innovation, and capacity building. Chapter 9 discussed the role of promoting innovations. In this section, some of the interesting initiatives in education, training, and knowledge networking have been highlighted.

CRCLR, Berlin is a Think-and-Do-Tank with a mission is to catalyze the transition towards a circular economy. The CRCLR team has created a unique, open space to explore creative community-based solutions to systemic global problems. It is the place to be for all things circular in Berlin. In 2019, the programs at CLCLR consisted of workshops, dinners, lessons, exhibitions, performances, concerts, markets, community meetups to screenings.

Sitra in Finland developed a circular economy and entrepreneurship game called Circula[43] for vocational educational institutions and supplementary training on the circular economy for teachers. This project was part of the

Circular economy teaching for all levels of education package implemented between 2017 and 2019. Sitra's vision was that all graduates should understand what the circular economy means from the point of view of their work and day-to-day life, and what decisions and actions they can take to promote the circular economy. The aim was to increase expertise in the circular economy in Finland by extensively developing circular economy training, materials, concepts, and co-operation from different points of view for all levels of education. More than 50 educational institutions, organizations and businesses took part in the package. With the Circula game workshops, training and events, the project involved more than 1700 teachers, students, and representatives of other stakeholders.

Ellen MacArthur Foundation in partnership with TU Delft has taken a lead in several online courses on circular economy. A webpage[44] and a compilation[45] is available that list one of the major courses offered. In July 2019, Ellen MacArthur Foundation launched a new circular economy learning hub[46].

Beyond providing an educational service, encouraging industry professionals to attend circular economy workshops or training can also contribute to building a circular "community of practice" and enabling networks of circularity champions within different industries. Communities of practice can be particularly valuable for helping smaller scale businesses to participate in circular economy practices and keep economic value within local communities.

Ellen MacArthur Foundation developed Circular Economy 100 (CE100) platform in 2013. CE 100 was launched with the aim of bringing together a network of 100 leading companies globally to facilitate development and commitment to new circular economy projects[47]. Members range from corporates, universities, city and government authorities, and affiliate networks, to emerging innovators, and include brands such as Google, Cisco, Coca-Cola, eBay, Apple, Novelis, IBM, and others.

The initiative serves as a collaborative network of businesses, innovators, cities, and governments, universities and thought leaders who work together to accelerate adoption of circular economy practices and processes that maximize the use of resources.

In 2016, the Ellen MacArthur Foundation announced the creation of a USA chapter of the CE100 program. The CE100 USA provides a national pre-competitive innovation platform addressing the specifics of the USA market. North America-based and focused organizations have access to unique collaboration, capacity building, networking as well as research and insight opportunities, to help them achieve their circular economy ambitions. By leveraging the network, members have access to insights to help them overcome local challenges and to explore circular opportunities which they might not be able to capture in isolation[48].

Platform for Accelerating Circular Economy (PACE)[49] is another recently launched platform. PACE was created to connect leaders who are committed to creating a circular economy. The platform catalyzes leadership from CEOs, government ministers, and the heads of civil society organizations who have a clear vision and the power to make things happen. PACE works with leading thinkers, researchers, and practitioners.

There are also examples of informal networks established on social networking platforms. Some of these platforms have been extremely popular and provide an opportunity to network. Box 11.12 presents an example of the Circular Economy Club that has recently embarked into a certificate program in circular economy.

BOX 11.12 THE CIRCULAR ECONOMY CLUB[50]

The Circular Economy Club (CEC) is the largest international network of circular economy professionals and organizations. The network is a not-for-profit organization which is open for anyone to join, free of cost.

They aim to bring the circular economy to cities worldwide, by building strong local networks who design and implement circular local initiatives.

Those goals are achieved by the CEC online platform and three main programs:

- CEC Chapters program – for circular leaders to bring the circular economy to live in their cities, universities, hubs and companies.
- CEC Mentors program – for members with expert skills to give free advice to the most promising circular talent.
- CEC Global events – for members to work collaboratively to solve local and global challenges.

CEC was formally established in London by Anna Tari, who realized that there were several great circular initiatives but lacked visibility, right tools, funding, and connections in order to have an impact. CEC was set up to bridge this gap by establishing strong connections amongst the circular economy community and sharing best practices.

CEC is managed voluntarily by a team of 37 volunteers based all over the world. Club members volunteer as CEC Mentors to give free mentorship to the youngest members. Alongside, CEC Organizers bring CEC activities to their cities, universities, and organizations.

Latest achievements of the club include:

- 7800 members in 150 countries.
- 260 CEC local Chapters in 110 countries.

- The Global Environmental Education Partnership issued a case study on the Circular Economy Club's organizers program.
- Winner or Finalist and Highly commended of awards and accolades: United Nations World Young Champion of the Earth, 30under30 North American Association of Environmental Education, Global Good Awards UK, Circular Awards from the World Economic Forum.

The Circular Economy Institute was founded in October 2020. The Institute is an agency that provides seminars, training, webcasts, and publications to allow alumni and other stakeholders to stay current on developments in the circular economy in sectors such as fashion and tourism. Currently it has 120 training sessions and operates in 30+ countries with its headquarters based in USA. Training is targeted for:

- Designers – Incorporating circularity when designing products, services, or any element of the built environment.
- Managers – Creating circular value chains for corporations, SMEs, big cities and small towns and logistics providers.
- Consultants – Developing circular initiatives in any industry, from fashion and tourism to the food and the automotive sector.
- Employers – find top talent with the highest degree of expertise in the field through the alumni search feature on the website.

ACTIVITY 11.4 DESIGNING AN AWARENESS PROGRAM FOR INTRODUCING CIRCULAR ECONOMY

Design an awareness program for introducing circular economy to the following:

- Politicians
- Policy makers and regulators
- Investors
- Community leaders.

Draft the contents and the pedagogy you would use.

11.6 Circular Living – A Way of Life

Chapter 4 described the "Rs" such as Refuse and Reduce with examples. Chapter 5 discussed product life extension and how consumers should move from "throw away" culture and show preference to the refurbished products. Chapter 6 argued that waste to landfill would decrease if the consumers were made aware about opportunities to recycle and recover resources from the waste streams.

Despite the several initiatives taken in circular economy, on average, European households use materials only once. The analysis of sectors has also found a significant economic wastage and there are opportunities for increased utilization of stocked resources. For example, the average European car is parked 92% of the time; 31% of Europe's food is wasted. The average European office is used only 35 to 50% of the time, even during working hours. The use of cycles is also short, and the average manufactured asset (excluding buildings) lasts only nine years[51].

Behavioral change is also needed to ensure that the consumers read and follow the instructions given while using the product. A study in the UK has revealed that more than half (58%) of British men did not know how to properly use a washing machine because they find the panel of the machine 'confusing'. According to this research, 16-to-24-year olds are most reluctant to do their own laundry, the most popular excuse being not knowing what buttons to press (40%). So even the best washing machines with good water and energy efficiency are available, but they may not get optimally used! Very few read the book of instructions (that is written in six languages) or ask for a demo of the washing machine. The consumers need to change their casual behavior and make serious effort to understand manufacturer's instructions and implications[52]. Indeed, this behavioral change will reduce resource consumption and increase the life span of the product.

The theory of behavioral change and communication plays an important role in sensitizing the consumers in these directions. This subject is of great interest today to product designers, manufacturers and brand owners, local municipal bodies, and the governments. Viability of several business models also depends on the behavioral change. Experts working in this field are finding opportunities to provide consulting service on how to influence consumer behaviors, e.g., exploring reuse options, selling second-hand, returning instead of stockpiling, and recycling instead of discarding.

Policies on sustainable procurement and local sourcing can make a difference to bring in behavioral change. It has been found that displaying expected product lifespan and ecolabels with assurance on traceability on products can influence purchase decisions. Communicating circular economy and its importance has however remained a challenge. For sustained results, circular living should be the way of life. Box 11.13 presents a case of sound material society in Japan. Government of Japan has now repositioned

BOX 11.13 SOUND MATERIAL SOCIETY IN JAPAN

The implementation of the concept of circular economy in Japan has followed a top-down approach, using legislation and enforcement. The motivation towards circularity was driven by lack of landfill space due to topography and limited domestic metal and mineral resources.

Japan became the first country to enact a comprehensive legislation to push circular economy with a societal goal, Law for effective utilization of recyclables was enacted in 1991. In 2007, 98% of Japan's metals were recycled and just 5% of its waste were landfilled. An enforced consumer's responsibility for returning electrical equipment resulted in the recovery of about 74% – 89% of the materials. Japan has also been the first country to demonstrate decoupling between economic growth and generation of wastes/emissions[53].

Transition to circular economy in Japan has been characterized by effective collaboration between consumers and manufacturers. The government developed an all-inclusive legal framework consisting several laws that reinforced the Japanese culture of frugal and sustainable living. In addition to legislation, the approach covered the following:

- Creation of educational courses on awareness of environmental issues in schools, companies, and communities.
- Provision of recycling laboratories in schools.
- Creation of enterprises' circular trading markets.
- Provision of incentives, enhancing public collaboration, and creating customer-friendly collection of old appliances.
- Building waste recycling facilities.

Integration of societal pursuits, business interest and education based on culture have thus been responsible for the sound material cycle society.

The communities responded positively to such a comprehensive approach. They complied with source separation of recyclables, did prompt payments of recycling fees, and exercised their rights as consumers. Manufacturers' used more recycled materials, produced long-lasting products, and designed products for repair, reuse, and recycling. Circular economy became a lifestyle and moved beyond economic interest to be inclusive with a social cause.

Today however the interest as well as consciousness of the new generation regarding a sound material-cycle society seem to be tapering. Percentage of those interested in waste management related issues for example is showing a decline. On the other hand, fun-filled 3R activities, such as cleanup after soccer games and other events, sports-like cleanup competitions, and cleanup by idols are fast catching attention. Factors responsible for change have been frequent occurrence of large-scale disasters, delays in responses, change in people's perspective (from material wealth to spiritual wealth) and shortages of human resources for waste recycling. It is necessary therefore that the concept of sound material-cycle society (now visioned as sustainable society) needs to be reemphasized using different channels of communication.

the goal towards sound material society to sustainable society. The 17 SDGs described in the next section provide a guidance in this direction.

ACTIVITY 11.5 DISCUSSION ON BEHAVIORAL CHANGE IN CIRCULAR ECONOMY

Behavioral change towards circular economy has always been a daunting task. Research on examples where a business or an enterprise included a campaign on promoting behavior change in the circular business strategy. Did it make a difference? What were the challenges faced and how were they overcome?

11.7 Sustainable Development Goals and Circular Economy

SDGs envision to eradicate poverty. These 17 goals and 169 targets were passed by UN General Assembly in September 2015 where all the nations have politically agreed to "heal the planet". The 17 SDGs are defined as "ambitions and transformational" as they apply globally. The SDGs are built on the Millennium Development Goals (MDGs) (2000) and attempt to cover all the global issues. The SDGs aim to fill the gaps that MDGs could not meet.

The SDGs are aspirational, ambitious, and transformational. They underline that each country has the primary responsibility for ensuring economic and social development. Each country needs to set the national goals based on the global ambitions while not ignoring national circumstances. Circular economy action plans thus contribute significantly to the attainment of SDGs. Box 11.14 shows a mapping between circular economy and the SDGs.

Many of the targets listed under the SDGs provide input to the national circular economy action plans and an overarching guidance.

BOX 11.14 MAPPING CIRCULAR ECONOMY WITH SUSTAINABLE DEVELOPMENT GOALS

UN SDG	CE Linkages
	Adoption of circular economy practices, such as, repair, reuse, refurbish, recycle, remanufacture etc. or the 12Rs described in Chapter 4 help to improve waste and resource management, reduce these health impacts, and can lead to generation of employment; that indirectly contributes to poverty reduction.
	Adoption of circular economy practices, such as, reducing the amount of waste generated in the food system, reuse of food, repurpose and utilize by products and food waste, and nutrient recycling help build sustainable food production systems and ultimately help end hunger.
	Poor waste management practices, such as open burning of waste and uncontrolled dumping, cause serious health impacts, particularly those living close to waste sites. Regulating recycling and recycled content in the products further help to reduce risks to the consumers.
	Circular Economy practices such as water harvesting and purification, sustainable sanitization, waste-water treatment for reuse and recycling, nutrient recovery, biogas systems etc. can help increase access to safe drinking water, increase water-use efficiency, reduce pollution, and improve water quality.
	Circular economy emphasizes on using renewable energy, bio-based or-fully recyclable input materials to replace single life-cycle inputs.
	New circular business models are a major potential source of increased resource effectiveness and efficiency, waste valorization and green jobs. Several studies have found circular economy implementation to be a multi-trillion opportunity globally.
	While circular economy practices will contribute directly to retrofitting industries to make them more resilient and sustainable, achieving targets under this goal is also important for transitioning to a circular economy. This includes new infrastructure, such as for renewable energy, circular water and waste/resource management, reverse logistics, support to research and innovation as well ensuring access to suitable financing.

Circular economy policies that integrate the informal waste management sector would be beneficial to those poor and vulnerable (especially women) since they would be protected by laws and safety measures in their jobs.

With increase in the number of world's population living in cities, a transition to a circular economy is imperative for reducing cities' resource and environmental impacts. Also, circular economy principles such as modular, adaptable, and flexible building design, can help enable access to housing for low-income groups.

Circular economy practices are all about decoupling economic activity from resource use and associated environmental and social impacts, which is also very much at the heart of this goal. Importantly, this goal is an important enabler for achieving most of the other SDGs, making the indirect impact of circular economy practices even more profound.

Research by organizations such as Ellen MacArthur and Circle Economy has found that to meet the 2050 goals of the Paris Agreement to combat climate change, every nation must inculcate circular economy practices, especially to improve material efficiency.

Preventing waste generation and leakages from land-based activities through circular economy practices will directly reduce waste entering the oceans, especially plastics. This also includes recovery of nutrients from waste-water streams before entering oceans. Additionally, circular economy contribution to tackling climate change will indirectly reduce ocean acidification.

At the core of circular economy practices is the aim to restore natural capital. This involves adopting sustainable and regenerative agricultural and agroforestry practices that embrace and protect biodiversity and returning biological material back to soils as nutrients – practices which are fundamental for restoring terrestrial ecosystems.

Introducing a circular economy in a globalized world requires a global coalition for action that is both diverse and inclusive. Businesses, governments, NGOs, and academics have to collectively boost capacity and capability to better serve societal needs through circular practices.

11.8 The COVID-19 Pandemic

In late 2019, the world was hit with a global health crisis caused by the infectious Coronavirus, which is popularly known as the COVID-19 pandemic. As of November 2020, the WHO reported above 10 million confirmed cases of the Coronavirus disease, which included more than 1.2 million deaths globally[54].

Due to such unprecedented health, economic and social crisis, meeting the various targets under SDGs has become more challenging. Figure 11.5 shows a contraction in meeting the SDGs due to COVID-19 pandemic.

Increased use of single-use protection devices and medical equipment such as masks, and bodysuits have put a pressure on the recycling and disposal infrastructure for managing plastic and COVID-19 contaminated medical waste. Overall, volumes of packaging have already witnessed a steep rise and so the generation of packaging waste. Plastic streams to the material flows are expected to rise, as nearly 40% of packaging consists of plastic and adhesives. Packaging firms need to innovate on sustainable packaging, despite the rise in single-use plastic. Use of robotics and artificial intelligence in sorting, cleaning, and sanitizing waste streams will help to ensure that recycling is safe and there is less leakage of plastic waste streams to the oceans.

Extensive use of sanitizers where chemicals used are not controlled are expected to lead to ecological risks. The pandemic has also resulted in a greater emphasis on the issues of gender inequalities towards women where questions of unpaid care work and domestic violence have been highlighted. In some countries, labor laws have been relaxed to support the businesses. This could become a concern in meeting the codes of conduct in the supply chains.

The consumption patterns during the COVID-19 pandemic have changed and even if the pandemic is brought under control, some consumption patterns may take longer time to return to the normal and some changed patterns may continue. Work from home practice, intensive use e-commerce platforms and increased digitalization in transaction are some examples that have set the "new normal". The production and delivery of services will need to adjust to the changed consumption patterns.

Innovation in circular economy could help handling the adverse impacts of COVID-19 pandemic. Contact-less health care and use of less plastic in medical equipment have been a solution to reduce use of plastic. Some producers of protective and medical equipment are using machines developed by Precious Plastic[56], an open-source hardware plastic recycling initiative, to turn recycled plastic into face shields and masks using special recycling machines. Several European economies are now using these machines, capable of producing protective masks 75 times faster than a 3D printer, to supplement declining supplies. Since the machines expose plastics to temperatures

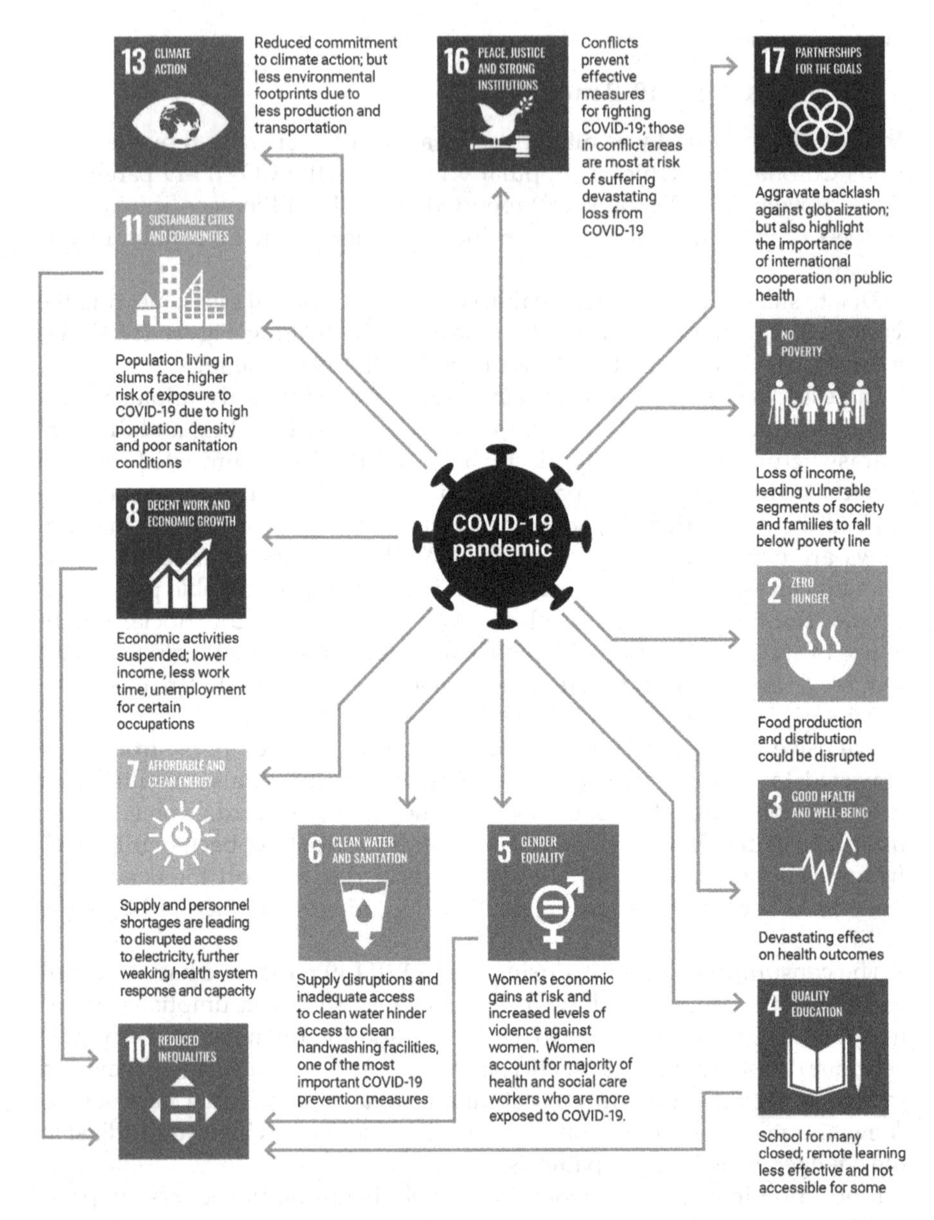

FIGURE 11.5
COVID-19 pandemic impact on the 17 UN SDGs[55].

over 200 degrees Celsius, the recycled plastics are sterilized in the process, adding the benefit of cleansing these used and shredded plastics for reuse later.

Nike is another example of firms using this principle to full effect in the fight against the pandemic. The prominent shoe manufacturer has redirected recycled material, earmarked to produce new Nike Air soles, into the production of personal protective gear.

Batelle, a non-profit institute engaged in scientific research, developed a way to decontaminate N95 masks using vaporized hydrogen peroxide. The technology received authorization for use from the US Food and Drug Administration in March 2020[57]. It is now being used in several hospitals in the United States. Through this novel method, masks can be decontaminated for use 20 times over before their quality and safety is compromised, and thus it provides a longer-lasting alternative to single-use products.

Other innovations include COVID-19 resistant fabrics. Many fashion companies engaged in textiles have launched their fabrics processed with "HieQ viroblock" made in Switzerland[58] that imparts anti-bacterial and anti-viral properties. Brazillian textile company *Dalila Têxtil* and Italian *Albini Group* have created fabrics capable of destroying the protective outer layer of COVID-19. Use of copper foils and silver coatings has however increased in packaging for anti-microbial/anti-viral activity, laser treatment is used for metal surfaces to make their texture virus resistant, especially in the furniture industry.

Practicing Sustainable Public Procurement (SPP) will play an important role to influence both consumption and production patterns. For example, only safe and biodegradable sanitizers and biodegradable and recyclable masks may be preferred, especially when procured on a scale for organizations.

A direct positive impact of the COVID-19 pandemic has been a boom in the digital sector offering services powered by technologies such as Blockchain, Internet of Things (IoT) and Artificial Intelligence (AI). Its application in sorting, recycling, aggregation and reverse logistic are expected to rise and boost business in circular economy.

Knowledge networks will need to manage actively that innovations in these areas are shared to benefit developing economies. The financial packages offered to help COVID-19 pandemic affected business sectors should allocate funds for circular economy action plans. Increased use of renewable energy, smart low carbon, or net zero buildings, green and safe public transport and sustainable food and agriculture systems should be on the agenda to provide financial support and attract private sector investments.

Many countries have come up with financial stimulus packages to restore economy and importantly the jobs. Budgets have also been set aside to spruce the medical infrastructure. arising from the COVID-19 crisis, to shift towards green economic growth. The situation is like the one faced in 2008-2009 that was a global economic crisis. UNEP's "green economy" initiative (described in Chapter 2) was considered as a strategy to reboot the global economy for green growth. Perhaps, a similar effort will need to be made leveraging on circular economy on a global scale. The linkages between SDGs and circular economy, articulated in Box 11.14 show circular economy could offer a solution.

Welsh Government and WRAP Cymru[59] realized that many organizations are facing significant challenges resulting from COVID-19 pandemic. In response, they reoriented the GBP 6.5 million Circular Economy Fund. In addition to funding the use of recycled materials in products, WRAP Cymru is now able to support preparation for re-use, refurbishment, and re-manufacturing activities in Wales. There are other funds who are re-examining their investment objects and aligning for an advantage by supporting business in circular economy.

An alliance of 180 European politicians, business leaders and environmental activists have urged that investments are directed towards the shaping of a "new European economic model: more resilient, more protective, more sovereign, and more inclusive". Over 100 investors, representing EUR 11.9 trillion in assets either managed or advised, are keen that a green recovery is delivered. These calls are taking place when investments and policy adjustments are happening for global economic recovery both in the short-term and the long-term. With around USD 10 trillion in economic stimulus being unveiled by governments all around the world, there is an unprecedented opportunity to "move away from unmitigated growth at all costs and the old fossil fuel economy, towards a lasting balance between people, prosperity, and planetary boundaries[60].

Overall, the consequences of COVID-19 pandemic are going to increase the global GHG emissions. Sharing of economy or collaborative consumption has come into a lens. Private transportation may be a preferred option over public transportation systems and car sharing could get low priority due to COVID-19 risks. In the year 2020, the 26th session of the Conference of the Parties (CoP 26) to the UNFCCC was scheduled to take place with the key issue for discussion set as natural solutions to climate change and countries expected to propose new commitments to lower their emissions, in line with the Paris Agreement. The international community was also due to set out a framework for better management of chemicals and waste. However, this has been postponed to 2021 due to the COVID-19 pandemic. Circular economy could be a win-win solution to reduce and slow down the GHG emissions. Climate change is however not yet mainstream in circular economy.

Experience of witnessing the COVID-19 pandemic has taught us the need to prepare resilient and regenerative societies. Practicing circular economy could be one of the potential strategies to achieve this goal. Mainstreaming circular economy (i.e., making circular living as a way of life) supported by policy tools, business models and financing could help in combating climate change – a risk that looms with a much larger magnitude compared to COVID-19 pandemic. There is hope that circular economy will receive much more attention during and post COVID-19 pandemic with a promise towards a sustainable, prosperous, inclusive, and resilient future.

11.9 Key Takeaways

- Metrics in circular economy is a need but also a challenge. Challenges include absence of a uniform methodology, lack of clarity on estimation of impacts, non-availability, and poor quality of data. The Circularity Gap Assessment report of 2020 does not show a promising picture on the transition to the circular economy.
- Regional circular economy will play an important role in the update of circular economy. Systematic methodologies need to be followed to develop regional and national scale circular action plans. The framework CIRCULAR suggested could help set guiding and operational principles to develop a strategic approach. Case study provided on development of India's circular economy action plan could help.
- Circular economy in the supply chains will provide a great opportunity to the corporates, especially with global operations. Here, following codes of conduct supported by traceability will help.
- Behavior change towards "circular living" is however extremely crucial and here circular economy needs to be promoted at local or neighborhood level through collective and collaborative actions. Education, training, and knowledge networks will play an important role. Circular economy is still misunderstood as recycling.
- Digital platforms and disruptive innovations will lead circular economy, especially in the times of pandemics such as COVID-19.
- Circular economy is intimately linked to the Sustainable Development Goals. It has a tremendous role to play to respond to the challenges like the COVID-19 pandemic to help towards a sustainable, prosperous, inclusive, and resilient future.

ADDITIONAL READING

1. **Pathways to a circular economy in cities and regions report**
   This document is a policy brief addressed to policy makers from European cities and regions. Cities and regions are uniquely positioned to accelerate Europe towards a more circular economy and harness economic, environmental, and social benefits related to this transition. ESPON, Interact, Interreg Europe, and URBACT call on cities and regions to act today and remember these four simple takeaways.

Source: European Commission. n.d. *Pathways to A Circular Economy In Cities And Regions.* <https://urbact.eu/sites/default/files/policy_brief_on_circular_economy.pdf> [Accessed 13 November 2020].

2. **The Circular Economy in Cities and Regions: Synthesis Report**
   Cities and regions play a fundamental role in the transition from a linear to a circular economy, as they are responsible for key policies in local public services such as transport, solid waste, water and energy that affect citizens' well-being, economic growth and environmental quality. This synthesis report builds on the findings from 51 cities and regions contributing to the OECD Survey on the Circular Economy in Cities and Regions and on lessons learnt from the OECD Policy Dialogues on the circular economy carried out in Groningen (Netherlands), Umeå (Sweden), Valladolid (Spain) and on-going in Glasgow (United Kingdom), Granada (Spain), and Ireland. The report provides a compendium of circular economy good practices, obstacles and opportunities, analyzed through the lens of its 3Ps analytical framework (people, policies and places). It concludes with policy recommendations, a Checklist for Action and a Scoreboard to self-assess the existence and level of implementation of enabling governance conditions to foster the transition towards the circular economy in cities and regions.

Source: OECD. 2020. *The Circular Economy in Cities and Regions: Synthesis Report, OECD Urban Studies.* Paris: OECD Publishing. https://doi.org/10.1787/10ac6ae4-en.

3. **Zero Waste Scotland Website**
   Zero Waste Scotland is working in partnership with organizations in five areas, as part of its Circular Cities and Regions programme, to deliver a tailored program of business engagement to identify and exploit the key sectors and businesses for circular growth. Visit this webpage to learn more about the program's key activities and progress.

Source: Zero Waste Scotland. n.d. *Circular Economy In Cities And Regions.* <https://www.zerowastescotland.org.uk/circular-economy/cities-and-regions> [Accessed 13 November 2020].

4. **Circular Cities Project**
   Circular Economy in Cities focuses on opportunities in three key urban systems – buildings, mobility, and products – and

looks at how city governments can work to enable a circular economy transition. The project addresses questions such as:

- Vision: What will the implementation of circular economy principles in cities look like?
- Factsheets: What benefits can a circular economy transition in key urban systems bring to cities?
- Policy levers: What can urban policymakers do to accelerate this transition?
- Case studies: What examples are there of urban policymakers already putting this into action?
- Other networks & resources: What are other organisations doing on the topic of circular economy and cities?

Visit the website to learn more.

Source: Ellen Macarthur Foundation. n.d. *Circular Cities.* <https://www.ellenmacarthurfoundation.org/our-work/activities/circular-economy-in-cities> [Accessed 13 November 2020].

5. **Government of Japan on Ecotowns**
   Visit this source to learn more about the Ecotowns initiative in Japan. This link takes you to the map of the Ecotowns in Japan.

Source: Ministry of Environment. n.d. *Eco Town Program.* <https://www.env.go.jp/en/recycle/manage/eco_town/index.html> [Accessed 13 November 2020].

6. **Accelerating Action for a Sustainable and Circular Garment and Footwear Industry: Which Role For Transparency And Traceability Of Value Chains?**
   This study shows that transparency and traceability are a key driver of sustainability and must be a collaborative effort. It looks into the key requirements and components of robust transparency and traceability framework and provides a series of recommendations on possible measures that public authorities could devise to create a conducive environment and sustain the implementation of such a framework at the industry level.

Source: UNECE. 2020. *Accelerating Action For A Sustainable And Circular Garment and Footwear Industry: Which Role For Transparency and Traceability of Value Chains?* <http://www.unece.org/fileadmin/DAM/trade/Publications/ECE_TRADE_449-AcceleratingTanspRraceabilityTextile.pdf> [Accessed 13 November 2020].

7. **Profiled Universities**
   This program is an initiative by the Ellen Macarthur Foundation. By showcasing circular economy related teaching and research globally, this program aims to enable collaborative ventures and knowledge exchange across academia, policy makers, and businesses outside the Foundation's formal programs.

Source: Ellen Macarthur Foundation. n.d. *Circular Economy Profiled Universities.* <https://www.ellenmacarthurfoundation.org/our-work/activities/universities/profiled-universities> [Accessed 13 November 2020].

8. **European Circular Economy Stakeholder Platform**
   A joint initiative by the European Commission and the European Economic and Social Committee, the European Circular Economy Stakeholder Platform brings together stakeholders active in the broad field of the circular economy in Europe. You can join the online discussion forum for topical conversations. You can submit your inspiring (or more challenging) stories, publications, circular economy events, existing network or platform, via the Contribute page.

Source: European Union. n.d. European Circular Economy Stakeholder Platform. <https://circulareconomy.europa.eu/platform/en/sector/education-and-skills> [Accessed 13 November 2020].

9. **Circular Economy Teaching for all Levels of Education**
   Sitra developed and tested circular economy learning materials and courses at all educational levels in Finland. Over 70,000 children and young all around Finland studied the circular economy in 2018-2019. Visit this source to learn more.

Source: Sitra. n.d. *Circular Economy Teaching for All Levels of Education.* <https://www.sitra.fi/en/projects/circular-economy-teaching-levels-education/#for-teachers.> [Accessed 13 November 2020].

10. **Breaking the Barriers to the Circular Economy**
    For this research, a survey with 153 businesses, 55 government officials and expert interviews with forty-seven thought leaders on the circular economy from businesses, governments, academia, and NGOs have been carried out. Two types of barriers emerged as main barriers. Firstly, there are the cultural barriers of lacking consumer interest and awareness as well as

a hesitant company culture. This finding is at odds with claims that the circular economy concept is hyped; rather, the concept may be a niche discussion among sustainable development professionals. Secondly, market barriers emerged as a core category of barriers, particularly low virgin material prices and high upfront investments costs for circular business models.

Source: Deloitte. n.d. *Breaking the Barriers To The Circular Economy.* <https://circulareconomy.europa.eu/platform/sites/default/files/171106_white_paper_breaking_the_barriers_to_the_circular_economy_white_paper_vweb-14021.pdf> [Accessed 13 November 2020].

11. **Circular economy indicators: What do they measure?**
    To understand what indicators used in CE measure specifically, the authors propose a classification framework to categorize indicators according to reasoning on what (CE strategies) and how (measurement scope). Despite different types, CE strategies can be grouped according to their attempt to preserve functions, products, components, materials, or embodied energy; additionally, indicators can measure the linear economy as a reference scenario. The measurement scope shows how indicators account for technological cycles with or without a Life Cycle Thinking (LCT) approach; or their effects on environmental, social, or economic dimensions.

Source: Moraga, G., Huysveld, S., Mathieux, F., Blengini, G., Alaerts, L., Van Acker, K., de Meester, S. and Dewulf, J., 2019. Circular economy indicators: What do they measure? *Resources, Conservation and Recycling*, 146, pp.452–461.

12. **The Circular Economy Business Model: Examining Consumers' Acceptance of Recycled Goods**
    The circular economy strategy supports the transformation of the *linear consumption* model into a *closed-production model* to achieve economic sustainability, with the consumers' acceptance of circular products being one of the major challenges. Further, one important aspect of product circularity remains unexplored, such as the consumers' purchase intention of recycled circular goods. In this context, the present study proposes and tests a conceptual model on consumers' acceptance of recycled goods through PLS Structural Equation Modeling (PLS-SEM), based on the data obtained from 312 respondents. Results indicate that the positive image of circular products

is the most important driver of consumers' acceptance, followed by the product's perceived safety. This study provides an empirical foundation for the important role of consumers in circular economy business models through the examination of consumers' acceptance of recycled goods.

Source: Calvo-Porral, C. and Lévy-Mangin, J., 2020. The Circular Economy Business Model: Examining Consumers' Acceptance of Recycled Goods. *Administrative Sciences*, 10(2), p.28.

13. **Design for Circular Behaviour: Considering Users in a Circular Economy**
    Circular design strategies have tended to focus on the physical aspects of a product (e.g., disassembly, material selection), but the design of products and services can also have an influence on user behavior and, to date, this aspect of circular design has not been fully explored. This project aims to define what key user behaviors are required for circular business models to work and to outline how design can enable these 'circular behaviors'. This research project consists of a literature review, case study analysis and expert interviews with practitioners. A theoretical framework for designing products and services to encourage circular behavior is developed. This work provides an initial step towards a better understanding of the user's role in the transition to a circular economy as well as a preliminary model for how design for behavior change strategies could be implemented in this context.

Source: Wastling, T., Charnley, F. and Moreno, M., 2018. Design for Circular Behaviour: Considering Users in a Circular Economy. *Sustainability*, 10(6), p.1743.

14. **The Circular Economy: A Transformative Covid-19 Recovery Strategy**
    With the Covid-19 pandemic revealing the vulnerability of global systems to protect the environment, health, and economy, many voices from governments, businesses, and civil society have been calling for a response to the devastating impacts of the pandemic that is inclusive and does not turn attention away from other global challenges. These calls are taking place at a pivotal time since investments and policy actions will determine the direction of economic recovery both in the short-term and the long-term. The pandemic may also be reconfiguring the roles of state and market actors for years to

come. Building on the past ten years of research carried out on the circular economy, the Ellen MacArthur Foundation highlights in this paper how policymakers can help pave the way towards a resilient recovery.

Source: Ellen Macarthur Foundation. 2020. *The Circular Economy: A Transformative Covid-19 Recovery Strategy.* <https://www.ellenmacarthurfoundation.org/assets/downloads/The-circular-economy-a-transformative-Covid19-recovery-strategy.pdf> [Accessed 13 November 2020].

Notes

1. Ellen Macarthur Foundation. 2015. *Circularity Indicators- An Approach To Measuring Circularity.* <https://www.ellenmacarthurfoundation.org/assets/downloads/insight/Circularity-Indicators_Project-Overview_May2015.pdf> [Accessed 13 November 2020].
2. Ellen Macarthur Foundation. n.d. *Circulytics - Measuring Circularity.* <https://www.ellenmacarthurfoundation.org/resources/apply/circulytics-measuring-circularity> [Accessed 13 November 2020].
3. Trayak - Sustainability Solutions. n.d. *About Us.* <https://trayak.com/> [Accessed 13 November 2020].
4. U.S. Chamber of Commerce Foundation. n.d. *Measuring Circular Economy.* <https://www.uschamberfoundation.org/circular-economy-toolbox/about-circularity/measuring-circular-economy> [Accessed 13 November 2020].
5. Institut National de l'Economie Circulaire and Entreprises pour l'Environnement. 2019. *Circular Economy Indicators For Businesses.* <https://figbc.fi/wp-content/uploads/sites/4/2020/05/circular-economy-indicators-for-businesses-february-2019.pdf> [Accessed 17 November 2020].
6. TERI. 2018. *Circular Economy: A Business Imperative For India.* <http://wsds.teriin.org/2018/files/teri-yesbank-circular-economy-report.pdf> [Accessed 17 November 2020].
7. WBCSD. 2018. *Circular Metrics Landscape Analysis.* <https://docs.wbcsd.org/2018/06/Circular_Metrics-Landscape_analysis.pdf> [Accessed 13 November 2020].
8. World Economic Forum. 2018. *Harnessing The Fourth Industrial Revolution For The Circular Economy- Consumer Electronics And Plastics Packaging.* <http://www3.weforum.org/docs/WEF_Harnessing_4IR_Circular_Economy_report_2018.pdf> [Accessed 13 November 2020].
9. Circle Economy. n.d. *About.* <https://www.circle-economy.com/> [Accessed 13 November 2020].
10. Global Reporting Initiative. 2020. *GRI 306: Waste 2020.* <https://www.globalreporting.org/standards/media/2595/gri-waste-leaflet.pdf> [Accessed 13 November 2020].

11. Ellen Macarthur Foundation. n.d. *Circular Cities*. <https://www.ellenmacarthur foundation.org/our-work/activities/circular-economy-in-cities> [Accessed 17 November 2020].
12. Global Environment Centre Foundation. 2005. *Eco-Towns in Japan -Implications And Lessons For Developing Countries And Cities* <https://wedocs.unep.org/bitstream/handle/20.500.11822/8481/Eco_Towns_in_Japan.pdf?sequence=3&%3BisAllowed=> [Accessed 13 November 2020].
13. WBCSD Regional Network Partner. n.d. *The Japanese Eco-Town Programme*. <http://www.govsgocircular.com/cases/the-japanese-eco-town-programme/> [Accessed 14 November 2020].
14. Eco-Industrial Park (EIP) Development in Viet Nam: Results and Key Insights from UNIDO's EIP Project (2014–2019) (DOI: 10.3390/su11174667)
15. OECD. 2020. The Circular Economy in Cities and Regions: Synthesis Report, OECD Urban Studies, OECD Publishing, Paris, https://doi.org/10.1787/10ac6ae4-en.
16. Ellen Macarthur Foundation. 2015. *DELIVERING THE CIRCULAR ECONOMY A TOOLKIT FOR POLICYMAKERS*. <https://www.ellenmacarthurfoundation.org/resources/apply/toolkit-for-policymakers> [Accessed 13 November 2020].
17. Circle Economy. n.d. *The 7 Key Elements Of The Circular Economy*. <https://www.circle-economy.com/circular-economy/7-key-elements> [Accessed 13 November 2020].
18. NITI Aayog and EU Delegation to India. 2019. *Resource Efficiency and Circular Economy: Current Status and Way Forward* < https://www.eu-rei.com/pdf/publication/NA_EU_Status%20Paper%20&%20Way%20Forward_Jan%202019.pdf> [Accessed 14 December 2020].
19. India Brand Equity Foundation. 2017. *Indian Steel Industry Report*. <https://www.ibef.org/industry/steel-presentation> [Accessed 13 November 2020].
20. Integrated Steel Plants include covers all the steps of primary steel production starting from iron making to steel making and product rolling.
21. Ministry of Steel. 2020. *Existing Steel Based Industries in The Country*. <http://pib.nic.in/newsite/PrintRelease.aspx?relid=77494> [Accessed 13 November 2020].
22. Niti Aayog. 2018. *Strategy Paper on Resource Efficiency in Steel Sector Through Recycling of Scrap & Slag*. <https://niti.gov.in/writereaddata/files/RE_Steel_Scrap_Slag-FinalR4-28092018.pdf> [Accessed 13 November 2020].
23. Green Spec. n.d. *Steel Production & Environmental Impact*. <http://www.greenspec.co.uk/building-design/steel-products-and-environmental-impact/> [Accessed 13 November 2020].
24. Central Pollution Control Board. 2016. *Final Document on Revised Classification of Industrial Sectors Under Red, Orange, Green and White Categories*. <http://moef.gov.in/wp-content/uploads/2017/07/Latest_118_Final_Directions.pdf> [Accessed 13 November 2020].
25. Sail. n.d. *Climate Change*. <https://www.sail.co.in/sites/default/files/Climate_Change.pdf> [Accessed 13 November 2020].
26. Government Of India Ministry Of Mines Indian Bureau Of Mines. 2018. *Indian Minerals Yearbook 2017*. <http://ibm.nic.in/writereaddata/files/03202018150040Slag_Iron_Steel_AR_2017.pdf> [Accessed 13 November 2020].
27. H&M Group. n.d. *Circularity and Our Value Chain*. <https://hmgroup.com/sustainability/circular-and-climate-positive/circularity-and-our-value-chain.html> [Accessed 13 November 2020].

28. Ethical Trading Initiative. n.d. *ETI Base Code.* <https://www.ethicaltrade.org/eti-base-code> [Accessed 13 November 2020].
29. SGS. n.d. *SA 8000 - Social Accountability Certification.* <https://www.sgs.com/en/sustainability/social-sustainability/audit-certification-and-verification/sa-8000-certification-social-accountability> [Accessed 13 November 2020].
30. Responsible Business. n.d. *Code of Conduct 6.0.* <http://www.responsiblebusiness.org/code-of-conduct/> [Accessed 13 November 2020].
31. The Global Language of Business. 2017. *Global Traceability Standard.* <https://www.gs1.org/sites/default/files/docs/traceability/GS1_Global_Traceability_Standard_i2.pdf> [Accessed 13 November 2020].
32. OECD. 2017. *OECD Due Diligence Guidance for Responsible Supply Chains In The Garment And Footwear Sector.* <https://mneguidelines.oecd.org/OECD-Due-Diligence-Guidance-Garment-Footwear.pdf> [Accessed 13 November 2020].
33. Green Car Congress. 2020. *BASF Introduces Blockchain Pilot Project To Improve Circular Economy And Traceability Of Recycled Plastics.* <https://www.greencarcongress.com/2020/02/20200216-basf.html> [Accessed 13 November 2020].
34. Everledger. 2020. *Why Traceability Matters In Recycling And The Circular Economy.* <https://www.everledger.io/why-traceability-matters-in-recycling-and-the-circular-economy/> [Accessed 13 November 2020].
35. Optel Group. n.d. *About.* <https://www.optelgroup.com/> [Accessed 13 November 2020].
36. Hunter, J., 2015. *What'S A Product Passport And Can It Work For Businesses?.*Make Resources Count. <http://makeresourcescount.eu/whats-a-product-passport-and-can-it-work-for-businesses/> [Accessed 14 December 2020].
37. European Commission. 2013. *Eco-Innovation Action Plan.* <https://ec.europa.eu/environment/ecoap/about-eco-innovation/policies-matters/eu/20130708_european-resource-efficiency-platform-pushes-for-product-passports_en> [Accessed 14 December 2020].
38. The International EPD® System. n.d. *Using EPD.* <https://www.environdec.com/> [Accessed 14 December 2020].
39. Ellen Macarthur Foundation. n.d. *Using Product Passports to Improve the Recovery And Reuse Of Shipping Steel.* <https://www.ellenmacarthurfoundation.org/case-studies/using-product-passports-to-improve-the-recovery-and-reuse-of-shipping-steel> [Accessed 14 December 2020].
40. Federal Ministry for the Environment, Nature Conservation and Nuclear Safety. n.d. *Environmental Digital Agenda.* <https://www.bmu.de/digitalagenda/> [Accessed 14 December 2020].
41. Pettit, M., 2020. *Germany'S New Climate-Friendly Digital Agenda Puts Sustainability and Digitalisation Front and Centre.* RESET. <https://en.reset.org/blog/germany%E2%80%99s-new-climate-friendly-digital-agenda-puts-sustainability-and-digitalisation-front-and-> [Accessed 14 December 2020].
42. Federal Ministry for the Environment, Nature Conservation and Nuclear Safety. n.d. *BMU Digital Policy Agenda For The Environment: Digital Product Passport.* <https://www.bmu.de/en/service/haeufige-fragen-faq/details-cluster/bmu-digital-policy-agenda-for-the-environment-digital-product-passport/> [Accessed 14 December 2020].
43. Sitra. n.d. *Circular Economy Teaching For All Levels Of Education.* <https://www.sitra.fi/en/projects/circular-economy-teaching-levels-education/> [Accessed 13 November 2020].

44. Ellen Macarthur Foundation. n.d. *Circular Economy Courses.* <https://www.ellenmacarthurfoundation.org/resources/learn/courses> [Accessed 13 November 2020].
45. Ellen Macarthur Foundation. 2018. *A Global Snapshot Of Circular Economy Learning Offerings In Higher Education.* <https://www.ellenmacarthurfoundation.org/assets/downloads/Global-Snapshot-19.10.18.pdf> [Accessed 13 November 2020].
46. Ellen Macarthur Foundation. 2019. *New Circular Economy Learning Hub Is Launched.* <https://www.ellenmacarthurfoundation.org/news/new-circular-economy-learning-hub-launches> [Accessed 13 November 2020].
47. Ellen Macarthur Foundation. 2013. *Foundation Launches Global Business Alliance: The Circular Economy 100.* <https://www.ellenmacarthurfoundation.org/news/foundation-launches-global-business-alliance-the-circular-economy-100#:~:text=Foundation%20launches%20global%20business%20alliance%3A%20the%20Circular%20Economy%20100,-February%2006%2C%202013&text=The%20Ellen%20MacArthur%20Foundation%20has,to%20new%20circular%20economy%20projects.> [Accessed 17 November 2020].
48. Ellen MacArthur Foundation. 2016. *The Ellen Macarthur Foundation'S Circular Economy 100 Programme Launches Its USA Network.* <https://bit.ly/3puRvhA> [Accessed 13 November 2020].
49. PACE. n.d. *Home.* <https://pacecircular.org/> [Accessed 13 November 2020].
50. Circular Economy Club. n.d. *About CEC.* <https://www.circulareconomyclub.com/gd-home/about-cec/> [Accessed 13 November 2020].
51. Ellen Macarthur Foundation. 2015. *Towards A Circular Economy: Business Rationale for an Accelerated Transition.* <https://www.ellenmacarthurfoundation.org/assets/downloads/publications/TCE_Ellen-MacArthur-Foundation_26-Nov-2015.pdf> [Accessed 13 November 2020].
52. Kirkova, D., 2013. *Half Of Men Can't Use A Washing Machine Properly and A Quarter Can't Even Figure Out How To Switch It On.* Mail Online. <http://www.dailymail.co.uk/femail/article-2340216/Half-men-use-washing-machine-properly-quarter-figure-switch-on.html> [Accessed 13 November 2020].
53. Ogunmakinde, 2019. A Review of Circular Economy Development Models in China, Germany and Japan. *Recycling*, 4(3), p.27.
54. WHO, 2020. *WHO Coronavirus Disease (COVID-19) Dashboard.* <https://COVID-19.who.int/> [Accessed 14 November 2020].
55. United Nations, 2020. *SHARED RESPONSIBILITY, GLOBAL SOLIDARITY: Responding to the Socio-Economic Impacts of COVID-19.* [ebook] United Nations. Available at: <https://unsdg.un.org/sites/default/files/2020-03/SG-Report-Socio-Economic-Impact-of-Covid19.pdf > [Accessed 18 November 2020].
56. Precious Plastic. n.d. *About.* <https://preciousplastic.com/> [Accessed 13 November 2020].
57. Battelle. 2020. *Battelle Develops System To Decontaminate Personal Protective Equipment To Meet Growing Demand During COVID-19 Crisis.* <https://www.battelle.org/newsroom/press-releases/press-releases-detail/battelle-develops-system-to-decontaminate-personal-protective-equipment-to-meet-growing-demand-during-covid-19-crisis> [Accessed 17 November 2020].

58. HEIQ Viroblock. n.d. *Technologies*. <https://heiq.com/technologies/heiq-viroblock/?no_popup=1> [Accessed 13 November 2020].
59. WRAP Cymru. n.d. *About Us*. <https://www.wrapcymru.org.uk/> [Accessed 13 November 2020].
60. Ellen Macarthur Foundation. 2020. *The Circular Economy: A Transformative Covid-19 Recovery Strategy*. <https://www.ellenmacarthurfoundation.org/assets/downloads/The-circular-economy-a-transformative-Covid19-recovery-strategy.pdf> [Accessed 13 November 2020].

Index

Note: *Italicized* page numbers indicate figures, **bold** indicate tables.

C

E

J

S

V

Y

Z

Printed in the United States
by Baker & Taylor Publisher Services